Bioenergy Crops
A Sustainable Means of Phytoremediation

Editors

Jos T. Puthur
Plant Physiology and Biochemistry Division
Department of Botany, University of Calicut
Kerala, India

Om Parkash Dhankher
Stockbridge School of Agriculture
University of Massachusetts
Amherst, U.S.A.

CRC Press
Taylor & Francis Group
Boca Raton London New York

CRC Press is an imprint of the
Taylor & Francis Group, an **informa** business
A SCIENCE PUBLISHERS BOOK

Cover credits: Image provided by Dr. Edappayil Janeeshma, Department of Botany, University of Calicut, Kerala, India. It has not been taken from any copyrighted source.

First edition published 2022
by CRC Press
6000 Broken Sound Parkway NW, Suite 300, Boca Raton, FL 33487-2742

and by CRC Press
2 Park Square, Milton Park, Abingdon, Oxon, OX14 4RN

ISBN: 978-0-367-48913-7 (hbk)
ISBN: 978-1-032-26033-4 (pbk)
ISBN: 978-1-003-04352-2 (ebk)

DOI: 10.1201/9781003043522

Typeset in Times New Roman
by Innovative Processors

Preface

Accumulation of heavy metals and organic xenobiotics has emerged as a big threat to living organisms. Extensive application of agrochemicals, mismanagement of organic wastes, industrial waste depositions, mining, and smelting are the different anthropogenic actions causing the accumulation of these toxic compounds in the environment. "Phytoremediation" was introduced as a strategy to remove or detoxify these compounds by the utilization of plants. However, growing hyperaccumulators in contaminated lands is a time-consuming and costly process. This calls for the application of bioenergy crops in the field of phytoremediation. Bioenergy crops are non-food crops cultivated for the production of biofuels, which generally needs low-cost for maintenance. The selection of plant candidates, which can play the dual role of phytoremediation and bioenergy production, is a task, and current researchers are investigating to identify apt plant candidates, which can make this process eco-friendly as well as economically viable. The main motive of writing this book was to address these issues. The present book includes 14 chapters dealing with different aspects of bioenergy production and phytoremediation that helps to update the current status of research in this field.

Organic pollutants and heavy metals induce different metabolic changes in plants that aid in tolerating the toxic effect of these contaminants. Different plants have different strategies to overcome the oxidative stress induced by the contaminants. It is essential to understand the mechanisms operational in plants, algae, and microorganisms with respect to the tolerance of heavy metals and organic pollutants. Chapter 1 of the book is a comprehensive approach towards phytoremediation and bioenergy production. It discusses the importance of bioenergy production and phytoremediation. Chapters 2 and 3 deal with the sources and impacts of organic and inorganic contaminants. Moreover, chapter 3 explains how microorganisms can enhance the tolerance level of plants by modifying the metabolic pathways as well as chelate the metal ions. Chapters 4 and 5 suggest important bioenergy plant candidates with phytoremediation potential, including *Jatropha curcas, Ricinus communis*, and *Eichhornia crassipes.* Algae can be considered as a good source of biofuel and fertilizer with heavy metal accumulation potential, and this is discussed in chapter 6. Wastewater treatment is another crisis faced by us, and utilization of the potentials of algae is one of the best strategies to overcome this. Chapter 7 explains the importance of aquatic plants with phytoremediation potential. The importance of constructed wetlands is detailed in chapter 8. Chapter 9 discusses the processes behind the conversion of waste materials to bioenergy, where the microbial fuel cells and anaerobic digestion are explained. The major actions taken by the governments of different countries for land reclamation are discussed in Chapter 10. The importance of the development of new industries for land clearing is explained in the same chapter. Along with the benefits, it is significant to discuss the challenges in utilizing bioenergy crops for phytoremediation, and this topic is dealt in Chapter 11. Chapters 12, 13, and 14 focus on the case studies, and explain the phytoextraction potential of willow plants, hydrocarbon generation, and bioelectricity production.

The book, therefore, comprises a unique combination of chapters on various aspects and will provide a comprehensive view of the utilization of bioenergy crops in phytoremediation. This book will guide the graduate, post-graduate and doctoral students as well as researchers to know the latest updates in the field of phytoremediation. Moreover, it gives a clear idea of utilizing the knowledge to develop different industries and job opportunities in the same field.

It gives immense pleasure to place this book in the hands of the scientific community working in the area of phytoremediation. The book could not been published without the substantial contribution of innovative ideas from enthusiastic researchers and the support of the publishers. Special thanks also due to Dr. Edappayil Janeeshma, Department of Botany, University of Calicut, who put a lot of labor in the timely publication of this book. The support rendered by other research scholars in Plant Physiology Division, Department of Botany, University of Calicut is also gratefully acknowledged.

Dr. Jos T. Puthur
Dr. Om Parkash Dhankher

Contents

CHAPTER

1

Bioenergy Plants: A Sustainable Solution for Heavy Metal Phytoremediation

P.P. Sameena[1], Nair G. Sarath[1], Louis Noble[1], M.S. Amritha[1], Om P. Dhankher[2] and Jos T. Puthur[1]*

[1] Plant Physiology and Biochemistry Division, Department of Botany, University of Calicut, Calicut University P.O., Kerala – 673635, India

[2] Stockbridge School of Agriculture, University of Massachusetts, Amherst, U.S.A.

1.1 Introduction

Abiotic stresses such as salinity, temperature, heavy metals and drought are the major problems in the agricultural field, harming crop yield and productivity (Zandalinas et al. 2020). Of the abiotic stressors, heavy metal contamination in arable lands is becoming a serious environmental problem. They are toxic to all forms of life at higher concentrations due to the complexation of these metal ions within the cell components (Hong et al. 2020). Because of rapid industrialization, the atmosphere is contaminated with several heavy metals. They are highly persistent, non-biodegradable, entering into the food chain, and create biological toxicity (Sall et al. 2020). The majority of the aquatic and terrestrial plants can accumulate and bio-concentrate these heavy metals in minute quantities. In contrast, some of the plants show extreme tolerance towards toxic heavy metal ions and have evolved different strategies to cope up with the metals; these groups of plants are termed as hyperaccumulators (Awa and Hadibarata 2020).

The land that is not appropriate for crop production because of low rainfall, poor soil fertility, and toxic pollutants is termed as marginal land (Miyake and Bargiel 2017). These areas are characterized by marginal, economic, and agronomic potential for food production, which may be classified as waste or degraded lands (Pancaldi and Trindade 2020). Degraded lands are the non-productive lands due to unsustainable use, and wastelands are the areas with environmental or physical constraints for farming. These lands are characterized by luxuriant growth of exotic

Corresponding author: jtputhur@yahoo.com

plants, which can grow abundantly without much water and nutrient supply (Blanco-Canqui 2016).

According to Milman (2015), one-third of the agricultural lands were lost in the past forty years because of soil pollution and/or erosion. The FAO reports (2018) showed that about 3 million sites are potentially polluted because of the industrial effluents, agricultural chemicals, petroleum-based products, and fumes generated from the burning of transportation fuels in the European Union. Approximately 10 million sites face soil pollution globally, of which 50% were contributed by heavy metals and metalloids (EPMC 2014). Therefore, it is essential to restore these contaminated lands for the efficient cultivation of food crops.

Various heavy metal remediation techniques have been employed in these contaminated lands, including the physico-chemical and biological remediation mechanisms (Liu et al. 2018). The methods such as chemical precipitation, adsorption, landfilling, flushing and ion exchange are the physico-chemical remediation methods, which result in irreversible changes in the soil characteristics and are also not cost-effective (Akhtar et al. 2020). Phytoremediation is a nature-friendly, cost-effective, economically feasible, and widely accepted technique in which the contaminants present in the water and soil are immobilized in the plant system (Yan et al. 2020). Therefore, cleaning the heavy metal contaminated lands using biomass-based crops can be an effective technique to rectify two broad problems, such as metal contamination and energy crisis (Pulighe et al. 2019, Pancaldi and Trindade 2020).

Bioenergy indicates the energy recovered from organic matter or biomass. Nowadays, fossil fuels contribute mainly towards the global energy demands. Bioenergy is an efficient alternative and can reduce carbon dioxide emissions and play a significant role in replacing petroleum-based fuels (Wu et al. 2018). Thus, the dedicated bioenergy plants, which have improved adaptation to the heavy metal polluted lands, can be effectively used for the coupled phytoremediation and bioenergy production (Sameena and Puthur 2021). In the current literature, various phytoremediation mechanisms functional in the bioenergy plants, production of biofuels from phytoremediation biomass, the quality assessment of biofuels, involvement of microbial system for bioenergy production and phytomining are taken into consideration and extensive analysis has been made to detail the aspects of the coupled role of phytoremediation and bioenergy production.

1.2 Mechanisms of Phytoremediation

Plants with phytoremediation potential utilize different strategies for metal decontamination. It includes (i) phytoextraction, (ii) phytostabilization, (iii) phytovolatilization, (iv) phytodegradation, and (v) rhizodegradation (Fig. 1.1). The mechanisms underlined in each plant vary from species to species. Among these mechanisms, phytoextraction and phytostabilization are two important strategies for successful phytoremediation (Sruthi et al. 2017, Sarath et al. 2020). The plants which are easily grown on the contaminated land and their associated microbial flora overcome the contaminant toxicity by immobilizing, degrading, or sequestering the toxicant in the growth habit and make it less toxic (Ibañez et al. 2016, Sameena and Puthur 2021).

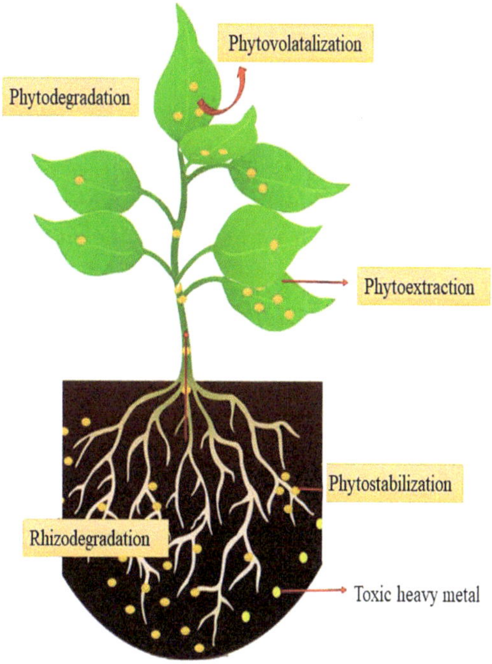

Figure 1.1: Various mechanisms adopted by plants for heavy metal remediation.

1.2.1 Phytostabilization

The plants with phytostabilization potential stabilize the toxic metal or the environmental contaminant in the root or the rhizosphere by reducing their mobility to aerial parts and minimizing the pollutant's toxicity. So, it cannot enter into another food chain and thus reduces the chance of passing it into different organisms (García-Sánchez et al. 2018, Shackira and Puthur 2019). Phytostabilization is an excellent strategy in preventing groundwater contamination because plants immobilize the metal ions in their roots and do not allow them to percolate into the deeper soil. The enzymes produced by the plant roots into the rhizosphere and associated microbial flora actively participate, reduce the toxicant's bioavailability, and convert them to fewer toxic forms (Janeeshma and Puthur 2019, Awa and Hadibarata 2020). This method mainly inactivates or stabilizes the metal in the soil but does not remove the metal from the source (either soil or water). The success of the phytostabilization is dependent on several factors, including the microbial flora of the rhizosphere, soil texture and other soil characteristics, potential of plant root exudates to immobilize the metal, root's cell wall binding capacity, chelation of toxicant in the plant part by using plant-specific metal-chelating proteins and sequestration in the cell vacuoles. These processes occur in plants through precipitation, sorption or complexation (Chaignon et al. 2002, Ghosh and Singh 2005).

Since the absorbed metal is immobilized in the root system and cannot enter into the shoot system, the phytostabilization mechanisms can be effectively utilized for the cultivation of edible bioenergy plants such as *Glycine max* (soybean), *Helianthus annuus* (sunflower), *Saccharum officinarum* (sugarcane), *Sorghum bicolour* (sorghum) and *Zea mays* (maize) (Vermerris 2011, Li et al. 2016, Rojjanateeranaj et al. 2017, Hunce et al. 2019). In maize, after harvesting the seeds, the remaining biomass is used to produce bioethanol, which is most common and widely used for bioenergy (Sigua et al. 2019). These plants show a high level of tolerance towards heavy metals and can be efficiently cultivated in marginal lands. The significant advantage of using edible plants with the phytostabilization potential for bioenergy production is that the remaining biomass after harvesting the edible portion can be effectively used to produce bioenergy.

1.2.2 Phytoextraction

The accumulation of the toxicant in the plant body part is carried out by absorbing through the roots from the contaminated land. The absorbed toxic metal or material cannot alter the normal physiology of plants. The accumulation is mainly in the above-ground vegetative parts such as stem or leaves. These potent accumulators which can withstand a high level of metal contamination are called metallophytes or hyperaccumulators (Chandra et al. 2016). The hyperaccumulators flourish in the polluted land and produce high biomass in harsh conditions. They extend their root system into deeper soil and extract toxicants, translocate into the upper parts, and accumulate in a significantly less toxic form (Lei et al. 2018).

The desirable quality of the plants used for the phytoextraction includes

(i) The plants with high biomass production
(ii) Higher metal accumulation potential
(iii) Fast root establishment
(iv) Easy to grow, handle and process
(v) Capability to tolerate the other harsh soil conditions such as soil pH, salinity, soil structure and water content.

The remediation of the metal-contaminated land is a tedious process. The hyperaccumulator plants are planted on the contaminated site following the nature of the metal present and based on the tolerance potential of the plant. These plants flourish in the areas of contamination and produce high biomass than normal plants. After attaining particular biomass, the plant parts are harvested. The harvested plant parts can either be used to extract the metal or used for other purposes like bioenergy, biogas or biofuels (Cunningham and Ow 1996). Since the non-edible bioenergy plants such as *Arundo donax*, *Jatropha curcas*, *Miscanthus species*, *Ricinus communis* and *Salix viminalis* have an effective phytoremediation mechanism operational in them, they can be very well utilized for phytoremediation purpose. Martín et al. (2020) revealed that *J. curcas* accumulates a higher amount of toxic As and Fe in their body within 90 days of growth in the metal-containing medium. The translocation factors showed a value of more than 1 in all the treatments. The plant's capacity to restore the soil to normal and further utilization of the harvested plant for bioenergy production plays a vital role in the sustainable management of pollutants.

1.3 Demand for Bioenergy: Current Scenario

Due to the increase in world population, increased consumption level, increase in natural resources' utilization in industries, urbanization, etc., land pollution is increasing day by day. Increased consumption causes the depletion of naturally available fuel as well (Kharas 2010). Among them, petroleum products are heavily utilized for transportation and as a machinery fuel source because of their high heat emitting property, accessibility and quality burning attributes; however, its natural source is exhausting day by day (Ghobadian et al. 2009). Over-consumption of petrol-based fuel and ecological concern has led to investigating an alternative source (Hassan and Kalam 2013). In the transportation sector, as an alternative to petroleum fuel, biofuels can be used, reducing the environmental concerns caused by petroleum fuels. In the current scenario of increased environmental pollution, biofuel reduces greenhouse gas emissions, which are typically produced and ejected into the atmosphere due to petroleum fuel combustion (Hassan and Kalam 2013).

Another concern that arises due to land pollution can also be rectified by using energy plants for phytoremediation. In contaminated soils with various impurities, energy plants that are fast-growing, deep-rooted, and capable of producing high biomass can be used for phytoremediation. These energy plants can be used as a good source of biofuel (Sameena and Puthur 2021). Thus, a combination of phyto-management and biofuel production becomes a relevant necessity of the modern era. The promising bioenergy plants utilized to recover various heavy metal contaminated lands in different parts of the world are summarized in Table 1.1.

1.4 Production Techniques behind Bioenergy Generation from Plant Biomass Utilized for Phytoremediation

Biofuels have advanced from the first to fourth era and they are primarily classified based on the feedstock and production technologies (Kaur et al. 2019). The two most usually used biofuels are biodiesel and bioethanol, which are obtained predominantly from vegetable oils, seeds and lignocellulosic biomass. Biodiesel can be used to substitute diesel, and bioethanol can substitute petroleum-based fuel. Typical biodiesel feedstock comes from plant oils like rapeseed, soybean, sunflower, palm and some other non-consumable oils like mahua, neem, karanja, and jatropha (Nordborg et al. 2014). The production of biofuels from microalgae, cellulolytic bacteria and other microorganisms with fast growth rate and enhanced CO_2 fixation capability constitutes the third era of biofuel (Kaur et al. 2019).

Biofuel production techniques can be classified into four:

(i) First-generation biofuel technology for the production of biodiesel and bioethanol (Leung et al. 2010, Dias et al. 2013),
(ii) Technology of second-generation biofuels by converting lignocellulosic biomass (Humbird et al. 2011),
(iii) Technology of third-generation biofuels by algae processing (Leite et al. 2013), and

Table 1.1: The promising bioenergy plants utilized for the recovery of various heavy metals from contaminated lands in different parts of the world

Sl. No.	Bioenergy plants	Type of marginal land and location	Heavy metals/ metalloids	Soil amendments used	Biomass/energy yield	References
1	*Arundo donax* (Giant reed)	Caffaro Area, Landriano, Italy	Cd, Cu, Hg, Ni, Pb and Zn	None	3.23±1.33 tons dry weight ha^{-1}	Danelli et al. (2021)
2	*Helianthus annuus* (Sun flower)	Trace element contaminated mining soil, Murcia, Spain	As, Cd, Cu, Pb and Zn	Solid fraction of pig slurry and paper mill sludge	134-154 ml g^{-1} biogas	Hunce et al. (2019)
3	*Jatropha curcas* (Physic nut)	Santa Maria mine site, Zimapan, Hidalgo, Mexico	Cd, Cu, Pb and Zn	Biochar and a mycorrhizal fungus	2 g $plant^{-1}$	Gonzalez-Chavez et al. (2017)
4	*Populus* spp. (Poplar)	Industrial zone of Boom near Antwerp, Belgium	Al, As, Cu, Cd, Co, Fe, Mn, Ni, Pb and Zn	None	18.1 mg ha^{-1}	Laureysens et al. (2005)
5	*Ricinus communis* (Castor)	DDT contaminated cotton plantations, Cixi county of Zhejiang Province, China	Cd	None	8.3 g pot^{-1}	Huang et al. (2011)
6	*Salix alba* (Willow)	Lowland part of Poland	Cd, Cu, Hg, Pb and Zn	None	6.81 kg yr^{-1} $shrub^{-1}$	Mleczek et al. (2010)
7	*Silybum marianum* (Milk thistle)	Trace element contaminated mining soil, Murcia, Spain	As, Cd, Cu, Pb and Zn	Solid fraction of pig slurry and paper mill sludge	194-223 ml g^{-1} biogas	Hunce et al. (2019)
8	*Vertiveria zizanioides* (Vetiver)	Pb/Zn mine area, Lechang, Guangdong Province, China	Cd, Pb and Zn	N:P:K (1:1:1) fertilizer and EDTA	30 t ha^{-1}	Zhuang et al. (2007)
9	*Zea mays* (Maize)	Land near smelting industry, Flanders, Belgium	Cd, Pb and Zn	None	33,000–46,000 kWh $ha^{-1}yr^{-1}$	Meers et al. (2010)

(iv) Fourth-generation biofuels technology by involving the genetically modified algal processing (Ale et al. 2019).

The production technologies used for the first-generation biofuel production are the biodiesel generation from trans- esterification process and bioethanol generation by fermentation process. Compared to this, processing of lignocellulosic biomass is more complicated, and it involves thermo-chemical and biological conversion processes, which leads to the production of second-generation biofuel. Generation of biofuel from algae comprises oil extraction as well as trans-esterification process, which represents the third-generation biofuels. Biofuel production at fourth generation is similar to that of third generation; however, it is an advanced technique which involves the conversion of genetically engineered plant biomass into biofuel *viz.* genetically modified microbes such as algae.

1.5 Involvement of Microbial System in Bioenergy Production from Biomass Utilized for Phytoremediation

Cereal straw, sugarcane and agriculture residues are considered ligno-cellulosic biomass and are made up of lignin, pectin cellulose and hemicellulose. It is widely used as the raw material for the generation of biofuel (Bhatia et al. 2016). Microbes are well efficient to convert this ligno-cellulosic biomass and ferment these into lipids, biogas and alcohols. The use of microbes in converting biomass to biofuel is considered one of the sustainable, environmentally friendly and economic approaches for the generation of biofuel (Bhatia et al. 2017a). The ligno-cellulosic biomass can be used for the generation of different types of biofuels using microbial fermentation. Butanol, isobutanol, lipid, ethanol, etc., are the biofuels produced as the results of bacterial fermentation. *Clostridium acetobutylicum* generates butanol by fermenting beech wood, which is well known for its accumulation of heavy metals from contaminated soil; similarly, corn stover used for phytoremediation is converted to lipid and isobutanol by the bacterial fermentation of *Cryptococcus* and *Clostridium,* respectively (Gong et al. 2014, Tippkotter et al. 2014, Evangelou et al. 2015, Wei et al. 2015). Many more bacterial strains are known for their fermentation property in converting phyto-remediating biomass into biofuel and are listed in Table 1.2.

Rhodococcus spp., *Rhodosporidium* spp., and *Cryptococcus* spp. are the microbes that have been extensively used for biodiesel production. Similarly, *Candida* spp., *Saccharomyces* spp. and *Kluyveromyces* spp. help in the production of bio-alcohol and *Clostridium* spp. as well as *Thermoanaerobacterium* spp. help in biogas production (Oberoi et al. 2010, Fei et al. 2016, Jiang et al. 2016, Bhatia et al. 2017b) (Fig. 1.2). For the production of biodiesel and lipid, bacterial strains such as *Rhodococcus* spp. and *Yarrowia lipolytica* were reported to utilize sugarcane bagasse as the carbon source for the generation of lipids. Also, engineered strains of *E. coli* are well known for the production of biodiesel. Engineered *E. coli* strain was able to produce FAME (fatty acid methyl esters) by utilizing *Panicum virgatum* (switchgrass)

Table 1.2: Examples for various kinds of biofuels generated by the microbial fermentation of ligno-cellulosic materials

Sl. no.	Raw materials	Strains	Products	References
1	Beech wood	*Clostridium acetobutylicum*	Butanol	Tippkötter et al. (2014)
2	Cellobiose	*Geobacillus thermoglucosidasius*	Isobutanol	Lin et al. (2014)
3	Corn stalk	*Thermoanaerobacterium thermosaccharolyticum* strain DD32	Hydrogen	Sheng et al. (2015)
4	Corn stover	*Cryptococcus curvatus* ATCC 20509	Lipid	Gong et al. (2014)
5	Corn stover	*Clostridium cellulolyticum*	Isobutanol	Li et al. (2014)
6	Fruit pulp	*Rhodosporidium kratochvilovae* HIMPA1	Lipid	Patel et al. (2015)
7	Jatropha hull	*Clostridium butyricum*	Hydrogen	Jiang et al. (2016)
8	Miscanthus	*Saccharomyces cerevisiae* CHY1011	Ethanol	Kang et al. (2015)
9	Oil palm	*Rhodococcus* sp. YHY01	FAEE	Bhatia et al. (2017b)
10	Oil palm trunk sap	*Clostridium acetobutylicum* DSM 1731	Butanol	Gottumukkala et al. (2013)
11	Pine	*Rhodococcusopacus* DSM 1069	Lipid	Wei et al. (2015)
12	Rice straw	*Candida tropicalis* ATCC 13803	Ethanol	Oberoi et al. (2010)
13	Rice straw	*Clostridium sporogenes* BE01	Butanol	Komonkiat and Cheirsilp (2013)
14	Sugarcane bagasse	*Trichosporon fermentans*	Lipid	Tsigie et al. (2011)
15	Sugarcane bagasse	*Candida shehatae* NCIM 3501	Ethanol	Chandel et al. (2007)
16	Sugarcane bagasse	*Clostridium butyricum*	Hydrogen	Pattra et al. (2008)
17	Sweetgum	*Rhodococcusopacus* PD630	Lipid	Wei et al. (2015)
18	Wheat bran	*Paecilomyces variotii*	Ethanol	Zerva et al. (2014)
19	Wheat straw	*Cryptococcus curvatus*	Lipid	Yu et al. (2011)
20	Yellow poplar	*Enterobacter aerogenes*	2,3-Butanediol	Joo et al. (2016)

as a carbon source (Bokinsky et al. 2011, Tsigie et al. 2011). Successfully engineered strains of *R. opacus* with over-expression of endoxylanases *xylA* and *xylB* were also able to generate lipid from corn stover (Kurosawa et al. 2014). Ethanol, propanol, butanol, isobutanol, etc., can be synthesized by bacterial fermentation.

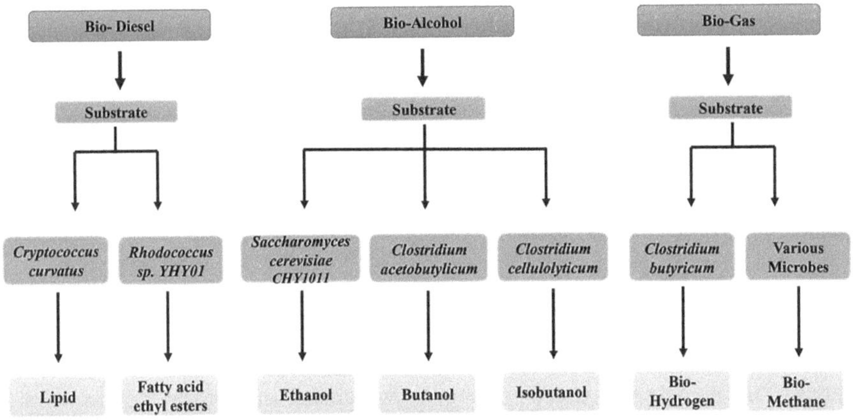

Figure 1.2: Microbes involved in biofuels production and the generation of various byproducts.

Genetically modified *Saccharomyces cerevisiae* having cell surface expression is well known for producing ethanol from rice straw. Co-culture of *Saccharomyces* and *Clostridium* strains are also reported to produce ethanol from cellulose (Zuroff et al. 2013). Butanol is another biofuel that is used as a substitute for gasoline. Different microbes and feedstocks are involved in the production of butanol. *Clostridium* spp. and *E. coli* are reported to generate butanol from oil palm trunk and switchgrass simultaneously (Bokinsky et al. 2011, Komonkiat and Cheirsilp 2013). *Geobacillus thermoglucosidasius* is used to produce isobutanol by utilizing cellobiose as a carbon source, which is not as volatile as that of ethanol (Lin et al. 2014).

Along with these biofuels, biogas such as biohydrogen and biomethane hold their position of being a renewable energy source and as a substitute for fossil fuel (Amigun et al. 2008). Fermentation of sugarcane bagasse by *Clostridium butyricum* generates high quality biohydrogen. As *Clostridium* strains face difficulty in producing biohydrogen at elevated temperatures, *Thermoanaerobacterium thermosaccharolyticum,* a thermophilic strain, is generally used to conduct the process of fermentation at high temperatures (Islam et al. 2006). Anaerobic fermentation of organic materials leads to the production of biomethane. It is a highly complex and complicated process which involves various steps such as by the breakdown of polymeric materials into methane and carbon dioxide. It further involves different group of microorganisms to complete rest of the biochemical process such as hydrolysis, acetogenesis, acidogenesis, and methanation (Chandra et al. 2012).

Because of the recalcitrant nature of lignocellulosic materials, microbes are unable to use them. Therefore, various pretreatment methods such as physical, chemical and biological methods are used to digest the same and release free sugars (Kumar and Sharma 2017). In addition to sugar, pretreatment prompts the creation of other toxic byproducts such as organic acid, furanaldehyde and phenolic compounds, which have the potential to affect microbial growth and fermentation (Allen et al. 2010). This drawback of using microbes in biofuel generation can be rectified using detoxification techniques. In the current era, the ongoing advances in molecular level studies and synthetic biology help in creating microbial biosynthetic pathways to generate biofuel raw materials such as alcohols, esters, alkanes, and alkenes from the isoprenoid and unsaturated fatty acid biosynthetic pathways. Similarly, industrial fermentation can be promptly applied to the microbial-based generation of biofuels. Advancement in science allows producing biofuels with the help of various microorganisms in bioreactors (Peralta-Yahya and Keasling 2010).

1.6 Recovery of Heavy Metals from the Biomass: Phytomining

The biomass harvested after successful phytoremediation is a rich source of heavy metals and, at the same time, potentially hazardous. Therefore, the recovery of the metals present in the biomass turns to be vital concerning phytoremediation (Mohanty 2016). The biomass processing should be carried out with utmost care to prevent metal re-entry to the environment and the food chain (Kidd et al. 2018). Such recovered heavy metals can be brought out for some economic benefits by effectively using these metals, which will be added advantage over bioenergy production (Jiang et al. 2015).

After harvesting the plant used for phytoremediation, the first step is to reduce the volume and weight of harvested biomass and maximize the density for easier handling, safer disposal and increasing the energetic value (Jasinskas et al. 2020). There are several crop disposal methods such as ashing, compaction, composting, incineration, pyrolysis, etc., of which incineration or gasification is the most practically viable (Mukherjee et al. 2020). This process can reduce the biomass weight and leaching of heavy metals as compared to other means of handling the biomass. By preparing the biomass into pellets, the volume of the material can be reduced to 10 times (Jasinskas et al. 2020). After incineration, heavy metals remain at the bottom ash called 'bio-ore', which can finally be extracted by phytomining (Chaney et al. 2018).

Phytomining is an eco-friendly method for the recovery of the soluble metals uptaken by the plant biomass. This process can generate revenue, depending upon the metal present in the 'bio-ore' (Patra and Mohanty 2013). The recovery methods are complicated and include hydrometallurgical processes, ion exchange, flotation, magnetic field, electrolysis, bio-electrochemical procedures, etc. (Elekes 2014). Phytomining can even be a substitute for environmentally destructive mining practices if undertaken at a large scale by growing high biomass yielding plants requiring less favorable environmental conditions such as reduced nutrient content, less water utilization, etc.

1.7 Strengths and Weakness of Bioenergy Production on Marginal Lands

The coupled approach of phytoremediation and subsequent bioenergy production from the phytoremediation biomass would be a significant advancement concerning sustainability to optimize the environmental and socio-economic issues (Bauddh et al. 2017). This integrated phytoremediation and bioenergy production will also help to restore the degraded lands and polluted water, reduce the greenhouse gas emissions and associated global climate change (Singh et al. 2019). The sustainable production of biofuels promotes efficiency in land management practices, which optimize land, water, and solar energy, thereby guarantee food and fuel security, maximize carbon storage, and reduce greenhouse gas emissions (Acharya and Perez-Pena 2020). Even though bioenergy dramatically contributes to maintaining energy security, the production of biomass feedstocks from marginal lands may lead to the presence of metals in the final energy product (Wu et al. 2018). Therefore, proper management of the feedstock is essential to prevent the presence of metal in energy products.

1.8 Conclusion

Bioenergy can be regarded as an efficient and sustainable solution for the ever-increasing demands for the energy supply. The depletion of arable lands due to the mismanaged farming methods and the energy crisis can be overcome by the sustainable interdisciplinary practices in phyto-technology for large-scale bioenergy production. Recent advancement in science helps in the large-scale production of biofuels by using bioreactors with various microorganisms. Inadequacies in the current phyto-management technologies need to be evaluated to increase the effectiveness of biofuel systems, and market-based frameworks need to be designed to enhance the sustainable production of bioenergy.

Acknowledgments

PPS, LN and MSA gratefully acknowledge the financial assistance from the University Grants Commission (UGC) and Council of Scientific and Industrial Research (CSIR), India, through JRF fellowships. NGS and JTP acknowledge the financial assistance provided by the Kerala State Council for Science, Technology and Environment in the form of KSCSTE Research Grant (KSCSTE/5179/2017-SRSLS). The authors extend their sincere thanks to the Department of Science and Technology (DST), Government of India, for granting funds under Fund for Improvement of S&T Infrastructure (FIST) programme (SR/FST/LSI-532/2012).

References

Acharya, R.N. and R. Perez-Pena. 2020. Role of comparative advantage in biofuel policy adoption in Latin America. Sustainability 12: 1411.

Akhtar, F.Z., K.M. Archana, V.G. Krishnaswamy and R. Rajagopal. 2020. Remediation of heavy metals (Cr, Zn) using physical, chemical and biological methods: A novel approach. SN Appl. Sci. 2: 267.

Ale, S., P.V. Femeena, S. Mehan and R. Cibin. (2019). Environmental impacts of bioenergy crop production and benefits of multifunctional bioenergy systems. pp. 195-217. *In*: Bioenergy with Carbon Capture and Storage. Academic Press.

Allen, S.A., W. Clark, J.M. McCaffery, Z. Cai, A. Lanctot, P.J. Slininger, Z.L. Liu and S.W. Gorsich. 2010. Furfural induces reactive oxygen species accumulation and cellular damage in *Saccharomyces cerevisiae*. Biotechnol. Biofuel. 3: 1-10.

Amigun, B., R. Sigamoney and H. von Blottnitz. 2008. Commercialisation of biofuel industry in Africa: a review. Renew. Sust. Energ. Rev. 12: 690-711.

Awa, S.H. and T. Hadibarata. 2020. Removal of heavy metals in contaminated soil by phytoremediation mechanism: A review. Water Air Soil Pollut. 231: 1-15.

Bauddh, K., S.B. Bhaskar and J. Korstad. 2017. Phytoremediation Potential of Bioenergy Plants. Springer, Singapore. doi:10.1007/978-981-10-3084-0.

Bhatia, S.K., B.R. Lee, G. Sathiyanarayanan, H.S. Song, J. Kim, J.M. Jeon, J.J. Yoon, J. Ahn, K. Park and Y.H. Yang. 2016. Biomass-derived molecules modulate the behavior of *Streptomyces coelicolor* for antibiotic production. 3 Biotech. 6: 1-8.

Bhatia, S.K., S.H. Kim, J.J. Yoon and Y.H Yang. 2017a. Current status and strategies for second generation biofuel production using microbial systems. Energ. Convers. Manage. 148: 1142-1156.

Bhatia, S.K., J. Kim, H.S. Song, H.J. Kim, J.M. Jeon and G. Sathiyanarayanan. 2017b. Microbial biodiesel production from oil palm biomass hydrolysate using marine *Rhodococcus* sp. YHY01. Biortech. 233: 99-109.

Blanco-Canqui, H. 2016. Growing dedicated energy crops on marginal lands and ecosystem services. Soil Sci. Soc. Am. J. 80: 845-858.

Bokinsky, G., P.P. Peralta-Yahya, A. George, B.M. Holmes, E.J. Steen, J. Dietrich, T.S. Lee, D. Tullman-Ercek, C.A. Voigt, B.A. Simmons and J.D. Keasling. 2011. Synthesis of three advanced biofuels from ionic liquid-pretreated switchgrass using engineered *Escherichia coli*. Proc. Natl. Acad. Sci. 108: 19949-19954.

Chaignon, V., D. Di Malta and P. Hinsinger. 2002. Fe-deficiency increases Cu acquisition by wheat cropped in a Cu-contaminated vineyard soil. New Phytol. 154: 121-130.

Chandel, A.K., R.K. Kapoor, A. Singh and R.C. Kuhad. 2007. Detoxification of sugarcane bagasse hydrolysate improves ethanol production by *Candida shehatae* NCIM 3501. Bioresour. Technol. 98: 1947-1950.

Chandra, R., H. Takeuchi and T. Hasegawa. 2012. Methane production from lignocellulosic agricultural crop wastes: A review in context to second generation of biofuel production. Renew. Sust. Energ. Rev. 16: 1462-1476.

Chandra, R., W. Cho and H. Kang. 2016. Phytoextraction potential of four poplar hybrids under greenhouse conditions. Forest Sci. Technol. 12: 199-206.

Chaney, R.L., A.J.M. Baker and J.L. Morel. 2018. The long road to developing agromining/phytomining. pp. 1-22. *In*: A. Van der Ent, G. Echevarria, A.J.M. Baker, J.L. Morel [eds.]. Agromining: Farming for Metals. Springer, Cham.

Cunningham, S.D. and D.W. Ow. 1996. Promises and prospects of phytoremediation. Plant Physiol. 110: 715-719.

Danelli, T., A. Sepulcri, G. Masetti, F. Colombo, S. Sangiorgio, E. Cassani, S. Anelli, F. Adani and R. Pilu. 2021. *Arundo donax* L. biomass production in a polluted area: Effects of two harvest timings on heavy metals uptake. Appl. Sci. 11: 1147.

Dias, M.O., T.L. Junqueira, C.E.V. Rossell, R. Maciel Filho and A. Bonomi. 2013. Evaluation of process configurations for second generation integrated with first generation bioethanol production from sugarcane. Fuel Process. Technol. 109: 84-89.

Elekes, C.C. 2014. Eco-technological solutions for the remediation of polluted soil and heavy metal recovery. pp. 309-335. *In*: M.C.H. Soriano [ed.]. Environmental Risk Assessment of Soil Contamination. InTech, Rijeka, Croatia.

Environmental Protection Ministry of China (EPMC). 2014. National survey report of soil contamination status of China. Environmental Protection Ministry of China, Beijing, China.

Evangelou, M.W., E.G. Papazoglou, B.H. Robinson and R. Schulin. (2015). Phytomanagement: Phytoremediation and the production of biomass for economic revenue on contaminated land. pp. 115-132. *In*: A.A. Ansari, S.S. Gill, R. Lanza and G.R. Newman [eds.]. Phytoremediation. Springer, Cham.

Fei, Q., M. O'Brien, R. Nelson, X. Chen, A. Lowell and N. Dowe. 2016. Enhanced lipid production by *Rhodosporidium toruloides* using different fed-batch feeding strategies with lignocellulosic hydrolysate as the sole carbon source. Biotechnol. Biofuels 9: 1-12.

Food and Agricultural Organization of United States (FAO). 2018. The state of the food security and nutrition in the world, Building climate resilience for food security and nutrition.

García-Sánchez, M., Z. Košnář, F. Mercl, E. Aranda and P. Tlustoš. 2018. A comparative study to evaluate natural attenuation, mycoaugmentation, phytoremediation, and microbial-assisted phytoremediation strategies for the bioremediation of an aged PAH-polluted soil. Ecotoxicol. Environ. Saf. 147: 165-174.

Ghobadian, B., T. Yusaf, G. Najafi and M. Khatamifar. 2009. Diesterol: An environment-friendly IC engine fuel. Renew. Energy 34: 335-342.

Ghosh, M. and S.P. Singh. 2005. A review on phytoremediation of heavy metals and utilization of its by products. Asian J. Energy Environ. 6: 1-18.

Gong, Z., H. Shen, X. Yang, Q. Wang, H. Xie and Z.K. Zhao. 2014. Lipid production from corn stover by the oleaginous yeast *Cryptococcus curvatus*. Biotechnol. Biofuel. 7: 1-9.

González-Chávez, M.D.C.A., R. Carrillo-González, M.I. Hernandez Godinez and S. Evangelista Lozano (2017). Jatropha curcas and assisted phytoremediation of a mine tailing with biochar and a mycorrhizal fungus. Int. J. Phytoremediation 19: 174-182.

Gottumukkala, L.D., B. Parameswaran, S.K. Valappil, K. Mathiyazhakan, A. Pandey and R.K. Sukumaran. 2013. Biobutanol production from rice straw by a non-acetone producing *Clostridium sporogenes* BE01. Bioresour. Technol. 145: 182-187.

Hassan, M.H. and M.A. Kalam. 2013. An overview of biofuel as a renewable energy source: Development and challenges. Procedia Eng. 56: 39-53.

Hong, Y.J., W. Liao, Z.F. Yan, Y.C Bai, C.L. Feng, Z.X. Xu and D.Y. Xu. 2020. Progress in the research of the toxicity effect mechanisms of heavy metals on freshwater organisms and their water quality criteria in China. J. Chem. 2020: 9010348

Huang, H., N. Yu, L. Wang, D.K. Gupta, Z. He, K. Wang, Z. Zhu, X. Yan, T. Li and X. Yang. 2011. The phytoremediation potential of bioenergy crop *Ricinus communis* for DDTs and cadmium co-contaminated soil. Bioresour. Technol. 102: 11034-11038.

Humbird, D., R. Davis, L. Tao, C. Kinchin, D. Hsu, A. Aden, P. Schoen, J. Lukas, B. Olthof, M. Worley, D. Sexton and D. Dudgeon. 2011. Process design and economics for biochemical conversion of lignocellulosic biomass to ethanol: Dilute-acid pretreatment and enzymatic hydrolysis of corn stover (No. NREL/TP-5100-47764). National Renewable Energy Laboratory (NREL), U.S. Department of Energy, Golden, Colorado, United States.

Hunce, S.Y., R. Clemente and M.P. Bernal. 2019. Energy production potential of phytoremediation plant biomass: *Helianthus annuus* and *Silybum marianum*. Ind. Crop. Prod. 135: 206-216.

Ibañez, S., M. Talano, O. Ontañon, J. Suman, M.I. Medina, T. Macek and E. Agostini. 2016. Transgenic plants and hairy roots: Exploiting the potential of plant species to remediate contaminants. New Biotechnol. 33: 625-635.

Islam, R., N. Cicek, R. Sparling and D. Levin. 2006. Effect of substrate loading on hydrogen production during anaerobic fermentation by *Clostridium thermocellum* 27405. Appl. Microbiol. Biotechnol. 72: 576-583.

Janeeshma, E and J.T. Puthur. 2019. Direct and indirect influence of arbuscular mycorrhizae on enhancing metal tolerance of plants. Arch. Microbiol. 202: 1-16.

Jasinskas, A., D. Streikus and T. Vonžodas. 2020. Fibrous hemp (Felina 32, USO 31, Finola) and fibrous nettle processing and usage of pressed biofuel for energy purposes. Renew. Energy 149: 11-21.

Jiang, D., Z. Fang, S.X. Chin, X.F. Tian and T.C. Su. 2016. Biohydrogen production from hydrolysates of selected tropical biomass wastes with *Clostridium butyricum*. Sci. Report 6: 1-11.

Jiang, Y., M. Lei, L. Duan and P. Longhurst. 2015. Integrating phytoremediation with biomass valorisation and critical element recovery: A UK contaminated land perspective. Biomass Bioenerg. 83: 328-339.

Joo, J., S.J. Lee, H.Y. Yoo, Y. Kim, M. Jang, J. Lee, S.O. Han, S.W. Kim and C. Park. 2016. Improved fermentation of lignocellulosic hydrolysates to 2,3-butanediol through investigation of effects of inhibitory compounds by *Enterobacter aerogenes*. Chem. Eng. J. 306: 916-924.

Kang, K.E., D.P. Chung, Y. Kim, B.W. Chung and G.W Choi. 2015. High-titer ethanol production from simultaneous saccharification and fermentation using a continuous feeding system. Fuel 145: 18-24.

Kaur, M., M. Kumar, D. Singh, S. Sachdeva and S.K. Puri. 2019. A sustainable biorefinery approach for efficient conversion of aquatic weeds into bioethanol and biomethane. Energy Convers. Manag. 187: 133-147.

Kharas, H. 2010. The emerging middle class in developing countries. OECD Development Centre Working Papers, 285, OECD Publishing, Paris.

Kidd, P.S., A. Bani, E. Benizri, C. Gonnelli, C. Hazotte, J. Kisser, M. Konstantinou, T. Kuppens, D. Kyrkas, B. Laubie, R. Malina, J.L. Morel, H. Olcay, T. Pardo, M.N. Pons, Á. Prieto-Fernández, M. Puschenreiter, C. Quintela-Sabarís, C. Ridard, B. Rodríguez-Garrido, T. Rosenkranz, P. Rozpądek, R. Saad, F. Selvi, M.O. Simonnot, A. Tognacchini, K.Turnau, R. Ważny, N. Witters and G. Echevarria. 2018. Developing sustainable agromining systems in agricultural ultramafic soils for nickel recovery. Front. Environ. Sci. 6. doi:10.3389/fenvs.2018.00044.

Komonkiat, I. and B. Cheirsilp. 2013. Felled oil palm trunk as a renewable source for biobutanol production by *Clostridium* spp. Bioresour. Technol. 146: 200-207.

Kumar, A.K. and S. Sharma. 2017. Recent updates on different methods of pretreatment of lignocellulosic feedstocks: A review. Bioresour. Bioprocess. 4: 1-19.

Kurosawa, K., S.J. Wewetzer and A.J. Sinskey. 2014. Triacylglycerol production from corn stover using a xylose-fermenting *Rhodococcus opacus* strain for lignocellulosic biofuels. J. Microb. Biochem. Technol. 6: 254-259.

Laureysens, I., L. De Temmerman, T. Hastir, M. Van Gysel and R. Ceulemans. 2005. Clonal variation in heavy metal accumulation and biomass production in a poplar coppice culture. II: Vertical distribution and phytoextraction potential. Environ. Pollut. 133: 541-551.

Lei, M., X. Wan, G. Guo, J. Yang and T. Chen. 2018. Phytoextraction of arsenic-contaminated soil with *Pteris vittata* in Henan Province, China: Comprehensive evaluation of remediation efficiency correcting for atmospheric depositions. Environ. Sci. Pollut. Res. 25: 124-131.

Leite, G.B., A.E. Abdelaziz and P.C. Hallenbeck. 2013. Algal biofuels: Challenges and opportunities. Bioresour. Technol. 145: 134-141.

Leung, D.Y., X. Wu and M.K.H. Leung. 2010. A review on biodiesel production using catalyzed transesterification. Appl. Energy 87: 1083-1095.

Li, Y., Q. Wang, L. Wang, L.Y. He and X.F. Sheng. 2016. Increased growth and root Cu accumulation of *Sorghum sudanense* by endophytic *Enterobacter* sp. K3-2: Implications for *Sorghum sudanense* biomass production and phytostabilization. Ecotoxicol. Environ. Saf. 124: 163-168.

Li, Y., T. Xu, T.J. Tschaplinski, N.L. Engle, Y. Yang, D.E. Graham, Z. He and J. Zhou. 2014. Improvement of cellulose catabolism in *Clostridium cellulolyticum* by sporulation abolishment and carbon alleviation. Biotechnol. Biofuels 7: 1-13.

Lin, P.P., K.S. Rabe, J.L. Takasumi, M. Kadisch, F.H. Arnold and J.C. Liao. 2014. Isobutanol production at elevated temperatures in thermophilic *Geobacillus thermoglucosidasius*. Metab. Eng. 24: 1-8.

Liu, L., W. Li, W. Song and M. Guo. 2018. Remediation techniques for heavy metal-contaminated soils: Principles and applicability. Sci. Total Environ. 633: 206-219.

Martín, J.F.G., M.D.C. González Caro, M.D.C. López Barrera, M. Torres García, D. Barbin and P. Álvarez Mateos. 2020. Metal accumulation by *Jatropha curcas* L. adult plants grown on heavy metal-contaminated soil. Plants 9: 418.

Meers, E., S. Van Slycken, K. Adriaensen, A. Ruttens, J. Vangronsveld, G.D. Laing, N. Witters, T. Thewys and F.M.G. Tack. 2010. The use of bio-energy crops (*Zea mays*) for 'phytoattenuation' of heavy metals on moderately contaminated soils: A field experiment. Chemosphere 78: 35-41.

Milman, O. 2015. Earth has lost a third of arable land in past 40 years, scientists say. The Guardian 2: 12.

Miyake, S. and D. Bargiel. 2017. 'Underutilised' agricultural land: Its definitions, potential use for future biomass production and its environmental implications. pp. 626. *In:* EGU General Assembly, EGU2017, Proceedings from the Conference, Vienna, Austria.

Mleczek, M., P. Rutkowski, I. Rissmann, Z. Kaczmarek, P. Golinski, K. Szentner, K. Strażyńska and A. Stachowiak. 2010. Biomass productivity and phytoremediation potential of *Salix alba* and *Salix viminalis*. Biomass Bioenerg. 34: 1410-1418.

Mohanty, M. 2016. Post-harvest management of phytoremediation technology. J. Environ. Anal. Toxicol. 6: 398.

Mukherjee, C., J. Denney, E.G. Mbonimpa, J. Slagley and R. Bhowmik. 2020. A review on municipal solid waste-to-energy trends in the USA. Renew. Sustain. Energ. Rev. 119: 109512.

Nordborg, M., C. Cederberg and G. Berndes. 2014. Modeling potential freshwater ecotoxicity impacts due to pesticide use in biofuel feedstock production: The cases of maize, rapeseed, salix, soybean, sugar cane, and wheat. Environ. Sci. Technol. 48: 11379-11388.

Oberoi, H.S., P.V. Vadlani, K. Brijwani, V.K. Bhargav and R.T. Patil. 2010. Enhanced ethanol production via fermentation of rice straw with hydrolysate-adapted *Candida tropicalis* ATCC 13803. Process Biochem. 45: 1299-1306.

Pancaldi, F. and L.M. Trindade. 2020. Marginal lands to grow novel bio-based crops: A plant breeding perspective. Front. Plant Sci. 11: 227.

Patel, A., D.K. Sindhu, N. Arora, R.P. Singh, V. Pruthi and P.A. Pruthi. 2015. Biodiesel production from non-edible lignocellulosic biomass of *Cassia fistula* L. fruit pulp using oleaginous yeast *Rhodosporidium kratochvilovae* HIMPA1. Bioresour. Technol. 197: 91- 98.

Patra, H.K. and M. Mohanty. 2013. Phytomining: An innovative post phytoremediation management technology. The Ecoscan 3: 15-20.

Pattra, S., S. Sangyoka, M. Boonmee and A. Reungsang. 2008. Bio-hydrogen production from the fermentation of sugarcane bagasse hydrolysate by *Clostridium butyricum*. Int. J. Hydrog. Energy 33: 5256-5265.

Peralta-Yahya, P.P. and J.D. Keasling. 2010. Advanced biofuel production in microbes. Biotechnol. J. 5: 147-162.

Pulighe, G., G. Bonati, M. Colangeli, M.M. Morese, L. Traverso, F. Lupia, C. Khawaja, R. Janssen and F. Fava. 2019. Ongoing and emerging issues for sustainable bioenergy production on marginal lands in the Mediterranean regions. Renew. Sustain Energy Rev. 103: 58-70.

Rojjanateeranaj, P., C. Sangthong and B. Prapagdee. 2017. Enhanced cadmium phytoremediation of *Glycine max* L. through bioaugmentation of cadmium-resistant bacteria assisted by biostimulation. Chemosphere 185: 764-771.

Sall, M.L., A.K.D. Diaw, D. Gningue-Sall, S.E. Aaron, and J.J. Aaron. 2020. Toxic heavy metals: Impact on the environment and human health, and treatment with conducting organic polymers – A review. Environ. Sci. Pollut. Res. 27: 29927-29942.

Sameena, P.P. and J.T. Puthur. 2021. Heavy metal phytoremediation by bioenergy plants and associated tolerance mechanisms. Soil Sediment Contam. Int. J. 30: 253-274.

Sarath, N.G., P. Sruthi, A.M. Shackira and J.T. Puthur. 2020. Heavy metal remediation in wetlands: Mangroves as potential candidates. pp. 1-27. *In*: M.N. Grigore [ed.]. Handbook of Halophytes: From Molecules to Ecosystems towards Biosaline Agriculture. Springer International Publishing, Switzerland.

Shackira A.M. and J.T. Puthur. 2019. Phytostabilization of heavy metals: Understanding of principles and practices. pp. 263-282. *In*: S. Srivastava, A. Srivastava and P. Suprasanna [eds.]. Plant-Metal Interactions. Springer Nature, Switzerland.

Sheng, T., L. Gao, L. Zhao, W. Liu and A. Wang. 2015. Direct hydrogen production from lignocellulose by the newly isolated *Thermoanaerobacterium thermosaccharolyticum* strain DD32. RSC Advanc. 5: 99781-99788.

Sigua, G.C., J.M. Novak, D.W. Watts, J.A. Ippolito, T.F. Ducey, M.G. Johnson and K.A. Spokas. 2019. Phytostabilization of Zn and Cd in mine soil using corn in combination with biochars and manure-based compost. Environ. 6: 69.

Singh, R., A.B. Jha, A.N. Misra and P. Sharma. 2019. Differential responses of growth, photosynthesis, oxidative stress, metals accumulation and *NRAMP* genes in contrasting *Ricinus communis* genotypes under arsenic stress. Environ. Sci. Pollut. Res. 26: 31166-31177.

Sruthi, P., A.M. Shackira and J.T. Puthur. 2017. Heavy metal detoxification mechanisms in halophytes: An overview. Wetl. Ecol. Manag. 25: 129-148.

Tippkötter, N., A.M. Duwe, S. Wiesen, T. Sieker and R. Ulber. 2014. Enzymatic hydrolysis of beech wood lignocellulose at high solid contents and its utilization as substrate for the production of biobutanol and dicarboxylic acids. Bioresour. Technol. 167: 447-455.

Tsigie, Y.A., C.Y. Wang, C.T. Truong and Y.H. Ju. 2011. Lipid production from *Yarrowia lipolytica* Po 1 g grown in sugarcane bagasse hydrolysate. Bioresour. Technol. 102: 9216-9222.

Vermerris, W. 2011. Survey of genomics approaches to improve bioenergy traits in maize, sorghum and sugarcane free access. J. Integr. Plant Biol. 53: 105-119.

Wei, Z., G. Zeng, F. Huang, M. Kosa, Q. Sun, X. Meng, D. Huang and A.J. Ragauskas. 2015. Microbial lipid production by oleaginous *Rhodococci* cultured in lignocellulosic autohydrolysates. Appl. Microbiol. Biotechnol. 99: 7369-7377.

Wu, Y., F. Zhao, S. Liu, L. Wang, L. Qiu, G. Alaxandrov and V. Jothiprakash. 2018. Bioenergy production and environmental impacts. Geosci. Lett. 5: 14.

Yan, A., Y. Wang, S.N. Tan, M.L.M. Yusof, S. Ghosh and Z. Chen. 2020. Phytoremediation: A promising approach for revegetation of heavy metal-polluted land. Front. Plant Sci. 11. https://doi.org/10.3389/fpls.2020.00359

Yu, X., Y. Zheng, K.M. Dorgan and S. Chen. 2011. Oil production by oleaginous yeasts using the hydrolysate from pretreatment of wheat straw with dilute sulfuric acid. Bioresour. Technol. 102: 6134-6140.

Zandalinas, S.I., Y. Fichman, A.R. Devireddy, S. Sengupta, R.K. Azad and R. Mittler. 2020. Systemic signaling during abiotic stress combination in plants. Proc. Natl. Acad. Sci. 117: 13810-13820.

Zerva, A., A.L. Savvides, E.A. Katsifas, A.D. Karagouni and D.G. Hatzinikolaou. 2014. Evaluation of *Paecilomyces variotii* potential in bioethanol production from lignocellulose through consolidated bioprocessing. Bioresour. Technol. 162: 294-299.

Zhuang, P., Q.W. Yang, H.B. Wang and W.S Shu. 2007. Phytoextraction of heavy metals by eight plant species in the field. Water Air Soil Pollut. 184: 235-242.

Zuroff, T.R., S.B. Xiques and W.R. Curtis. 2013. Consortia-mediated bioprocessing of cellulose to ethanol with a symbiotic *Clostridium phytofermentans*/yeast co-culture. Biotechnol. Biofuel. 6: 1-12.

CHAPTER

2

Organic Contaminants and Phytoremediation: A Critical Appraisal

Rogimon P. Thomas[1]*, Joby Paul[2], Vinod V.[3] and Kannan V. Manian[4]

[1] Department of Botany, CMS College Kottayam (Autonomous), Kerala – 686001, India
[2] Department of Botany, St. Thomas College (Autonomous), Thrissur, Kerala, India
[3] College of Medicine, University of Florida, USA
[4] University of Rochester Medical Center, Rochester, New York 14642

2.1 Introduction

Agricultural and other industrial activities are the major reasons for polluting air, soils, sediments and water in neighboring environments. Organic and inorganic fertilizers, pesticides, herbicides and insecticides, organic matter such as animal wastes and decaying plant material, irrigation residues like salts and trace metals and microorganisms are regarded as chief agricultural contaminants (Ansari et al. 2018). To protect agricultural land, stored grain, flower gardens and exterminate the pests transmitting perilous infectious disease, pesticides are used widely. Several human acute and chronic illnesses have been associated with the influence of pesticide. The debasement of water, soil and air quality by nutrients arising from agriculture is a consternating environmental issue. The rapid increase in the use of fertilizers and pesticides has amplified the adverse environmental consequences (Andrews et al. 2004, Mostafalou and Abdollahi 2012, Borah 2020).

Water in agricultural environments is affected by a large assortment of contaminants including pesticides, nutrients and sediments. Drinking water may become contaminated with pesticides and herbicides through agricultural run-off, leaching of organics through the soil, accidental spills and incorrect disposal (Sharma and Bhattacharya 2016). The recognition of vestige quantities of organic contaminants and other pollutants in groundwater has aroused severe public scare. The incidence of organic contaminants in water may generate noxious chemicals all through disinfection. Groundwater system can be contaminated by organic contaminants by infiltration and through the interaction of surface water. In addition, due to the disposal of large amounts of manure and slurries from the animal rearing

Corresponding author: rogimon@cmscollege.ac.in

units, problems also arise. While considering the groundwater systems in India, it has been found that mainly two herbicides, Alachlor and Atrazine, and one insecticide, Malathion, are found in significant levels as pollutants. Moreover, nitrate and phosphate fertilizers have also been detected in groundwater systems of India in significant amounts (Duttagupta et al. 2018).

Soil pollution, due to polycyclic aromatic hydrocarbons (PAHs), organochlorine pesticides (OCPs) and polychlorinated biphenyls (PCBs), is a global problem in both developed and developing countries (Sosa et al. 2017). Soil contaminants enter the food chain through producers at the first trophic level and may affect the structure of an ecosystem. The redundant and ever increasing use of pesticides, insecticides, synthetic fertilizers, unfavorable and harmful irrigation practices and also other industrial activities have led to the deterioration of soil which is the most essential requirement for human life.

2.2 Organic Contaminants: An Overview

Organic contaminants may be specified as carbon-based chemicals, e.g. organic solvents, pesticides, petroleum-based wastage, and gas or liquid phase volatile compounds (Gama et al. 2012). Organic contaminants are expected to cause adverse impacts on the environment and they include herbicides, pesticides, fungicides and plant and animal tissues (Hassaan and El Nemr 2020). Trace levels of organic contaminant residues present in the soil, water, air, and sometimes food may create harmful effects for human and environmental health (Kookana 2010). Organic contaminants and wastes are mainly framed by carbon, hydrogen, and potentially other elements (Wilcox 2005). Volatile organic compounds (VOCs), comprising gas or liquid phase compounds, are present in the form of solvents, petroleum compounds, chemical precursors and intermediates. Solid organic compounds comprise sludges, still bottoms, resins, chemicals, waxes, paper, plastic, wood and foodstuffs (Wilcox et al. 2009). Organic contaminants are important pollutants in wastewaters and they include dye, humic substances, phenolic compounds, petroleum, surfactants, pesticides, and pharmaceuticals. The contaminants resulting from the decay of organic matter and pharmaceuticals such as antibiotics, humic substances such as humic acid, fulvic acid, or humin are most abundant in farm wastewaters (Dougherty et al. 2010). The emerging contaminants (Table 2.1) such as pharmaceuticals, steroids, antibiotics and antibiotic-resistance genes of bacteria have potential to enter the environment and cause known or suspected adverse ecological or human health effects (Stuart and Lapworth 2013, Juliano and Magrini 2017, Montes-Grajales et al. 2017, Xing et al. 2018).

2.3 Organic Contaminants in Agriculture

Fertilizers, either organic or inorganic, are regarded as indispensable sources of nutrients to the agriculture ecosystem. These fertilizers, besides supplying essential nutrients and acting as soil conditioners, pose potential pollution risk in agriculture and might contain a significant amount of contaminants (Rasmi et al. 2020). Indiscriminate and long-term use of fertilizer and manures, improper handling and

Table 2.1: Emerging organic contaminants

Pharmaceuticals		Personal care products	Pesticides	Industrial chemicals
Caffeine	Fexofenadine	Methylparaben	Atrazine	Benzotriazole
Benzotriazole	Lignocaine	Propylparaben,	Simazine	Methylbenzotriazole
Methylbenzotriazole	Losartan	Isopropylparaben,	Metaldehyde	Triethylphosphate
DEET	Metaxalone	Isobutylparaben,	Acetanilides	Metabolites
Gabapentin	Metoprolol	Pentylparaben	Dicamba	Atenolol acid
Metformin	Metronidazole	Triclosan	Flufenacet	N_4-Acetyl sulfamethoxazole
N_4-acetylsulfamethoxazole	Oxcarbazepine	Plastic microbeads	Uron	O-Desmethyl venlafaxine
Triethanolamine	Paracetamol			
Triethylphosphate	Ranitidine			
Citalopram	Sulfamethoxazole			
Flecainide	Tramadol			
Carbamazepine	Trimethoprim			
Cefalexin	Valsartan			
Citalopram	Venlafaxine			

Source: Ncube et al. 2012, Perez et al. 2012.

storage facilities often put the pristine terrestrial and aquatic ecosystems downstream and human health at risk. The high amount of heavy metals such as Chromium (Cr), Cadmium (Cd), Nickel (Ni) ,and Mercury (Hg) in the agriculture soil is often associated with the excess application of manures and fertilizers. Direct and long term effects have been produced by these contaminants on soil properties, such as the decline in soil organic carbon, the high buildup of salts, densification and compaction, surface crusting and imbalance of essential nutrients (Khan et al. 2018).

Organic chemicals have been deposited into the soil both naturally and anthropogenically, and many of the organic chemicals released into the air and water ultimately fetch up in the soil. Soil contamination is a serious environmental problem usually created by humans. In agricultural areas, farmers have been using an increasing amount of organic chemicals for the production of adequate quantities of agricultural products, but this leads to environmental pollution (Lynn 2013). PAHs, PCBs, polybrominated biphenyls, polychlorinated dibenzofurans, organophosphorus/carbamate insecticides, herbicides and organic fuels, particularly gasoline and diesel are the cases of organic pollutants generally seen in soils. Another source of soil pollution is the complex mixture of organic chemicals, metals and microorganisms in the effluent from septic systems, animal wastes and other sources of biowaste (Havugimana et al. 2017).

2.4 Risks Associated with Organic Contaminants

Perilous organic contaminants are chemicals having carbon as the major part. They are normally intact in the environment for a long period and eventually have been found to pile up in aquatic life and sediments. Organic chemicals can cause long-term health problems to humans by polluting the drinking water supplies. Many toxic organic chemicals such as organic solvents, pesticides, dioxins, PCBs, furans, and other nitrogen-containing derivatives are the main cause of water pollution (Luthy 2004). Unconventional disposal of industrial and household wastes and runoff of pesticides are regarded as the two major reasons for dangerous organic compounds in aquatic systems. Excessive use of insecticides, herbicides, fungicides, and rodenticides can result in toxic compounds carried by storm water runoff from agricultural lands, construction sites, and residential lawns to natural water sources (Agarwal et al. 2010).

The net enhancement of contaminants corresponding to that in the environment is identified and described by the process of bioaccumulation. Bioaccumulation is the net result of all uptake and loss processes, such as respiratory and dietary uptake, and loss by ejection, passive diffusion, metabolism, transfer to offspring and growth. Organic contaminants may dissolve and accumulate in organic phases of biotic life. An organism can adjust the absorbed blend of contaminant; some chemicals are maintained and retained, while others are eliminated from the body, resulting in no net accretion. In the food web, animals thereby show very different bioaccumulation of various chemicals, both in levels and in the relative composition (Cravedi 2003).

Persistent Organic Pollutants (POPs) are toxic substances composed of organic (carbon-based) chemical compounds and they are present in our food, soil, air and water, e.g. PCBs and DDT. They are chiefly the products and spin-offs from industrial processes, chemical manufacture and ensuing wastes. Wildlife and humans around the world carry amounts of POPs in their bodies that are at or near levels that can cause injury (Guo et al. 2019). They accumulate in the body fat of animals especially human and marine mammals and are passed from mother to fetus. The characteristics that make POPs noxious and dangerous are toxicity, persistence, resistance to normal processes that break down contaminants and ease of travel to great distances on wind and water systems. Even minor quantities of POPs can make for violent and needless disturbance in human and animal tissue, engendering nervous system damage, diseases of the immune system, reproductive and developmental disorders, and cancers (Pawelczyk 2013).

The most persistent bio-accumulative chemicals have been acknowledged for priority action. It was decided that the Stockholm Convention on Persistent Organic Pollutants will phase out and remove the use and manufacture of those chemicals besides new ones. The twelve targeted POPs consist of eight pesticides (aldrin, chlordane, DDT, dieldrin, endrin, heptachlor, mirex, and toxaphene), two types of industrial chemicals (polychlorinated biphenyls or PCBs and hexachlorobenzene), and two chemical families of chance by-products of the manufacture, use, and/or burning and combustion of chlorine and chlorine-containing materials (dioxins and furans) (Sah and Joshi 2011). These chemicals have reported endocrine-disrupting properties. Endocrine disrupting chemicals can be harmful at exceptionally small doses and create specific risks that alter the body's sensitive systems and lead to serious health problems. During prenatal life, endocrine disruptors can alter development and undermine the ability to learn, fight disease, and reproduce (Gregoraszczuk and Ptak 2013). Risk management and control of soil PAHs in agriculture soils are required to ensure the safety of the biocoenosis and human health (Fitzgerald and Wikoff 2014).

The volatile organic compound methyl tert-butyl ether (MTBE) causes several health problems such as leukaemia, thyroid glands and kidneys lymphoma and tumors in the testicles. By mimicking, interfering or blocking the function of hormones, POPs can alter the hormone homeostasis; moreover, POPs are hypothesized to modify the risk of breast cancer. POPs and breast cancer associations have been widely studied (Wielsoe 2017). Hg, Cr, As, Zn, Cd, Ur, Se, Ag, Au and Ni are hazardous heavy metals that not only contaminate the environment but also adversely affect the quality of the soil, crop production as well as public health (Ayangbenro and Babalola 2017). These pollutants are the main sources of degenerative diseases such as cancer, atherosclerosis, Alzheimer's disease and Parkinson's disease affecting humans (Muszynska and Hanus-Fajerska 2015).

2.5 Remediation of Organic Contaminants

Over the past few decades, organic pollutants have become a major threat to the ecosystem and human health due to its contamination of the water and soil. A few of these pollutants are highly toxic and may mount up to top trophic levels via food chain contamination. Integrated bio-chemical remediation technologies have potential uses in the remediation of environmental contamination by multiple contaminants. Various conventional physico-chemical techniques like composting, land forming have been used to combat this problem. These are all invasive, time-consuming, and lead to the generation of more toxic substances and even result in the emission of greenhouse gases. Therefore, biological remediation of these pollutants is a viable option. Bioremediation involves the use of living organisms to neutralize these pollutants (Dada et al. 2015).

The undesirable characteristics of organic compounds like toxicity and flammability arise from the structure of the compounds rather than the basic elements comprising them. They may be eliminated from contaminated media and destroyed in bulk by mechanisms that alter their structure through oxidation or thermal decomposition. PAHs can be sequestered from the soil through adsorption, volatilization, photolysis, and chemical degradation, but these are overpriced operations (Lu et al. 2014). The mobilization of a special organic contaminant is mostly influenced by physical and chemical features viz. volatility, miscibility in water, electric charge, density, size, and relations with neighboring environment system. In a soil ecosystem, important parameters such as pH, porosity, bulk density, microbial communities, nutrient content, and organic matter availability affect the transport and immobilization of a contaminant to a greater extent. Furthermore, the contaminant sequestered into below-ground biomass may pose a significant challenge for their removal and may prove a time-consuming and expensive process, restricting their field-scale feasibility (Singh et al. 2019).

Many methods are reported in the elimination of organic contaminants from the media (Kwon and Pignatello 2005, Wang and Gong 2010, Gidley et al. 2012, Tang et al. 2013, Guan et al. 2014, Lidy et al. 2014, Oliveira et al. 2017, Shen et al. 2019, Saxena et al. 2020, Ghosh and Acharyya 2020, Guo et al. 2021). Several bacterial, fungal and algal strains have been shown to degrade a wide variety of PAHs and the most commonly reported bacterial species include *Acinetobacter calcoaceticus*, *Alcaligens denitrificans*, *Mycobacterium* sp., *Pseudomonas putida*, *Pseudomonas fluorescens*, *Pseudomonas vesicularis*, *Pseudomonas cepacia*, *Rhodococcus* sp., *Corynebacterium renale*, *Moraxella* sp., *Bacillus cereus*, *Beijerinckia* sp., *Micrococcus* sp., *Pseudomonas paucimobilis* and *Sphingomonas* sp. (Jain et al. 2005). *P. putida* strain was engineered to increase the efficiency of degradation of naphthalene and salicylate (Cao et al. 2009). The bio-degradative pathways of hydrocarbon bacteria have also been reported from the genera Mycobacterium, Aeromonas, Rhodococcus, Staphylococcus, Streptococcus, Shigella, Alcaligenes, Acinetobacter, Escherichia, Klebsiella, Enterobacter and Bacillus (Joutey 2013).

2.6 Phytoremediation of Organic Contaminants

Plant-microbial degradation (Table 2.2) is considered as one of the prospective low-cost removal procedure of organic contaminants (Tokala et al. 2002). Plants have an array of cellular and physiological mechanisms that enable them capable of withstanding high quantities and concentrations of organic pollutants without exhibiting their toxic results; such plants often gather and change these organics into nontoxic or comparatively less toxic metabolites. Phytoremediation is the technology based on plants for extraction, sequestration, and/or degradation of environmental contaminants (Choudhary et al. 2017, Tripathi et al. 2020). The process of phytoremediation is a green and suitable alternative to widely

Table 2.2: Degradation of PAHs and heavy metals by plant-microbe interactions

Bacterial /Fungi strain	**Plants**	**Names of PAHs/Heavy metals**	**References**
Arthrobacter, Bacillus, Flavobacterium, Nocardia, Pseudomonas, Sphingomonas, Stenotrophomonas and Streptomyces	*Vigna unguiculata, Helianthus annus, Austrodanthonia caespitosa, Zea mays, Sorghum sudanense* and *Vetiveria zizanoides*	Phenanthrene, pyrene	Sivaram et al. (2020)
Sphingomonas sp.,*Pseudomonas fluorescens*	*Sedum alfredii*	Cd	Chen et al. (2019)
Pseudomonas aeruginosa, Bacillus sp.	*Rhizophora mangle*	Naphthalene, Anthracene, Benzo pyrene, Dibenzo anthracene, Acenaphthene	Sampaio et al. (2019)
Bacillus licheniformis, B. mojavensis	*Festuca arundinacea*	Naphthalene, phenanthrene, benzo(a)anthracene, Dibenzo [a,h]anthracene	Eskandary et al. (2017)
Vibrio sayamiensis	*Spartina maritima*	As, Cu, and Zn	Mesa et al. (2015)
Magnaporthe oyzae	*Oryza sativa*	As	Lakshmanan et al. (2015)
Acinetobacter sp., *Alcaligens* sp., *Listeria* sp., *Staphylococcus* sp., *Alcaligens* sp., and *Listeria* sp.	*Nymphaea pubescens, Typha* sp., *Juncus effusus,* and *Phragmites australis.*	Cu, Zn, Pb, Cd, and Fe	Kabeer et al. (2014)
Micromonospora sp., *Bacillus* sp., *Arthrobacter* sp., *Leifsonia* sp., and *Staphylococcus* sp.	*Brassica napus*	Na, Mg, K, Fe, Cu, Zn, Cd, and Pb	Croes et al. (2013)
Pseudomonas putida, Pseudomonas fluorescens, Chlorella vulgaris, Methylobacterium oryzae, Berknolderia sp., *Pseudomonas aeruginosa, Citrobacter* sp.,	*Viola baoshanensis, Sedum alfredii, Rumex crispus, Helianthus annus, Anthyllis vulneraria,* and *Festuca arvernensis,*	Naphthalene, anthracene, fluoranthene, pyrene, benzo(a) pyrene, As, Zn, Cu, Ni, Cr, Cd, Hg, and Pb	Wanil et al. (2012)

Pseudomonas fluorescens, *P. aeruginosa*, *Bacillus subtilis*, *Bacillus* sp., *Alcaligenes* sp., *Acinetobacter lwoffi*, and *Flavobacterium* sp.	*Cordia subcordata*, *Thespesia populnea*, *Prosopis pallida*, *Scaevola serica*, and *Medicago sativa*	Naphthalene	Das and Chandran (2011)
Acinetobacter sp., *Acinetobacter lwoffii*, *Actynomices* sp., *Actynomices viscosus*, *Agrobacterium radiobacter*, and *Alcaligenes faecalis*	*Sinapis alba*, *Lepidium sativum*, and *Sorghum saccharatum*	Benzo(a)pyrene	Coccia et al. (2009)
Mycobacterium sp., *Haemophilus* sp., *Rhodococcus* sp., *Paenibacillus* sp., *Agrobacterium* sp., *Burkholderia* sp., *Rhodococcus* sp. and *Mycobacterium* sp.	*Bouteloua gracilis*, *Cynodon dactylon*, *Festuca arundinacea*, *Festuca rubra*, and *Melilotus officinalis*	Benzo(a)pyrene, anthracene, benzo[b]fluoranthene, benzo(e)pyrene, fluoranthene, naphthalene, and phenanthrene	Haritash and Kaushik (2009)
Pseudomonas sp., *Rhizobium* sp., *Arthrobacter* sp., *Nocardia* sp., *Streptomyces* sp., and *Burkholderia* sp.	*Artemisia frigida*, *Banksia integrifolia*, *Triticum durum*, *Lupinus albus*, *Alternanthera ficoidea*, and *Avena barbata*	Phenanthrene and α-pyrene	Chaudhry et al. (2005)
Pseudomonas oleovorans, *Aquaspirillum* sp., *Flavobacterium indologenes*, *Pseudomonas* sp., and *Burkholderia* sp.	*Bromus carinatus*, *Elymus glaucus*, *Festuca ruba*, *Hordeum californicum*, and *Nassella pulchra*	Hexadecane, naphthalene, and phenanthrene	Siciliano et al. (2003)

accomplished physicochemical approaches (Jan and Parray 2016). The action of plant-based contaminant removal could be either inside the plant body or outside the plant body. Phytoremediation involves the facilitation of different biochemical and physiological mechanisms like absorption, accumulation, sequestration, transport, and degradation (Dixit et al. 2015). Several organic contaminants including PCBs, PAHs and halogenated hydrocarbons have been targeted for effective remediation by utilization of diverse plant groups (Isiuku and Ebere 2019). Plant-associated bacteria, such as endophytic bacteria and rhizospheric bacteria, were evaluated for their biodegradation potential of toxic organic compounds in contaminated soil for improving phytoremediation programs. Many experimental investigations have been conducted to develop genetically modified plants and endophytic bacterial strains harboring genes of interest displaying efficient contaminant degradation ability (Jabeen et al. 2009).

Depending on the compound structure, environmental factors, and plant genotypes, plants can remediate organic pollutants through volatilization, immobilization, transformation to different extents (even mineralization), or a combination of all. Usually, a range of phytoremediation technologies are employed to mitigate organic pollutants: (a) phytoextraction (phytoaccumulation, phytoabsorption, phytosequestration) refers to the amassing of soil contaminants within plant's tissues (Ali et al. 2013, Van der Ent et al. 2013); (b) phytodegradation/rhizodegradation deals with the transformation of toxic environmental contaminants into nontoxic forms using plants and associated microorganisms followed by either accumulation or secretion in the vicinity (Rascio and Navari-Izzo 2011, Wiszniewska et al. 2016); (c) phytovolatilization describes the phenomenon of pollutant uptake from contaminated sites and their release into atmosphere in volatile forms (Tak et al. 2013, Wiszniewska et al. 2016); (d) phytostabilization discusses the process of contaminant entrapment on a suitable matrix through adsorption (Blaylock et al. 1995); (e) phytodesalination refers to the use of halophytic plants for removal of excess salts from saline soils (Zorrig et al. 2012); and (f) phytofiltration could be in any of the three forms of rhizofiltration (the use of plant roots), blastofiltration (the use of seedlings) and caulofiltration (the use of excised plant shoots) (Mesjasz-Przybyłowicz et al. 2004, Rahman et al. 2016). *Axonopus compressus* (Bordoloi et al. 2012), *Cyperus brevifolius* (Basumatary et al. 2012), and *Cyperus rotundus* (Basumatary et al. 2013) have been identified as model plants showing optimum performance when planted in the hydrocarbon-contaminated soil.

Rhizoremediation is an *in-situ* remediation approach involving microorganisms for the biodegradation of organic pollutants and various other contaminants in the root zone (Mohan et al. 2006, Kong and Glick 2017). Rhizosphere and the root exudates provide a suitable niche for the microorganisms to grow and in turn microbes act as biocatalysts to neutralize the pollutants. The dangerous pollutants like PAHs, pesticides and herbicides can be converted to degradable compounds, while heavy metals such as Zn, Cu, Pb, tin (Sn), and Cd can be transformed to different oxidation state or organic complex (Dzantor 2007, Wenzel 2009).

Various mechanisms employed are: production of bio-surfactants which are amphiphilic molecules that solubilize hydrophobic contaminants in their core by their spherical form or lamellar micelles, and enhance their bacterial degradation to simple harmless compounds, production of metal-chelating siderophores for heavy metal acquisition, increased humification, biofilm and acid production (Oberai and Khanna 2018). One of the most promising technologies for remediation of contaminated environmental sites is the genetic engineering of endophytic and rhizospheric bacteria useful in plant-associated degradation of toxic compounds in the soil (Afzal et al. 2014).

Genetically engineered microbes such as *Escherichia coli* strain M109 and *Pseudomonas putida* containing the merA gene can be used to effectively eradicate Hg from contaminated soils and sediments (Azad et al. 2014). Transgenic plants exhibiting biodegradation capabilities of microorganisms bring the promise of an efficient and environmentally friendly technology for cleaning up polluted soils and water (Aken 2009, Prasad et al. 2021).

2.7 Role of Bioenergy Crops in Mitigation of Organic Contaminants

Bioenergy crops are defined as any plant material utilized to generate bioenergy. These crops have the capacity to produce large volume of biomass, high energy potential, and can be grown in marginal soils. Plant species evaluated as potential bioenergy crops has fast growth, tolerance to biotic and abiotic stresses, and low requirements for biological, chemical or physical pre-treatments. They need low inputs for the establishment, require low fossil fuel inputs, should be adjustable to marginal lands and offer high biomass and energy yield that is promised to help dilute global warming and fight Global Climate Change (GCC). Planting bioenergy crops in degraded soils is one of the hopeful agricultural choices for carbon sequestration. Agronomically, they should not be limited to the low proportional provision of dry matter to reproductive structures, long canopy duration, perennial growth, sterility to prevent escape, and low moisture content at harvest (Oliver et al. 2009).

Bioenergy crops consist of herbaceous grasses and woody perennials. They can be classified into five types viz. first (FGECs), second (SGECs), third-generation bioenergy crops (TGECs), dedicated energy crops (DECs) and halophytes. The first-generation bioenergy crops include oil palm, corn, sorghum, rapeseed and sugarcane, whereas the second-generation bioenergy crops are comprised of switchgrass (*Panicum virgatum*), reed canary grass (*Phalaris arundinacea*), alfalfa (*Medicago sativa*), Napier grass (*Pennisetum purpureum*), *Cynodon* spp, silver grass (*Miscanthus giganteus*), and *Pongamia* sp. The third-generation bioenergy crops comprise boreal plants, crassulacean acid metabolism (CAM) plants, eucalyptus and microalgae. Microalgal and seaweed mediated degradation of organic pollutants has been well studied (Kurade et al. 2016, Baghour 2017, 2019). Tarla et al. (2020) reported biodegradation of pesticides, microorganisms that degrade pesticides, bioremediation and phytoremediation of pesticide-contaminated soils, and soil amendments for pesticide remediation. Bioenergy halophytes include different species of *Acacia, Eucalyptus, Casuarina, Melaleuca, Prosopis, Rhizophora* and *Tamarix*. The dedicated energy crops include perennial herbaceous and woody plant species viz. giant miscanthus, switchgrass, jatropha and algae (Yadav et al. 2019).

Dedicated and promising energy crops must be cultivated on the contaminated lands for manifold profits. Several TGECs oleaginous crops can help to reduce greenhouse gas emissions by capturing carbon dioxide released from power plants or by generating biomass through photosynthesis (Peterson 2008).

Throughout the world, there are two key concerns such as escalating the chain of contaminated lands and bioenergy demands. Therefore, connecting phytoremediation with energy crops is essential for sustainable development. Research has shown that energy crops have increased soil stability, decreased surface water runoff, decreased transport of nutrients and sediment, and increased soil moisture, in comparison to traditional crops. They also require fewer fertilizers, herbicides and insecticides than traditional row crops; this reduction in herbicide and pesticide use reduces the potential for water pollution and other environmental problems. Another environmental benefit from the use of energy crops is a decrease in emissions. Unlike fossil fuels, plants grown for energy crops absorb the amount of CO_2 released during their combustion/ use. Use of inedible bioenergy crops for remediation of heavy metal-polluted sites has the benefit that biomass produced can be used to generate biofuel and utilizes lands which are incompatible for raising food crops. Energy crops may also shelter natural forests by catering an alternative supply of wood, which can be raised on-farm or pasture land that is no longer desirable for traditional row crops. They are beneficial in providing certain ecosystem services, including carbon sequestration, biodiversity enhancement, salinity mitigation, and enhancement of soil and water quality (Dipti and Priyanka 2013, Bauddh et al. 2017).

2.8 Conclusion

The globally increasing population demands the production of more goods and services to fulfil the increasing needs of human beings, which resulted in urbanization and industrialization, which in turn resulted in an energy crisis and quickened environmental pollution throughout the world. Organic contamination is a disquieting environmental issue that leads to the deterioration of immaculate natural resources. Organic contaminants such as volatile (VOCs) and semivolatile organic compounds (SVOCs) are acknowledged as mutagenic and carcinogenic and have the ability to negatively affect human health and the environment. Several physical and chemical methods have been deployed for the removal of organic contaminants from the sources. But, an approach that could amalgamate the two aspects, i.e. energy production and the removal of organic contaminants, can be achieved through phytoremediation. Phytoremediation, a sustainable, environment-friendly, and potentially cost-effective technology, can be used to decontaminate heavy metal-contaminated land. Plants can sequester carbon dioxide and absorb, degrade, and stabilize environmental pollutants such as heavy metals, poly-aromatic hydrocarbons, poly-aromatic biphenyls, radioactive materials, and other chemicals. The use of green plants for pollution mitigation and energy production will also deal with some other major global concerns like global climate change, ocean acidification, and land degradation through carbon sequestration reduced emissions of other greenhouse gases, restoration of degraded lands and water and more. The international and national literature on EOCs research was reviewed and it confirmed that they are released in the environment from a range of sources.

Bioenergy crops offer the only basis of substitute energy with the ability to trim down the use of fossil fuels. Energy crops perform as auxiliary crops for pollution control as they eliminate pesticides and surplus fertilizer from surface water before it pollutes groundwater or rivers. They can care for riverbank and water from erosion and chemical runoff and provide a filter system for contaminant removal. Bioenergy-driven restoration of degraded ecosystems can also increase terrestrial carbon sequestration due to large biomass production and root residues as well as slowing decomposition of soil organic materials. The booming employment of genetic engineering techniques together with the existing information on phytoremediation, plant physiology and rhizoremediation measures could be influential in gaining profound appreciations into the process of elimination of environmental contaminants. An improved refinement and understanding of the mechanisms of heavy metal intake, movement, accruement and tolerance will facilitate scientists to formulate effectual and productive transgenic bioenergy crops for alleviation of heavy metals in soil.

Acknowledgments

The authors are thankful to the Manager and the Principal, CMS College Kottayam (Autonomous), Kerala for facilities and support.

References

Afzal, M., Q.M. Khan and A. Sessitsch. 2014. Endophytic bacteria: Prospects and applications for the phytoremediation of organic pollutants. Chemosphere 117: 232-242.

Agarwal, A., R.S. Pandey and B. Sharma. 2010. Water pollution with special reference to pesticide contamination in India. Journal of Water Resource and Protection 02(05): 432-448.

Aken, B.V. 2009. Transgenic plants for enhanced phytoremediation of toxic explosives. Current Opinion in Biotechnology 20(2): 231-236.

Akram, M.S., N. Rashid and S. Basheer. 2021. Physiological and molecular basis of plants tolerance to linear halogenated hydrocarbons. pp. 591-602. *In*: Handbook of Bioremediation: Physiological, Molecular and Biotechnological Interventions.

Ali H., E. Khan and M.A. Sajad. 2013. Phytoremediation of heavy metals-concepts and applications. Chemosphere 91(7): 869-881.

Ali, N., T. Mehdi, R.N. Malik, S.A.M.A.S. Eqani, A. Kamal, A.C. Dirtu, H. Neels and A. Covaci. 2014. Levels and profile of several classes of organic contaminants in matched indoor dust and serum samples from occupational settings of Pakistan. Environmental Pollution 193: 269-276.

Andrews, S.S., D.L. Karlen and L.A. Cambardella. 2004. The soil management assessment framework: A quantitative soil quality evaluation method. Soil Science Society of America Journal, 68: 1945-1962.

Ansari, A.A., M. Naeem and S.S. Gill. 2018. Contaminants in agriculture: Threat to soil health and productivity. Agricultural Research & Technology 16(1): 555975.

Ayangbenro, A.S. and O.O. Babalola. 2017. A new strategy for heavy metal polluted environments: A review of microbial biosorbents. International Journal of Environmental Research and Public Health 14(1).

Azad, M.A.K., L. Amin and N.M. Sidik. 2014. Genetically engineered organisms for bioremediation of pollutants in contaminated sites. Chinese Science Bulletin 59: 703-714.

Baghour, M. 2017. Effect of seaweeds in phyto-remediation. pp. 47-83. *In*: Nabti, E. [ed.]. Biotechnological Applications of Seaweeds. Nova Science Publishers, New York.

Baghour, M. 2019. Algal degradation of organic pollutants. *In*: Martínez, L., O. Kharissova and B. Kharisov [eds.]. Handbook of Ecomaterials. Springer, Cham. https://doi.org/10.1007/978-3-319-68255-6_86

Basumatary, B., S. Bordoloi and H.P. Sarma. 2012. Crude oil-contaminated soil Phytoremediation by using *Cyperus brevifolius* (Rottb.) Hassk. Water Air Soil Pollution 223: 3373-3383.

Basumatary, B., R. Saikia, H.C. Das and S. Bordoloi. 2013. Field Note: Phytoremediation of petroleum sludge contaminated field using Sedge species, *Cyperus rotundus* (Linn.) and *Cyperus brevifolius* (Rottb.) Hassk. International Journal of Phytoremediation 15(9): 877-888.

Bauddh, K., B. Singh and J. Korstad. 2017. Phytoremediation Potential of Bioenergy Plants. Springer, Singapore.

Blaylock, M., B. Ensley, D. Salt, N. Kumar, V. Dushenkov and I. Raskin. 1995. Phytoremediation: A novel strategy for the removal of toxic metals from the environment using plants. *Biotechnology*, 13(7): 468-474.

Borah, P., M. Kumar and P. Devi. 2020. Chapter 2 – Types of inorganic pollutants: Metals/metalloids, acids, and organic forms. pp. 17-31. Pooja Devi, Pardeep Singh, Sushil Kumar Kansal [eds.]. Inorganic Pollutants in Water, Elsevier. Cambridge MA.

Bordoloi, S., B. Basumatary, R. Saikia and H.C. Das. 2012. *Axonopus compressus* (Sw.) P. Beauv. A native grass species for phytoremediation of hydrocarbon contaminated soil in Assam, India. Journal of Chemical Technology and Biotechnology, 87(9): 1335-1341.

Borgå, K. 2013.Ecotoxicology: Bioaccumulation. Earth Systems and Environmental Sciences. https://doi.org/10.1016/B978-0-12-409548-9.00765-X.

Cao, B., K. Nagarajan and C.L. Kai. 2009. Biodegradation of aromatic compounds: Current status and opportunities for biomolecular approaches. Applied Microbiology and Biotechnology 85: 207-228.

Chaudhry, Q., M. Blom-Zandstra, S.K. Gupta and E. Joner. 2005. Utilising the synergy between plants and rhizosphere microorganisms to enhance breakdown of organic pollutants in the environment. Environmental Science and Pollution Research, 12: 34-48.

Chen, B., Y. Zhang, M.T. Rafiq, K.Y. Khan, F. Pan, X. Yang and Y. Feng. 2014. Improvement of cadmium uptake and accumulation in *Sedum alfredii* by endophytic bacteria *Sphingomonas* SaMR12: Effects on plant growth and root exudates. Chemosphere 117C (1): 367-373.

Choudhary, D.K., A. Varma and N. Tuteja. 2017. Plant-Microbe Interaction: An Approach to Sustainable Agriculture. Springer; New Delhi, India.

Chowdhury, S., N. Khan, G.H. Kim, J. Harris, P. Longhurst and N.S. Bolan. 2016. Chapter 22 – Zeolite for nutrient stripping from farm effluents. pp. 569-589. *In*: M.N.V. Prasad and Kaimin Shih [eds.]. Environmental Materials and Waste. Academic Press. London.

Coccia, A.M., P.M.B. Gucci, I. Lacchetti, E. Beccaloni, R. Paradiso, M. Beccaloni and L. Musmeci. 2009. Hydrocarbon contaminated soil treated by bioremediation technology: Microbiological and toxicological preliminary findings. Environmental Biotechnology 5(2): 61-72.

Covaci, A., T. Geens, L. Roosens, N. Ali, N. Van den Eede, A.C. Ionas, G. Malarvannan and A.C. Dirtu. 2011. Human exposure and health risks to emerging organic contaminants. *In*:

Barceló, D. [ed.]. Emerging Organic Contaminants and Human Health. The Handbook of Environmental Chemistry, Vol. 20. Springer, Berlin, Heidelberg.

Cravedi, J.P. 2003. Residues of organic micropollutants in fish: Bioaccumulation conditions and safety for the consumer. Cahiers de Nutrition et de Diététique, 38(1): 45-52.

Croes, S., N. Weyens, J. Janssen, H. Vercampt, J.V. Colpaert, R. Carleer and J. Vangronsveld. 2013. Bacterial communities associated with *Brassica napus* L. grown on trace element contaminated and non-contaminated fields: A genotypic and phenotypic comparison. Microbial Biotechnology 6(4): 371-384.

Dada, E., K. Njoku, A. Osuntoki and M. Akinola. 2015. A review of current techniques of Physico-chemical and biological remediation of heavy metals polluted soil. Ethiopian Journal of Environmental Studies and Management, 8(5): 606.

Das, N. and P. Chandran. 2011. Microbial degradation of petroleum hydrocarbon contaminants: An overview. Biotechnology Research International. doi:10.4061/2011/941810.

Dipti and Priyanka. 2013. Bioenergy crops: An alternative energy. International Journal of Environmental Engineering and Management 4(3): 265-272.

Dixit, R., D. Malaviya, K. Pandiyan, U.B. Singh, A. Sahu, R. Shukla, B.P. Singh, J.P. Rai, P.K. Sharma, H. Lade and D. Paul. 2015. Bioremediation of heavy metals from soil and aquatic environment: An overview of principles and criteria of fundamental processes. Sustainability 7: 2189-2212.

Dougherty, J.A., P.W. Swarzenski, R.S. Dinicola and M. Reinhard. 2010. Occurrence of herbicides and pharmaceutical and personal care products in surface water and groundwater around Liberty Bay, Puget Sound, Washington. Journal of Environmental Quality 39(4): 1173-1180.

Duttagupta, S., A. Mukherjee, J. Bhattacharya and A. Bhattacharya. 2018. An overview of agricultural pollutants and organic contaminants in groundwater of India. Chapter: 15. pp. 1-9. *In*: Abhijit Mukherjee [ed.]. Groundwater of South Asia. Springer Nature, Singapore Pte Ltd.

Dzantor, E.K. 2007. Phytoremediation: The state of rhizosphere "engineering" for accelerated rhizodegradation of xenobiotic contaminants. Journal of Chemical Technology & Biotechnology 82: 228-232.

Ely, C.S. and B.F. Smets. 2017. Bacteria from wheat and cucurbit plant roots metabolize PAHs and aromatic root exudates: Implications for rhizodegradation. International Journal of Phytoremediation 19(10): 877-883.

Eskandary, S., A. Tahmourespour, M. Hoodaji and A. Abdollahi. 2017. The synergistic use of plant and isolated bacteria to clean up polycyclic aromatic hydrocarbons from contaminated soil. Journal of Environmental Health Science and Engineering 15(12).

Fitzgerald, L. and D.S. Wikoff. 2014. Persistent Organic Pollutants. Encyclopedia of Toxicology, Elsevier, New York.

Gama, S., J.A. Arnot and D. Mackay. 2012. Toxic organic chemicals. *In*: Meyers, R.A. [ed.]. Encyclopedia of Sustainability Science and Technology. Springer, New York, NY.

Geng, L., N. Junjie, G. Wenjiong, A. Xiangsheng and Z. Long. 2016. Ecological and health risk-based characterization of agricultural soils contaminated with polycyclic aromatic hydrocarbons in the vicinity of a chemical plant in China. Chemosphere 163: 461-470.

Ghosh, S. and M. Acharyya. 2020. Pyridine rich Novolac-based network as an effective adsorbent for removing azo dyes. Chemistry Select 5(34): 10727-10735.

Gidley, P.T., S. Kwon, A. Yakirevich, V.S. Magar and U. Ghosh. 2012. Advection dominated transport of polycyclic aromatic hydrocarbons in amended sediment caps. Environmental Science & Technology 46(9): 5032-5039.

Gregoraszczuk, E.L. and A. Ptak. 2013. Endocrine-disrupting chemicals: Some actions of POPs on female reproduction. International Journal of Endocrinology, https://doi.org/10.1155/2013/828532.

Guan, Y., X. Meng and D. Qiu. 2014. Hollow microsphere with mesoporous shell by pickering emulsion polymerization as a potential colloidal collector for organic contaminants in water. Langmuir 30(13): 3681-3686.

Guo, S., Z. Yang, H. Zhang, W. Yang, J. Li and K. Zhou. 2021. Enhanced photocatalytic degradation of organic contaminants over $CaFe_2O_4$ under visible LED light irradiation mediated by peroxymonosulfate. Journal of Materials Science and Technology, 62: 34-43.

Guo, W., B. Pan, S. Sakkiah, G. Yavas, W. Ge, W. Zou, W. Tong and H. Hong. 2019. Persistent organic pollutants in food: Contamination sources, health effects and detection methods. International Journal of Environmental Research and Public Health 16(22): 4361.

Haritash, A.K. and C.P. Kaushik. 2009. Biodegradation aspects of Polycyclic Aromatic Hydrocarbons (PAHs): A review. Journal of Hazardous Materials 169(1-3): 1-15.

Hassaan, M.A. and A. El Nemr. 2020. Pesticides pollution: Classifications, human health impact, extraction and treatment techniques. The Egyptian Journal of Aquatic Research, 46(3): 207-220.

Havugimana, E., B.S. Bhople, A. Kumar, E. Byiringiro, J. Pierremugabo and A. Kumar. 2017. Soil pollution – Major sources and types of soil pollutants. Environmental Science and Engineering, 11: 53-82.

Isiuku, B.O. and E.C. Ebere. 2019. Water pollution by heavy metal and organic pollutants: Brief review of sources, effects and progress on remediation with aquatic plants. Analytical methods in Environmental Chemistry Journal. 2(3): 5-38.

Jabeen, R., A. Ahmad and M. Iqbal. 2009. Phytoremediation of heavy metals: Physiological and molecular mechanisms. The Botanical Review 75: 339-364.

Jain, R.K., M. Kapur, S. Labana, B. Lal, P.M. Sarma, D. Bhattacharya and I.S. Thakur. 2005. Microbial diversity: Application of microorganisms for the biodegradation of xenobiotics. Current Science 89: 1-112.

Jan, S. and J.A. Parray. 2016. Approaches to Heavy Metal Tolerance in Plants. Springer; New Delhi, India.

Joutey, N.T., W. Bahafid, H. Sayel and N.E. Ghachtouli. 2013. Biodegradation: Involved microorganisms and genetically engineered microorganisms, biodegradation-life of science, Rolando Chamy and Francisca Rosenkranz, IntechOpen, doi: 10.5772/56194.

Juliano, C. and G.A. Magrini. 2017. Cosmetic ingredients as emerging pollutants of environmental and health concern. A mini-review. Cosmetics 4(2).

Kabeer, R., R. Varghese, V.M. Kannan, J.R. Thomas and S.V. Poulose. 2014. Rhizosphere bacterial diversity and heavy metal accumulation in *Nymphaea pubescens* in aid of phytoremediation potential. Journal of BioScience & Biotechnology 3(1): 89-95.

Khan, M.N., M. Mobin, Z.K. Abbas and S.A. Alamri. 2018. Fertilizers and their contaminants in soils, surface and groundwater. pp. 225-240. *In*: Dominick A. DellaSala and Michael I. Goldstein [eds.]. The Encyclopedia of the Anthropocene, Vol. 5, Oxford: Elsevier.

Kong, Z. and B.R. Glick. 2017. The role of plant growth-promoting bacteria in metal phytoremediation. Advances in Microbial Physiology 71: 97-132.

Kookana, R.S., S. Baskaran and R. Naidu. 1998. Pesticide fate and behaviour in Australian soils in relation to contamination and management of soil and water: A review. Australian Journal of Soil Research 36(5), 715-764.

Kookana, R.S. 2010. The role of biochar in modifying the environmental fate, bioavailability, and efficacy of pesticides in soils: A review. Australian Journal of Soil Research 48(7): 627-637.

Kurade, M.B., J.R. Kim, S.P. Govindwar and B.H. Jeon. 2016. Insights into microalgae mediated biodegradation of diazinon by *Chlorella vulgaris*: Microalgal tolerance to xenobiotic pollutants and metabolism. Algal Research 20: 126-134.

Kuroda, K. and T. Fukushi. 2008. Groundwater contamination in urban areas. *In*: Takizawa, S. [ed.]. Groundwater Management in Asian Cities. cSUR-UT Series: Library for Sustainable

Urban Regeneration. Vol 2. Springer, Tokyo. https://doi.org/10.1007/978-4-431-78399-2_7.

Kwon, S. and J.J. Pignatello. 2005. Effect of natural organic substances on the surface and adsorptive properties of environmental black carbon (char): Pseudo pore blockage by model lipid components and its implications for N_2-probed surface properties of natural sorbents. Environmental Science & Technology 39(20): 7932-7939.

Lakshmanan, V., D. Shantharaj, G. Li and A. Seyfferth. 2015. A natural rice rhizospheric bacterium abates arsenic accumulation in rice (*Oryza sativa* L.). Planta 242(4): 1037-1050.

Lapworth, D., N. Baran, M. Stuart and R. Ward. 2012. Emerging organic contaminants in groundwater: A review of sources, fate and occurrence. Environmental Pollution 163: 287-303.

Lu, M., Z. Zhang, J. Wang, M. Zhang, Y. Xu and X. Wu. 2014. Interaction of heavy metals and pyrene on their fates in soil and tall Fescue (*Festuca arundinacea*). Science & Technology 48(2): 1158-1165.

Luthy, R.G. 2004. Organic contaminants in the environment: Challenges for the water/environmental engineering community. *In*: National Research Council (US) Chemical Sciences Round Table; Norling, P., Wood-Black, F. and Masciangioli, T.M. [eds]. Water and Sustainable Development: Opportunities for the Chemical Sciences: A Workshop Report to the Chemical Sciences Roundtable, Washington (DC).

Lydy, M.J., P.F. Landrum, A.M.P. Oen, M. Allinson, F. Smedes, A.D. Harwood, H. Li, K.A. Maruya and J. Liu. 2014. Passive sampling methods for contaminated sediments: State of the science for organic contaminants. Integrated Environmental Assessment and Management 10(2): 167-178.

Lynn, B. 2013. The Effects of Organic Pollutants in Soil on Human Health. EGU General Assembly Conference Abstracts, Harvard.

Mesa, J., E. Mateos-Naranjo, M.A. Caviedes, S. Redondo-Gómez, E. Pajuelo and I.D. Rodríguez-Llorente. 2015. Endophytic cultivable bacteria of the metal bioaccumulator *Spartina maritima* improve plant growth but not metal uptake in polluted marshes soils. Frontiers in Microbiology. 6(1450).

Mesjasz-Przybyłowicz, J., M. Nakonieczny, P. Migula, M. Augustyniak, M. Tarnawska, W. Reimold, C. Koeberl, W. Przybylowicz and E. Glowacka. 2004. Uptake of cadmium, lead nickel and zinc from soil and water solutions by the nickel hyperaccumulator *Berkheya coddii*. Acta Biologica Cracoviensia Series Botanica, 46: 75-85.

Mohan, S.V., T. Kisa, T. Ohkuma, R.A. Kanaly and Y. Shimizu. 2006. Bioremediation technologies for treatment of PAH contaminated soil and strategies to enhance process efficiency. Review on Environmental Science and Biotechnology 5: 347-374.

Montes-Grajales, D., M. Fennix-Agudelo and W. Miranda-Castro. 2017. Occurrence of personal care products as emerging chemicals of concern in water resources: A review. Science of the Total Environment 595(94): 601-614.

Mostafalou S. and M. Abdollahi. 2012. Concerns of environmental persistence of pesticides and human chronic diseases. Clinical and Experimental Pharmacology S5: e002.

Muszynska, E. and E. Hanus-Fajerska. 2015. Why are heavy metal hyperaccumulating plants so amazing? BioTechnologia. Journal of Biotechnology, Computational Biology and Bionanotechnology, 96: 265-271.

Ncube, B., J.F. Finnie and J. Van Staden. 2012. Quality from the field: The impact of environmental factors as quality determinants in medicinal plants. South African Journal of Botany, 82: 11-20.

Ncube, E.J., K. Voyi and H. du Preez. 2012. Implementing a protocol for selection and prioritisation of organic contaminants in the drinking water value chain: Case study of Rand Water, South Africa. Water SA 38(4): 487-504.

Oberai, M. and V. Khanna. 2018. Rhizoremediation – Plant microbe interactions in the removal of pollutants. International Journal of Current Microbiology and Applied Sciences, 7(1): 2280-2287.

Oliveira, F.R., A.K. Patel, D.P. Jaisi and S. Adhikari. 2017. Environmental application of biochar: Current status and perspectives. Bioresource Technology, 246.

Oliver, R.J., J.W. Finch and G. Taylor. 2009. Second generation bioenergy crops and climate change: A review of the effects of elevated atmospheric CO_2 and drought on water use and the implications for yield. GCB Bioenegy 1: 97-114.

Pawełczyk, A. 2013. Assessment of health risk associated with persistent organic pollutants in water. Environmental Monitoring and Assessment 185: 497-508.

Perez, F., M. Llora, M. Farre and D. Barcelo. 2012. Perfluorinated compounds in drinking water, food and human samples. Emerging organic contaminants and human health. The Handbook of Environmental Chemistry 20: 337-374.

Petersen, J.E. 2008. Energy production with agricultural biomass: Environmental implications and analytical challenges. European Review of Agricultural Economics, 1-24.

Prasad, M., P. Saraswat, A. Gupta and R. Ranjan. 2021. Molecular basis of plant-microbe interaction in remediating organic pollutants. pp. 603-623. *In*: Handbook of Bioremediation: Physiological, Molecular and Biotechnological Interventions.

Rahman, M.A., S.M. Reichman, L. De Filippis, S.B.T. Sany and H. Hasegawa. 2016. Phytoremediation of toxic metals in soils and wetlands: Concepts and applications. *In*: Hasegawa, H., Rahman, M.M., Rahman, I. [eds.]. Environmental Remediation Technologies for Metal-Contaminated Soils. Springer; Tokyo, Japan, 161-195.

Rascio, N. and F. Navari-Izzo. 2011. Heavy metal hyperaccumulating plants: How and why do they do it? And what makes them so interesting? Plant Science 180: 169-181.

Rashmi, I., T. Roy, K.S. Kartika, R. Pal, V. Coumar, S. Kala and K.C. Shinoji. 2020. Organic and inorganic fertilizer contaminants in agriculture: Impact on soil and water resources. *In*: Naeem, M., Ansari, A. and Gill, S. [eds]. Contaminants in Agriculture. Springer, Cham.

Sah, R.C. and K.R. Joshi. 2011. Fact Sheets of 22 Persistent Organic Pollutants (POPs) Under Stockholm Convention. Center for Public Health and Environmental Development (CEPHED), Lalitpur, Nepal.

Sampaio, C.J.S., J.R.B. de Souza, A.O. Damião, C. Thiago, T.C. Bahiense and M.R.A. Roque. 2019. Biodegradation of polycyclic aromatic hydrocarbons (PAHs) in a diesel oil contaminated mangrove by plant growth promoting rhizobacteria. Biotech 9: 115.

Saxena, R., M. Saxena and A. Lochab. 2020. Recent progress in nanomaterials for adsorptive removal of organic contaminants from wastewater. Chemistry Select 5(1).

Sharma, S. and A. Bhattacharya. 2016. Drinking water contamination and treatment techniques. Applied Water Science, 7: 1043-1067.

Shen, Z., J. Zhang, D. Hou, D.C.W. Tsang, Y.S. Ok and D.S. Alessi. 2019. Synthesis of MgO-coated corncob biochar and its application in lead stabilization in a soil washing residue. Environment International 122: 357-362.

Siciliano, S.D., J.J. Germida, K. Banks and C.W. Greer. 2003. Changes in microbial community composition and function during a polyaromatic hydrocarbon phytoremediation field trial. Applied and Environmental Microbiology 69(1): 483-489.

Singh, P., A. Kumar and A. Borthakur. 2019. Abatement of Environmental Pollutants: Trends and Strategies, Elsevier, USA.

Sivaram, A.K., S.R. Subashchandrabose, P. Logeshwaran, R. Lockington, R. Naidu and M. Megharaj. 2020. Rhizodegradation of PAHs differentially altered by C3 and C4 plants. Scientific Reports, 10: 16109.

Sosa, D., I. Hilber, R. Faure, N. Bartolomé, O. Fonseca, A. Keller, P. Schwab, A. Escobar and T.D. Bucheli. 2017. Polycyclic aromatic hydrocarbons and polychlorinated biphenyls in soils of Mayabeque, Cuba. Environmental Science and Pollution Research 24: 12860-12870.

Stuart, M. and D. Lapworth. 2013. Emerging organic contaminants in groundwater. *In*: Mukhopadhyay, S. and Mason, A. [eds.]. Smart Sensors for Real-Time Water Quality Monitoring. Smart Sensors, Measurement and Instrumentation, Vol. 4. Springer, Berlin, Heidelberg.

Sun, J., L. Pan, D. Tsang, Y. Zhan, L. Zhu and X. Li. 2018. Organic contamination and remediation in the agricultural soils of China: A critical review. The Science of the Total Environment, 615: 724-740.

Tak, H.I., F. Ahmad and O.O. Babalola. 2013. Advances in the application of plant growth – Promoting rhizobacteria in phytoremediation of heavy metals. Reviews of Environmental Contamination and Toxicology 223: 33-52.

Tang, J., W. Zhu, R. Kookana and A. Katayama. 2013. Characteristics of biochar and its application in remediation of contaminated soil. Journal of Biosciences and Bioengineering, 116(6).

Tarla, D.N., L.E. Erickson, G.M. Hettiarachchi, S.I. Amadi, M. Galkaduwa, L.C. Davis, A. Nurzhanova and V. Pidlisnyuk. 2020. Phytoremediation and bioremediation of pesticide-contaminated soil. Applied Sciences 10: 1217.

Tokala, R.K., J.L. Strap, C.M. Jung, D.L. Crawford, M.H. Salove, L.A. Deobald, J.F. Bailey and M.J. Morra. 2002. Novel plant-microbe rhizosphere interaction involving *Streptomyces lydicus* WYEC108 and the pea plant (*Pisum sativum*). Applied and Environmental Microbiology 68(5): 2161-2171.

Tripathi, S., V.P. Singh, P. Srivastava, R. Singh, R.S. Devi, A. Kumar and R. Bhadouria. 2020. Phytoremediation of organic pollutants: Current status and future directions. Chapter 4. pp. 81-105. *In*: Pardeep Singh, Ajay Kumar and Anwesha Borthakur [eds.]. Abatement of Environmental Pollutants Trends and Strategies. Elsevier, USA.

Van der Ent, A., A.J. Baker, R.D. Reeves, A.J. Pollard and H. Schat. 2013. Hyperaccumulators of metal and metalloid trace elements: Facts and fiction. Plant and Soil 362: 319-334.

Wang, X., P. Gong, T. Yao and K.C. Jones. 2010. Passive air sampling of organochlorine pesticides, polychlorinated biphenyls, and polybrominated diphenyl ethers across the Tibetan Plateau. Environmental Science & Technology 44(8): 2988-2993.

Wanil, S.H., G.S. Sanghera, H. Athokpam, J. Nongmaithem, R. Nongthongbam, B.S. Naorem and H.S. Athokpam. 2012. Phytoremediation: Curing soil problems with crops. African Journal of Agricultural Research, 7(28): 3991-4002.

Weber, S., S. Khan and J. Hollender. 2005. Human risk assessment of organic contaminants in reclaimed wastewater used for irrigation. Desalination 187: 724-735.

Wenzel, W.W. 2009. Rhizosphere processes and management in plant assisted bioremediation (phytoremediation) of soils. Plant Soil 321: 385-408.

Wielsoe, M., P. Kern and E.C. Bonefeld-Jorgensen. 2017. Serum levels of environmental pollutants is a risk factor for breast cancer in Inuit: A case control study. Environmental Health, 16: 56.

Wilcox, H.S., J.B. Wallace, J.L. Meyer and J.P. Benstead. 2005. Effects of labile carbon addition on a headwater stream food web. Limnology and Oceanography, 50(4): 1300-1312.

Wilcox, J.B. 2005. Electrical and thermal solutions. Chapter 9. pp. 203-211. *In*: Franklin J. Agardy and Nelson Leonard Nemerow [eds.]. Environmental Solutions. Elsevier Academic Press, USA.

Wilcox, J.D., J.M. Bahr, C.J. Hedman, J.D.C. Hemming, M.A.E. Barman and K.R. Bradbury. 2009. Removal of organic wastewater contaminants in septic systems using advanced treatment technologies. Journal of Environmental Quality 38(1): 149-156.

Wiszniewska, A., E. Hanus-Fajerska, E. Muszyńska and K. Ciarkowska. 2016. Natural organic amendments for improved phytoremediation of polluted soils: A review of recent progress. Pedosphere 26(1): 1-12.

Xing, Y., Y. Yu and Y. Men. 2018. Emerging investigators series: Occurrence and fate of emerging organic contaminants in wastewater treatment plants with an enhanced nitrification step. Environmental Science: Water Research and Technology, 4(10): 1412-1426.

Yadav, P., P. Priyanka, D. Kumar, A. Yadav and K. Yadav. 2019. Bioenergy crops: Recent advances and future outlook. *In*: Rastegari, A., A. Yadav and A. Gupta [eds.]. Prospects of Renewable Bioprocessing in Future Energy Systems. Biofuel and Biorefinery Technologies, Springer, Cham.

Zorrig, W., M. Rabhi, S. Ferchichi, A. Smaoui and C. Abdelly. 2012. Phytodesalination: A solution for salt-affected soils in arid and semi-arid regions. Journal of Arid Land, 22(1): 299-302.

CHAPTER

3

Bioremediation: Plants and Microbes for Restoration of Heavy Metal Contaminated Soils

Harsh Kumar[1]*, Shumailah Ishtiyaq[1], Mayank Varun[2], Paulo J.C. Favas[3,4], Clement O. Ogunkunle[5] and Manoj S. Paul[1]

[1] Department of Botany, St. John's College, Agra – 282 002, India
[2] Department of Botany, Hislop College, Nagpur – 440 001, India
[3] School of Life Sciences and the Environment, University of Tras-os-Montes e Alto Douro, Vila Real, Portugal
[4] Faculty of Sciences and Technology, MARE – Marine and Environmental Sciences Centre, University of Coimbra, Coimbra, Portugal
[5] Environmental Botany Unit, Department of Plant Biology, Faculty of Life Sciences, University of Ilorin, Ilorin, 240003, Nigeria

3.1 Introduction

The build-up of heavy metals (HMs) and metalloids within the environment and notably in soils and the subsequent transfer along the food chain to man is a matter of growing environmental concern. At present, HMs and metalloids such as As (first place), Pb (second place), Hg (third place), and Cd (seventh place) are currently on the top ten priority list of threatening substances as issued by the American Agency for the toxic substances and Disease Registry (ASTDR 2019).

Heavy metal(loid)s contamination of soil and wastewater is a serious environmental downside that has a detrimental effect on living organisms and is harmful or toxic even at low levels. The contamination of heavy metal(loid)s is becoming a grave issue with the increased growth of industries and phantom alteration of regular biogeochemical cycles. Heavy metal is a typical generic term used for a category of metal(loid)s with an atomic density >5 gcm^{-3} (Hawkes 1997). Heavy metal(loid)s include arsenic (As), cadmium (Cd), lead (Pb), cobalt (Co), nickel (Ni), iron (Fe), zinc (Zn), chromium (Cr), silver (Ag) and the platinum (Pt) group elements. Their toxicity to plants is dependent on plant type/species, the precise nature of the metal, and its concentration, pH, and soil composition.

Corresponding author: kumarharsh1911@gmail.com

Heavy metal(loid)s can strongly bound to the soil particle. Heavy metals cannot be degraded by biological or chemical activities as in the case of organic material. A pollutant is any environmental element that can result in undesired effects like harm living organisms, deteriorate life, and can cause death even when present in low concentrations. Heavy metals in soil can cause toxicological effects on plants and soil microbiota and thus consequently reduce the microbial population of soil and their enzymatic activities (Khan et al. 2010).

Some heavy metal(loid)s like Cd, As, Hg, Se, Pb, and Se are non-essential, as they do not play a vital role in the plant's physiology. Further to such non-essential HMs and metalloids, some HMs such as Zn, Fe, Mn, and Cu are crucial for proper growth and expansion. However, at high doses, certain metals cause some lethal consequences and are identified as environmental toxicants (Kumar et al. 2016). Due to their persistent and tenacious nature, these HMs accumulate in the environment and cause serious health problems. They get accumulated along the ecological food chain through direct uptake by the primary producers and enter the next successive ecological levels. Heavy metals are cytotoxic even at low levels and can induce malignancies in humans (Dixit et al. 2015).

3.2 Toxic Metals Sources and Their Impact on the Environment

A number of sources of toxic metals have been identified in the ecosystem such as natural sources, agricultural activities, domestic effluents, industrial activities, atmospheric sources, and traffic emissions.

3.2.1 Natural sources

The primary natural sources of serious metal(loid) pollutants within the soil are surface erosion, volcanic activities, urban/municipal runoffs, and suspended aerosols/particles. Volcanic eruptions have been reported to have a venturous effect on the environment, including the climate and human health of the exposed populations. Volcanic eruptions contain high levels of Cu, Pb, Mn, Hg, Ni, and toxic gases.

Several significant metals are present in inland coastal areas because of aerosols created in oceanic practices. Aerosols (fine mixture particles or water droplets within the air) carry a completely different form of contaminants, such as smoke clouds and HMs. The aerosols that contain significant amounts of HMs usually settle on leaf blades within a variety of fine particulates and might penetrate through stomata into the leaf mesophylls (Sardar et al. 2013).

3.2.2 Agricultural sources

Organic and inorganic fertilizers are the most prominent sources of HMs and metalloids pollution in farming soil including the dumping of urban and industrial wastes and atmospheric emission from motor vehicles and the burning of fossil fuels (Zhang 2006). Increased amounts of these heavy metals in soils will have a detrimental effect on plant physiological activities resulting in plant growth reductions, dry matter accumulation, and yield (Suciu et al. 2008).

3.2.3 Industrial sources

Metal processing and mining, electroplating, nuclear energy, and textile activities are some of the factors responsible for the pollution of HMs in soil. Metal processing and electroplating require the coating of thin protective layers into prepared metal surfaces by electrochemical processes. Metal mining and smelting operations are known to be a significant source of toxic metals. In areas where such activities take place, deposition of significant amounts of HMs in corresponding soil, and crop plants are very common (Wei et al. 2008).

3.2.4 Anthropogenic activities

Anthropogenic practices such as mining, waste sludge applications, improper waste management, and low land operations are the main reason for severe metal emissions (Ha et al. 2014). These practices result in the accumulation of several toxic metals into the environment and generating their accumulation in the soil (Glick 2003). Heavy metals are usually incorporated into the urban environment by urban waste, industrial effluents, vehicle pollution, building waste, and the widespread use of agrochemicals (Zhao et al. 2003). Selected HMs and their anthropogenic sources have been listed in Table 3.1.

Table 3.1: Anthropogenic origins of selected HMs contamination

Metal	Sources	References
Cd	Electroplating, plastic stabilizers, paints and pigments, combustion of cadmium-containing plastics, phosphate fertilizers	Pulford and Watson (2003)
As	Wood preservatives and pesticides	Thangavel and Subbhuraam (2004)
Pb	Emission from the burning of leaded fuel, manufacturing units of batteries, insecticides, and herbicides	Wuana and Okieimen (2011)
Ni	Industrial effluents, combustion of fossil fuels, medical instruments, batteries, mining, and electroplating	Tariq et al. (2006)
Hg	Release from Au–Ag mining, medical waste and, coal-burning	Rodrigues et al. (2012)
Cr	Steel companies, tanneries, fly ash	Khan et al. (2007)
Cu	Cu pesticides and fertilizers	Khan et al. (2007)

3.3 Heavy Metal Toxicity

Soil pollution by toxic metal(loid)s can disproportionate the microflora and microfauna of a place and could be detrimental for a healthy biodiversity. Generally, soils with high metal exposure show poor plant growth and surface quality, resulting in metal's leakage into water runoffs with subsequent absorption into the soil and water sediments (Lasat 2000). The most hazardous HM contaminants have been mentioned in Table 3.2, where the limits of tolerance are indicated.

Table 3.2: Concentration ranges and regulatory limits for toxic metals in soil

Metals	Soil concentration range[a] (mg kg^{-1})	Regulatory Limit[b] (mg kg^{-1})
As	0.10-02	20
Pb	1-6,900	600
Hg	< 0.01-800	270
Cr	0.05-3950	100
Cd	0.10-345	100
Zn	150-5000	1,500
Ni	10-1000	35-40

[a] Riley et al. 1992

[b] Non-residential direct contact soil cleanup criteria (NJDEP 1996).

3.3.1 Arsenic

Arsenic (As) is a silver-grey naturally-occurring metalloid with brittle crystalline characteristics and an atomic number of 33. It is non-essential, but it can accumulate in plants to lethal levels. As a consequence, it can enter the food chain and pose a health risk to humans. In the environment, arsenic may be mixed with oxygen (O_2), sulfur, and form inorganic As compounds. Inorganic As compounds are primarily used to protect the wood. Organic As compounds are used as pesticides, particularly in cotton plants. These metals in the environment are also associated with other elements (Au, Ag, and Sn in particular), and the production and processing of these ores have contributed to extensive arsenic contamination in mining regions around the world. Arsenic concentration has a range of <10 mg kg^{-1} (non-contaminated soils) to 30,000 mg kg^{-1} (contaminated soils) (Adriano 1986, Vaughan 1993). It has been observed that the inorganic types, arsenite (As^{3+}) and arsenate (As^{5+}), are the prevalent species occurring in most habitats; however, the organic species could also be present (Andrianisa et al. 2008). Organic and inorganic forms and the oxidation state of As can be influenced by the mineral composition, pH, and microbial activity of soil.

Arsenic toxicity in plants

As^{3+} and As^{5+} are toxic and they interrupt plant physiology in a different manner. It has been reported that the As^{3+} form is more soluble (5-10 folds) in water than As^{5+}. Methylated arsenic (+3) is perhaps more noxious than inorganic arsenic because it is more effective at inducing the breakdown of DNA (Vaclavikova et al. 2008). However, As^{5+} appears to be less harmful than As^{3+} because it is thermodynamically more steady due to its predominance under ideal circumstances and is the source of significant groundwater contaminant. Arsenate in the (As^{5+}) state is often known to be toxic and carcinogenic to humans (Yusof and Malek 2009). It interrupts oxidative phosphorylation and ATP production. In plants, transition of arsenate to arsenite results in the development of reactive oxygen species leading to damage to DNA, protein, and lipids. The high toxicity of arsenate is due to its chemical resemblance with phosphate. Arsenic reduces the germination percentage and chlorophyll content in *Triticum aestivum* (Chun-xi et al. 2007), root and shoot dry weight in *Oryza sativa* (Shri et al. 2009), decreased fresh weight in *Cicer arietinum* (Gunes et al. 2009), and *Arabidopsis thaliana* (Leterrier et al. 2012).

3.3.2 Cadmium

Cadmium (Cd) is a naturally present metal found in the periodic table (d-block) of the elements between zinc (Zn) and mercury (Hg), with chemical activity similar to Zn. Cd (II) is usually more stable in the positive valence of (+2) and exists in most of the natural aquatic environments in that state (Baes and Mesmer 1976). It was first discovered in Germany in 1817 by Stromeyer and Hermann. Cadmium is one of the major toxic contaminants released from mining or industrial activities and poses a major threat to humans. The average concentration in the Earth's crust is in the range of 0.1-0.5 ppm.

Cadmium is extensively used in the production of dyes, paints, photography, glass, and electroplating. In the 1970s, Cd was being used as a plating agent, pigment, stabilizer, and batteries. The heat, light, weathering, and alkali resistance of Cd, as well as its color, promoted its use as a pigment in various plastics, paints, textiles, rubber, glass, and enamels. In batteries, Cd is complexed with Ni, while in alloys, it is combined with metals like Cu and Zn (Nriagu et al. 1982). Other important sources of Cd, which particularly affect agricultural soils, are phosphate fertilizers, sewage sludge, and other organic waste (Adriano 1992).

Table 3.3 depicts HM pollution and its toxicological effects on humans. The absorption of Cd in soil and water triggered significant environmental and health problems (Salt et al. 1995a). It is poisonous at extremely low levels, and it has no known useful function in higher organisms. Itai-Itai is one of the chronic diseases caused by the consumption of rice, grown on Cd contaminated rice fields, which causes osteomalacia and osteoporosis (Kaji 2012). Long-run exposure in humans leads to renal dysfunction, characterized by tubular proteinuria.

Table 3.3: HM pollution and its toxicological effects on humans

Metals	Diseases	Symptoms	References
Hg	Minamata disease	Affects kidney, brain, and the development of the fetus	Alina et al. (2012)
As	Arsenicosis	Causes cancer of lungs, liver, bladder, pigmentation, and keratosis	Chowdhury et al. (2000)
Cd	Itai-Itai disease	Bone fracture, lung disorders, etc.	Nogawa et al. (1987)
Pb	Pb poisoning	Mental retardation, congenital disabilities, psychosis, paralysis hyperactivity, paralysis, weight loss, muscular weakness, brain damage	Martin and Griswold (2009)
Zn	Flu-like symptoms	Gastro-intestinal and respiratory damage as well as injuries to brain, heart, and kidney	Friberg et al. (1986)

Cadmium toxicity in plants

Cadmium is a major phytotoxic HM (Ogunkunle et al. 2020) and its sequestration at elevated concentrations not only results in toxicity symptoms but also causes structural and ultra-structural changes. The effects of higher levels of cadmium on plant species have been extensively studied with ecological and environmental implications that rely on the presence of Cd in soil or water. Plants growing in soil contaminated with Cd exhibit clear signs of injury expressed in chlorosis, leaf rolls, growth inhibition, and finally death (Mohanpuria 2007).

Cadmium exposure in higher plants results in stomatal closure. The rise in stomatal susceptibility was closely associated with a rise in the abscisic acid (ABA) level in leaves (Poschenrieder et al. 1989). Several studies have shown that the main sites of action of Cd are chlorophyll (Chl 'a' and chl 'b'), and carotenoids biosynthesis (Prasad 1995). Disruption of photosynthesis was observed in several plant species, like oilseed rape (*Brassica napus*) (Baryla et al. 2001), and mungbean (*Vigna radiata*) (Wahid et al. 2008), after both long and short-term Cd exposure.

The primary target of Cd mostly affects two main enzymes of CO_2 fixation, i.e. RuBP (ribulose-1,5-bisphosphate), and PEP (phosphoenolpyruvate) carboxylases. Cadmium ions reduce the activity of RuBP and disrupt its structure by replacing Mg ions, which are essential co-factors of carboxylation reactions, and also, Cd can transfer the activity of RuBP carboxylase to oxygenation reactions (Siedlecka et al. 1998). A high concentration of Cd in plants can increase the respiration and enzymatic activity of the tricarboxylic acid cycle (TCA) and other carbohydrate utilization pathways (Arisi et al. 2008). This rise in respiration was associated with increased demand for ATP, which compensates for deficits in photophosphorylation (Ernst 1980).

3.3.3 Nickel

Nickel (Ni) is a naturally occurring metallic element with an atomic number and atomic weight of 28 and 58.71, respectively. It is a silvery-white lustrous metal with a shiny appearance. It is the 5th most prevalent element that widely occurs in the Earth's crust and core. Nickel is one of the ubiquitous trace metal which can be released from both anthropogenic and natural sources (WHO 1991). Among heavy metals, nickel is classified as hard and ductile metal. Nickel is an important micronutrient for plant life as well as being a part of the enzyme urase, it is necessary for nitrogen metabolism in plants. Nickel and its components have several commercial and industrial uses. The advancement of industrialization has contributed to an increase in the emission of contaminants into the environment.

Nickel could find its way into the atmosphere through man-made activities, such as smelting, mining, vehicle emissions, fossil fuel combustion, household, urban and industrial wastes, and organic manures (Alloway 1995). Ni toxicity in plants has also become a global threat to sustainable agriculture. In comparison to other toxic metal(loid)s like arsenic, cadmium, copper, chromium, lead, and mercury, plant scientists have paid little attention to Ni due to its dual existence and advanced chemistry.

Nickel toxicity in plants

Toxicological effects of higher doses of Ni in plants have been observed by many researchers at various degrees such as suppression of mitotic activity, plant-water relationship, photosynthesis, reduced plant growth, inhibition of antioxidative enzyme activity, and nitrogen metabolism (Gajewska et al. 2009). Overall, it decreases plant productivity because of poor plant growth and reduced availability of nutrients in the soil. In addition to harmful growth effects, there may be alterations in plant morphology and physiology. Seregin and Kozhevnikova (2006) reported the reduction in mesophyll tissue and vessel diameter of vascular bundles upon exposure to 1 mM $NiSO_4$ in wheat. The decrease in chloroplast size, dysfunction of chloroplast ultrastructure, including the diminished numbers of grana and thylakoids and the changes in the membrane lipid composition were reported in *Cajanus cajan* and *Brassica oleracea* when plants were grown in the presence of $NiSO_4.7H_2O$ (10–20 g/m^3) (Molas 1997).

The principles of suppression of plant growth and production by nickel are not well known. Nickel can reduce the plasticity of plant cell wall possibly by specific binding to pectins and can increase the peroxidase activity and intercellular spaces of the plant cell wall. In wheat plants, Ni stress can instigate a significant reduction in ascorbate peroxidase (APX) and superoxide dismutase activity in the leaves of the plant. APX could exert an essential role in the elimination of H_2O_2 (hydrogen peroxide) from the Ni-stressed plants as the maximum value of APX corresponds with a decrease in H_2O_2 content. In contrast to the plasma membrane, the toxic metals also harm structural and functional damage to the thylakoid membrane by causing modifications and destruction of lipid content in the membrane.

3.3.4 Lead

Lead (Pb) is a primary group member of the carbon group with an atomic number of 82. It is a light, malleable, poor metal, and widely distributed. Lead is the 2nd most significant and toxic heavy metal; it adsorbs mainly in the top soil, while its concentration reduces as soil depth increases (de Abreu et al. 1998). It exists in the environment as dust, fumes, mist, and minerals. When it is absorbed into the soil, it is very hard to remove. It is listed by EPA in category B1, as a probable human carcinogen (Evangelou 1998). Considerably affected soil contains 400-800 mg Pb kg^{-1} soil. The average Pb concentrations ranged from 2-200 mg kg^{-1} soil, and some researchers have indicated that less polluted soils contain less than 100 mg kg^{-1}. High Pb content is present in 'soil A horizon', while dissolved Pb in soil water is just around 0.005-0.13% of total soil Pb, though bioavailable for uptake by plants (Watmough et al. 2004).

Lead toxicity in plants

A high Pb content in the soil can cause irregular morphology in several species of plants. Lead inhibits plant growth such as germination rate, seedling production, root-shoot elongation, lamellar organization in the chloroplast, chlorophyll, and cell division (Maestri et al. 2010). Lead toxicity can alter the cell membrane as Pb^{2+}

is physiologically identical to calcium (Srivastava and Gupta 1996). Increased Pb influences the production of chlorophyll pigments and Fe metabolism. A high concentration of Pb is associated with high stomatal resistance, CO_2 uptake inhibition, and reduced photosynthetic activity (Poskuta et al.1987). Lead inhibits seed germination in *Spartina alterniflora* (Morzck and Funicclli 1982), *Pinus halepensis* (Nakos 1979), *Soybean* (Huang et al. 1975), and *Barley* (Stiborova et al. 1987) plants. Lead also inhibited stem and root elongation in *Raphanus sativus* and *Allium* species (Gruenhage and Jager 1985).

3.3.5 Zinc

Zinc (Zn) is an element with an atomic number of 30. Zn is an essential metal that belongs to Group II of the Periodic Table and occupied the 24th position among elements that are abundant in the Earth's crust. Zinc is a transition metal that typically exists in nature in its divalent state (+2). It is known as one of the most essential minerals for living organisms. However, significantly increased concentration of Ni is toxic (Hambidge and Krebs 2007). Zn plays an important function in plants, the most relevant is its activity as a functional activator of a variety of enzymes (peptidases, dehydrogenases, phosphohydrolases, and proteinases) (Clarkson and Hanson 1980). Maximum Zn in the natural soil is 10-300 ppm (Tisdale et al. 1985). The clay contains three times more Zn than sandy soil. The major areas of Zn mining are Russia, Australia, the USA, Canada, and Peru. Global supply crosses seven million tonnes per year, and commercially exploitable reserves reach a hundred million tons.

It is the primary metal used to make American pennies and die casting in the automobile sector. Some commonly used Zn compounds are zinc oxide (used in water colors or paints and as a modulator in the rubber industry), zinc chloride (used in deodorants), zinc sulfide (used for luminescent paints), zinc pyrithione (used as an anti-dandruff shampoo agent), and zinc carbonate (used as dietary supplements). Zinc metal is used in a tablet and is known for having antioxidant properties, which also defend against premature aging of the skin.

Zn smelting operations, and industrial effluents are some of the primary sources of Zn contamination. It is the most versatile heavy metal since it exists as soluble compounds at acidic and neutral pH values. Zn is a critical component of living systems and is harmful only at extremely high concentrations (McIntyre 2003). It is also an essential component for the synthesis of protein in terrestrial organisms and the second most crucial metal after iron (Fe) in living organisms (Broadley et al. 2007).

Zinc toxicity in plants

Symptoms of Zn toxicity include lower yields, stunted growth, decrease in chlorophyll biosynthesis, chloroplast deterioration, and interruption in the intensity of fundamental physiological processes, i.e. transpiration, photosynthesis, respiration, and reproduction (Kholodova et al. 2005). Metal toxicity resulted in an increase in the metal supplied to the root that triggered cell organelles to disintegrate, membrane disruption, and condensation of chromatin material were major events during zinc toxicity (Sresty and Madhava 1999). Khurana and Chatterjee (2001) reported a reduction in biomass, seed weight, seed number, and soluble proteins in sunflower (*Helianthus annuus*) plants grown under Zn-laden soil. The high level of Zn can disrupt the Mg ions and OEC (oxygen-evolving complex) of PSII, thus impeding both photosystems (PS) and the electron transport system (ETS) in Zn-treated *Phaseolus vulgaris* plants (Van Assche and Clijsters 1986). A high concentration of Zn in the soil also inhibits many plant metabolic functions, results in leaf senescence, retarded growth, and chlorosis in younger leaves, which can spread to an older part of leaves after long-term exposure under Zn contaminated soils (Ebbs and Kochian 1997). Zinc toxicity also inhibits plastidial ATP synthesis in *Spinacia oleracea* (Teige et al. 1990).

3.4 Bioavailability of HMs in the Environment

In order to understand the accessibility and solubility, HMs in the environment are classified into two categories:

1. Bioavailable (soluble, non-sorbed, and mobile)
2. Non-bioavailable (precipitated, complexed, and non-mobile)

Bioavailability is a measure of the potential for the entry of metal into biological receptors. Most of the researches on metal bioavailability have been undertaken in the soil system as identifying the outcomes of metals in soil and sediments are difficult in order to determine the effects of metals on the biota, metal leaching to groundwater, and metal transfer up the food chain. In general, the transition can be influenced by different reactions, including ion exchange, adsorption, complexation with organic/inorganic ligands, and precipitation-dissolution (Morel 1997).

The inaccessible portion is adsorbed firmly to soil particles, or embedded into organic and inorganic compounds. Doelsch et al. (2008) evaluated fractionation patterns in soils with elevated Zn (104–242 mg kg^{-1}), Cr (106–175 mg kg^{-1}), Ni (89–310 mg kg^{-1}), and Cu (34–118 mg kg^{-1}) concentrations. They stated less than 5% of the total HM contents were present in the exchangeable fractions. Conversely, the residual fraction contained the most considerable bit of the studied heavy metals and ranged from 39–97.9% of the total Cu, Cr, Ni, and Zn concentrations. Similarly, Sánchez-Martin et al. (2007) studied the fractionation patterns of soils amended with heavy metals tainted sewage sludge and found that >60% of Ni, Cd, Cr, Pb, Cu, and Zn were present in the residual fraction.

In soil, metal's bioavailability and modification depend on the oxidation-reduction reactions, acidification, precipitation, and chelation capacity of the soil (Seneviratne et al. 2017). Plant roots naturally produce certain chemicals that alter the bioavailability of metals under stress conditions. These exudates form complexes with metal ions and provide nutrients for colonizing microbes. The toxicity of heavy metals increases ROS production and thus reduces the antioxidant systems (GR, SOD, CAT, APX, etc.), which protects the cell. If this situation continues, the structure and role of the organism will be disrupted and can eventually lead to cell death (Ishtiyaq et al. 2018, 2020).

3.5 Remediation Methods

3.5.1 Conventional methods

Over the last few decades, many methods have been used to eliminate hazardous metal(loid)s from soil and sediments, but most of them are expensive with low efficiency. Conventional methods for extracting HMs from an aqueous medium comprise ion exchange, membrane technologies, chemical precipitation, electrochemical treatment, filtration, and chemical oxidation or reduction. Chemical technologies produce massive volumetric sludge and lifting costs (Rakhshaee et al. 2009). Another major downside of conventional treatment technologies is the development of hazardous chemical sludge, and its disposal management is expensive and is not environmentally friendly (Fig. 3.1).

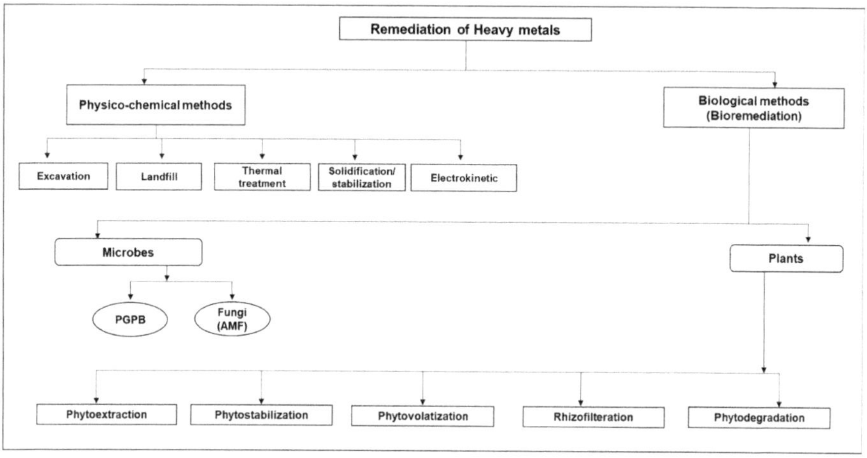

Figure 3.1: Schematic representation of HMs remediation methods.

Remediation methods based on thermal and chemical approaches are technically complicated and costly, and the valuable portion of the soil can also be weakened by these processes (Hinchman and Cristina 1997). Consequently, the elimination of hazardous heavy metals to an environmentally appropriate level in a cost-effective and environmentally safe manner is, therefore, of great importance. Excavation, capping, solidification, isolation, soil washing, and containment are some of the examples of physical remediation methods.

Conventionally, the rehabilitation of metal-polluted soils requires either onsite (*ex situ*) monitoring or excavation and proper disposal at the landfill site. This recycling strategy simply moves the contamination abroad along with the uncertainties involved in the transport of polluted soil and the transfer of pollutants from landfill sites to a nearby area. The cost can surpass \$300/ cubic yard making enormous disposal initiatives very expensive. One more limitation of this strategy is the physical constraints of the excavation equipment.

Capping technique is used to prevent water from infiltrating the soil but is site-specific (Mulligan et al. 2001). This technique is low-cost compared to other *in situ* remediation approaches. This mitigation method can only reduce the metal transfer rate in the soil, while the immobilization impact for heavy metal is minimum. Some chemical modifications (such as apatite, lime, zeolite, and rock phosphate) can also be added to the sand cap in order to improve the immobilization ability of the soil. This method does not remove contaminants from the soil, as the contaminants are still present in the soil; it merely limits their spread (USEPA 2002).

Solidification and stabilization (s/s) are two technologies that are closely related to the physical and chemical processes, which can reduce the stability of hazardous substances and pollutants in the environment. Solidification is a physical process in which the dirt soil sludge is transformed into a solid, monolithic substance that is more resistant to physical deterioration and sensitive to leaching than the untreated material. *Stabilization* typically involves the process that lowers the chances of a contaminant by converting it into a less reactive, immobile, and less harmful form. This method includes the insertion of binding materials to the soil profile to remove the pollutants in the soil matrix by a combination of chemical reactions, reduced surface area, and permeability (Evanko and Dzombak 1997). The drawbacks of solidification and stabilization remediation strategies are that they can be relatively expensive.

In *chemical extraction* process, the excavated soil is combined with a mixture of solvent or surfactant to eliminate the pollutants. This method can be applied to heavy metals as well as organic compounds. As this technique is costly, its implementation is not common (Dermont et al. 2008).

In the *thermal treatment* process, the soil is heated to a very extreme temperature, and volatile pollutants, primarily organics, are removed from the soil. The pollutants are either combusted or retrieved with solvent due to heat volatilization. This process can be effectively used to concentrate mercury (Hg) from soil (Stegmann 2001).

Soil washing is a commonly used method for the effective cleanup of soil, polluted with PCBs, PAHs, SVOCs (semi-volatile organic compounds), petroleum, pesticides, heavy metals, and other organic pollutants (Peters 1999).

Electrokinetic technique uses the effects of induced electric field to activate targeted contaminants and their displacement is facilitated by the use of suitable chemicals, that serve as complexing agents. This method can be used as an *in situ* tool, while for the treatment of excavated soil it is not that suitable. The key strength of this method is that it can be effectively used for the treatment of low permeable soils (Mulligan 2001).

3.5.2 Bioremediation: Mitigation strategies for removal of heavy metal contaminants

In addition to the chemical and physical methods, biological strategies have also been introduced as an important tool for the remediation of pollutants from water and soil. Bioremediation is the application of microorganisms and plant species to treat polluted/contaminated soils. It is a commonly accepted method of soil remediation as it is based on natural processes. It is a specific method that can utilize the flexibility of microorganisms and plants (Fig. 3.2).

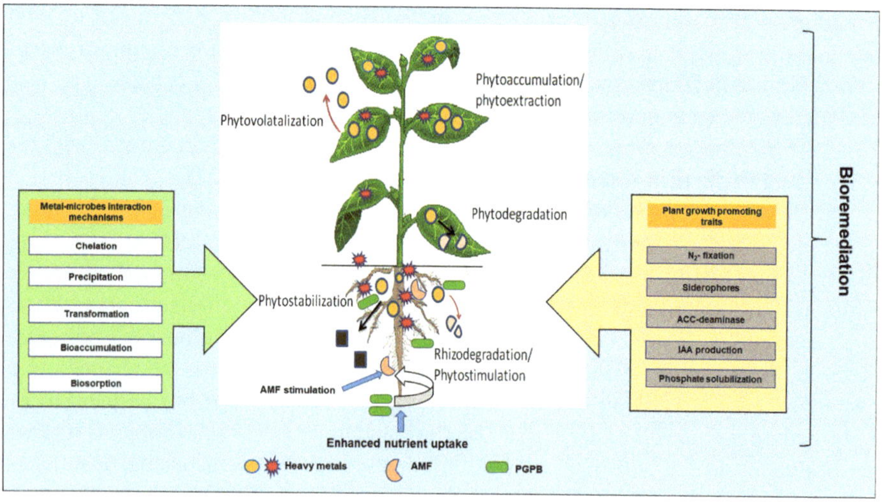

Figure 3.2: Mechanism of PGPB-mediated phytoremediation of HMs.

Phytoremediation

Phytoremediation involves the utilization of plant species that have fundamental and responsive processes for accepting or retaining high metal contents in their tissues. Phytoremediation can be used as an *in-situ* remediation technology for the remediation of soil and water contaminated with organic and inorganic contaminants (Glick 2003). Phytoremediation involves the exploitation of several plants referred to as phytoremediators, capable of extracting significant amounts of certain toxic HMs from the soil. These plants can store a significant concentration of HMs in their different tissues and organs.

This method illustrates the absorption processes of both organic and inorganic pollutants using phytoremediation technology. Remediation of organic pollutants can be done by phytodegradation, phytostabilization, rhizofiltration, and phytovolatilization. The mechanisms involving organic contaminants cannot be incorporated into the tissues of plants. Phytostabilization, phytovolatilization, rhizofiltration, and phytoextraction techniques/processes may be involved in the phytoremediation of inorganic pollutants.

The concept of employing plants to remove HMs from polluted soil was first conceived and implemented by Chaney (1983). Since then, the ongoing development of phytoremediation has mainly been influenced primarily by the spiraling costs correlated with conventional soil remediation strategies and the necessity to adopt the sustainable "green" method. Plants may be associated with the solar-driven pump method for the extraction and absorption of these elements from the contaminated growth matrix (Salt et al. 1995b). This technology has been regarded as a modern, highly promising technology for the elimination of contaminated sites. The major benefits and drawbacks of the phytoremediation technology have been presented in Table 3.4.

Table 3.4: Advantages and limitations of plant remediation technology

Advantages	Limitations
1. Low capital and maintenance costs and metal recycling are additional economic advantages	1. Slow process as compared to others, especially in hyperaccumulator plants
2. Relevant to a wide variety of inorganic pollutants	2. Restricted by depth, solubility, and accessibility of contaminants
3. Does not need costly equipment or highly trained staff	3. Limited to areas with low contaminated concentrations
4. It can be implemented *in situ*. It also reduces disturbances of soil and spread of contaminants/pollutants	4. Generated biomass of plant from phytoextraction needs to be properly disposed off as hazardous waste
5. Easy to apply and manage. Plants are a cheap and sustainable resource that is readily available	5. Implementation of unwanted or invasive plant species must be avoided (non-native species may affect the ecosystem)
6. Practically possible and publicly accepted. Less noisy than any other remediation process (Watt 2007).	6. Plants with metal accumulation may contaminate the food chain
7. Plant residue can also be used for phytomining.	7. Multiple cycles required to clean up the area

Depending on the pollutants, and site conditions, phytoremediation involves five major plant-based strategies with various modes of action to clean metal-polluted sites (Vamerali et al. 2010). Phytostabilization and phytoextraction are among the most effective methods of removing toxic metals from soils.

Phytoextraction

Phytoextraction is the process that involves the absorption of soil contaminants by plant and the accumulation of these contaminants in the harvestable portion of the plant (i.e. roots and shoots). Phytoextraction generally has the essential specifications, such as fast growth rate, high biomass yield, deep root structure, and the capacity to endure high levels of heavy metals. According to Kumar et al. (1995), phytoextraction is a continuous mitigation action that involves multiple crop cycles for the reduction of metal concentrations to admissible limits. One significant advantage of phytoextraction is that it can be used for the remediation of low to moderately polluted soils. However, also *Thlaspi caerulescens* and *Brassica juncea*, the strongest metal accumulators in field tests, will take 15 years of uninterrupted cultivation to mop up HMs from polluted sites.

Phytostabilization

It is a plant-based mitigation strategy that decreases the accessibility and potency of contaminants by plant roots in the rhizosphere for the elimination of inorganic contaminants. Marques et al. (2009) stated that this method of remediation is only applied where the soil is highly polluted and the use of extraction plants would take a very long time to be done and would therefore not be suitable. Phytostabilizing efficiency depends on the modification of the plant and soil used. Plants tend to stabilize the soil by their root systems, thereby avoiding erosion.

In a study of industrially-contaminated sites, Tam and Singh (2004) reported a high accumulation of Cr, Pb, and Zn in kangaroo grass (*Themeda australis*), ryegrass (*Lolium perenne*), and Kikuyu grass (*Pennisetum clandestinum*), respectively. They suggested that these species can be used for phytoremediation purposes. Such species may be ideal for clean up several Cd and Pb polluted sites such as shooting ranges, smelting sites, etc., where alternative remediation techniques may be impractical due to the size of the area or funds available for remediation. The micro-organisms present in the rhizosphere also play a significant role in phytoremediation. They help the plant to overcome the phytotoxicity by chaining the metal speciation, thus helping in the revegetation process (Van der et al. 1999).

Phytovolatilization

Phytovolatilization involves the removal of metals from the polluted environment and conversion of them to less harmful volatile forms, to facilitate the remediation of organic/inorganic contaminants. It is a special way of dealing with metal pollutants that may be present as gaseous species in the atmosphere, like selenium (Se) and mercury (Hg). Mercury and selenium are poisonous metals and there is some debate about whether volatilization of these metals into the environment is safe or not (Watanabe 1997). Despite the criticism, phytovolatilization is a promising technique for the remediation of As, Hg, and Se polluted soils. Volatilization of Se in the form of methyl selenate ($CH_3O_4Se^-$) was suggested as a significant mechanism for the elimination of selenium. *Brassica juncea* was recognized as an important plant for the extraction of selenium from the soil (Raskin et al. 1997).

Rhizofilteration

It is the utilization of plant root systems to remove (or adsorb) contaminants, primarily HMs, but could also be applied to organic contaminants, from aqueous solutions. This approach is well suitable for the rehabilitation of most metals (like Cu, Pb, Cr, Cd, and Ni), nutrients, and radionuclides (such as U, Sr, and Cs) in polluted water. Plant species successfully used for rhizofiltration are *Brassica* sp., *Helianthus* sp., (Dushenkov et al. 1995), *Lemna* sp., and *Thlaspi* sp. (Salt et al. 1995a), and *Eichhornia crassipes* (Zhu et al. 1999).

Phytodegradation or phytotransformation

This technique involves the destruction of contaminants by plants and soil microorganisms, which removes organic compounds. Plants pick up contaminants by the process of phytodegradation and reduce them to less toxic forms. This deterioration occurs in two ways, i.e. either by metabolic processes of the plant or through enzymes released by the plant. The process is suitable for removing organic contaminants like ethyl benzene, xylene, benzene, and chlorinated solvents (Zayed and Terry 2003).

Bioremediation cannot be achieved by plants alone. There is always a strong connection between the growth-promoting bacteria in the root system and plants which contribute to increased activity associated with soil remediation (also called micro-remediation), i.e. the use of microbes (Compant et al. 2010). Both methods are preferred for physical and chemical remediation due to their cost-effectiveness and lower adverse effects.

The method of bioremediation is divided into three phases. Initially, due to natural attenuation, contaminants are diminished by native bacteria without any human increase. Second, biostimulation is used to introduce nutrients and oxygen into the systems to increase their performance and enhance degradation, and in the last, microbes are added into systems during bioaugmentation. The main benefits and limitations of bioremediation techniques have been listed in Table 3.5.

Table 3.5: Advantages and limitations of bioremediation techniques

Bioremediation techniques	Examples	Advantages	Limitations
In situ	*In situ* bioremediation	Cost-effective	Environmental
	Bioaugmentation	Less damaging to the environment	Time monitoring difficulties
	Biosparging	Improves the efficiency of air sparging in the treatment of a broader range of petroleum hydrocarbons	Chemical, physical, and biological processes are not well understood.
	Bioventing	Bioremediation of volatile organic carbon pollutants	Time-consuming
	Biostimulation	Takes advantage of indigenous microorganisms that are well adapted to the environment	Extremely site-specific and requires immense scientific observation (Mohee and Mudhoo 2012)
Ex-situ	Land farming	Stimulate native biodegradable microorganisms and promote the aerobic degradation of pollutants (cost-efficient)	Space requirements
	Biopiles	Effective on organic constituents	Requires a large area for treatment
	Composting	Low-cost	Extended treatment time

Bioremediation cannot deteriorate toxic metals but can convert them from one oxidation state to another. In response to a change in oxidation state, these metals shift to less toxic, more or less water-soluble, and readily volatilized forms that can be easily removed from the environment (Garbisu and Alkorta 2003). Several microbes, especially growth-promoting bacteria like *Pseudomonas putida,* and *Enterobacter cloacae*, have been widely used to minimize Cr (+6) to the less reactive Cr (+3) (Garbisu et al. 1998).

The massive class of microorganisms involved in the removal of toxic metal includes bacteria (such as *Pseudomonas fluorescens, Bacillus* sp., *Citrobacter, Cyanobacteria, Rhodococcus, Enterobacter cloacae, Pseudomonas aeruginosa, Streptomyces* sp., *Alcaligenes, Sphinganonas,* and *Mycobacterium*) and fungi like *Rhodotorula mucilaginosa, Penicillium chrysogenum, Candida utilis,* and *Aspergillus terreus* (Ahirwar et al. 2016).

3.6 Plants and Microbes for Environmental Clean-up of HMs

Phytoremediation can be used to remediate various types of contaminants, namely pesticides, solvents, petroleum hydrocarbons, heavy metals, explosives, organic substances, etc. The metals chiefly targeted include Pb, As, Hg, U, Cr, Ni, Cd, Cs, etc. Unlike organic contaminants, heavy metals cannot be degraded by chemical or biological modifications (NRC 1997). Table 3.6 lists some phytoremediation studies related to various HMs and their tolerance mechanisms in various plants under abiotic stresses.

The addition of heavy metal-resistant bacteria is one of the most promising strategies used for phytoremediation; it could act on metals directly by accumulation, chelation, biosorption, precipitation, and transformation. A variety of bacteria growing in metalliferous soils tends to accumulate metals in plants in their harvestable part and have the ability to be used for remediation of heavy metal contaminated soil called plant-assisted bioremediation. Most soil bacteria withstand HM toxicity and play important roles in the mobilization of heavy metals (Wenzel 2009).

3.7 Mechanism of HM Uptake, Transport, and Accumulation in Plants

Plants developing in a heavy metal-contaminated matrix absorb HMs and effectively sequester them into various parts of the plant. Several reports have shown that the separation of toxic metals at the plant level may generally be separated into three groups. Chaney and Giordano (1977) classified Mn, Se, Zn, Cd, and Mo as easily moved to above-ground biomass; Ni, Co, and Cu belong to the intermediate group, while Cr, Pb, and Hg can be transported to the lowest extent (Alloway 1995). The accumulation of HMs involves four main key steps:

(i) **Mobilization and uptake:** In order to enhance the absorption rate of HMs, plants reduce the pH of their rhizosphere and release metal chelators such as organic acids (Clemens et al. 2002). This leads to the increased concentration gradient of HMs in the cell wall during the absorption of HMs. Heavy metals have been considered to remain in the negatively charged areas of the cell wall or distributed through an apoplastic pathway to the cortex (Greger 1999). In addition, a portion of HMs can enter the cytoplasm of plant root, and it is assumed to arise through:

Table 3.6: Heavy metals and their resistance mechanisms in plants

Plants	Heavy metal	Effect on plants	Tolerance mechanism	References
Atriplex canescens	Cu	Successfully grown in Cu mine tailings and displayed high Cu accumulation in leaves	Accumulation in roots (Phytostabilization)	Sabey et al. (1990)
Avicennia marina	Pb, Cu, and Zn	Increased guaiacol peroxidase (GPX) activity (in Pb only), decreased (in Cu and Zn)	Antioxidant system	Macfarlane and Burchett (2001)
Thlaspi caerulescens	Cd, Zn	Accumulate 50–160 mg kg^{-1} Cd and 13,000-19,000 mg kg^{-1} Zn	Phytoextraction	Zhao et al. (2003)
Atriplex lentiformis	Pb, Zn	Plants exhibit sufficient growth in moderate and high acidic Pb/Zn mine-tailings	Accumulation in roots (Phytostabilization)	Mendez et al. (2007)
Avicennia marina	Cu, Pb, Zn	Increases guaiacol peroxidase activity	Antioxidant system	Caregnato et al. (2008)
Atriplex halimus sub sp.*Schweinfurthii*	Cd	Accumulation of metal in the shoot	Accumulation in shoot	Nedjimi and Daoud (2009)
Sesuvium portulacastrum	Pb	Increased levels of Pb in root	Accumulation in roots (phytoextraction)	Zaier et al. (2010)
Eichhornia crassipes	As and Cu	Arsenic hyperaccumulation	Phytoextraction	Mokhtar et al. (2011)
Atriplex halimus	Cu, Pb, and Zn	Increased concentration of metals in root	Exclusion	Kachout et al. (2012)
Atriplex halimus	As, Cd, Pb, and Zn	Increase in biomass production	Phytostabilization	Clemente et al. (2012)
Atriplex lentiformis and *A. undulate*	Cd, Zn, and Pb	Maintain concentration of metals	Phytostabilization	Eissa (2015)
Chrysopogon zizanioides	Cu, Ni, and Zn	Metals' accumulation in roots and restricted translocation to the shoots	Phytoextraction	Melato et al. (2016)

(*Contd.*)

Table 3.6: (*Contd.*)

Plant	Heavy metal	Effect on plants	Tolerance mechanism	References
Oryza sativa	Cd	Secretion of organic acids by roots	Phytostabilization	Fu et al. (2017)
Arabidopsis	Hg	Increased accumulation of metal	Overexpression of ATP-binding cassette (ABC) transporter gene-PtABCC1	Sun et al. (2018)
Trigonella foenum-graecum	Cd, Cr, and Pb	Reduction in root and shoot length	Alteration in antioxidative defense	Alaraidh et al. (2018)

(a) *Passive transport:* This occurs when a significant amount of HMs present in the cell wall of the plant facilitates the mass diffusion of these metal ions into the cytoplasm via plasma membrane (Clemens et al. 2002).
(b) *Active transport (through plasma membrane carrier proteins):* This is necessary because the low concentration of metals cannot move freely through the lipid-soluble plasma membrane (Lasat 2002). Inorganic ion carriers often include Ca^{2+} and H^{+}ATPases channel proteins (Clemens et al. 2002).

(ii) **Root compartmentalization and chelation:** This involves the chelation of HMs to organic acids (e.g. histidine, citrate, malate), phytochelatins (PCs) form of thiol rich peptides, and metallothioneins (MTs) cysteine-rich compound once inside the root cytoplasm (Pilon-Smits and Pilon 2000). They are stored or transported to root and shoot vacuoles.
(iii) **Loading in xylem and transfer to shoots:** In order to reach the xylem, HMs are transported across the Casparian strip. This transport is a challenging process as Casparian strips are not entirely formed in all the differentiated parts of the root. Thus, these parts are primarily involved in the accumulation of HMs (Greger 1999). During the transport towards the xylem, HMs accumulation takes place in the cytoplasm of root cells and is later conveyed to the stele. The movement of HMs into the xylem tissue is a firmly regulated mechanism that is mediated by membrane carrier-proteins (Ghosh and Singh 2005).
(iv) **Sequestration and deposition in leaf tissues:** Xylem metals enter leaf tissues through the apoplastic pathway. Transfer of metals from xylem to leaf tissues takes place through membrane-binding proteins such as ABC-type vacuole transporters and P-type ATPase (Pilon-Smits 2000). Inside the plant cell, HM levels can also be controlled by aggregation in vacuoles.

3.8 Mechanism of PGPB in Plant Growth Promotion under Abiotic Stress

Plant beneficial bacteria that have a strong influence on the growth and development of plants are also named plant growth-promoting bacteria (PGPB). These bacteria act as a bio-control agent and encourage the development of plants indirectly by suppressing plant pathogens. The augmentation of metal-resistant bacteria is one of the most promising technique mainly used for the acceleration of phytoremediation, which might directly affect metals by chelation, precipitation, transformation, bioaccumulation, and absorption. Numerous soil bacteria resist toxic metals and play an essential part in the extraction of heavy metals (Wenzel 2009).

PGPBs promote plant growth through different mechanisms, like increased nutrient availability, suppression of pathogens by producing siderophores, bacterial and fungal antagonistic substances, nitrogen fixation, and phytohormone production. Many growth-promoting rhizospheric bacteria can naturally tolerate toxic metals and are thus used for phytoremediation of metal pollutants in the soil. Rhizospheric bacteria make essential nutrients available in soils to promote plant growth, generate plant hormones like IAA, protect the plant from diseases and remediate the soil from contaminants (Nehra and Choudhary 2015).

Rhizobacteria, which are active root colonizers, have gained significant attention in recent years. Siderophores pose growth benefits for both plants and microbes (Narendra et al. 2015). Many factors affect microorganisms to use or metabolize contaminants as substrates, including certain environmental factors such as available nitrogen, phosphorus sources, temperature, and pH, and therefore tend to have access to the rate and intensity of degradation (Fritsche and Hofrichter 2008).

The N_2 fixation is an important example of plant-microbe interaction, which is an endo-symbiotic relationship between roots of leguminous plants and microbes that can enzymatically convert gaseous N_2 (nitrogen) to NO_3^- (nitrate) and NO_2^- (nitrite). Over the last few years, common apprehensions about metal(loid)s pollution have been resolved by the adoption of legume-*Rhizobium* association, as a strategy of metal remediation from contaminated soil. This symbiotic pairing is a crucial part of the phytoremediation process (Pajuelo et al. 2011).

Nitrogen-fixing bacteria include *Mesorhizobium, Bradyrhizobium, Sinorhizobium, Rhizobium,* and *Azorhizobium,* which not only promote plant growth but also improve soil quality (Kong et al. 2015).

Growth-promoting bacteria live in the surrounding of the host plant where they strengthen the production of siderophore, 1-aminocyclopropane-1-carboxylate deaminase synthesis, phosphate solubilization, and which facilitates the plant to endure abiotic stress circumstances by decreasing ethylene levels, and improves hormone production (Ahemad and Kibret 2014). Most of the bacterial genera are endophytic microorganisms and include *Agrobacterium, Pseudomonas, Micrococcus, Azotobacter, Flavobacterium,* and *Azospirillium* (Bhattacharyya and Jha 2012). The metal-resistant microbes (PGPB) alleviate the adverse stress effects by altering the soil's physico-chemical properties, causing detoxification and soil removal. These PGPRs contain a variety of compounds that promote plant growth under stress conditions. It includes:

3.8.1 Siderophores

Siderophores are low molecular mass metal chelating agents produced by soil microbes and plants, especially under iron-limiting conditions (Ahmed and Holmström 2014). These metal-binding agents enhance the absorption of essential nutrients by plants that can support plant growth in a stressed environment. Iron has been one of the most important nutrients necessary for the biological process, although it is insufficient in soil. Siderophore forming PGPB may promote plant growth, directly by providing better mineral nutrition as well as indirectly impeding the expression of plant pathogens in the rhizosphere by restricting their accessibility of iron (Ma et al. 2011).

Improving Fe-nutrients by soil microorganisms in plants is necessary during exposure to severe external factors, like loss of important minerals or in the occurrence of excessive HMs that can restrict Fe-uptake. Siderophores could, therefore, make PGPR extremely helpful for the remediation of HMs in contaminated soils. The exact process through which siderophores contribute to the absorption of plant metals remains undefined (Rajkumar et al. 2010). However, certain soluble metals can be mobilized by bacterial siderophores; the efficacy of a siderophore-forming bacterium in mobilizing or immobilizing soil metals relies on the type of the metal, plant species, soil composition, concentration of the metal, type of siderophores secreted, and the potential of rhizosphere microbes to mobilize toxic metals.

3.8.2 Phosphate solubilization

It is a determining factor in growing plants since it is less mobile and readily available to plants under soil conditions than other major nutrients. Plants may take up the small amount of phosphorous present in the soil either as monobasic ($H_2PO_4^-$) or dibasic (HPO_4^{2-}) ions (Bhattacharyya and Jha 2012). Most of the PGPBs converting phosphorous to a soluble form for plants are known as phosphate-solubilizing bacteria (PSB). These PSB provide phosphorous to the plant under chronic conditions and increase plant activity by improving the fixation of biological nitrogen (N_2) and allowing certain trace elements by plant-growth-enhancing stimulants (Choudhary et al. 2017). In metal-polluted soils, PSB are one of the most important agents as they have the potential to solubilize stable and insoluble HMs by the sequestration of organic compounds, thus eventually enhance the bioavailability of metals for phytoextraction (Ahemad and Kibret 2014). Various microbial communities (particularly *Bacillus* spp and *Pseudomonas* spp) and fungi (*Aspergillus* spp and *Penicillium* spp) can develop phosphate to a soluble form, by releasing organic acids (Chunningham and Kuiack 1992).

3.8.3 ACC deaminase

ACC deaminase was first isolated from *Pseudomonas* sp. strain ACP and *Hansenula saturnus* (yeast) in 1978; thereafter, it has been identified in many other bacteria as well as in fungi. This enzyme regulates the production of ethylene by converting ACC into α-ketobutyrate and ammonia and can be used to protect plant pathogens from ethylene-generated stress (Glick et al. 1998). Most ACC deaminase-producing bacteria specifically stimulates plant growth by producing the 1-aminocyclopropane-1-carboxylate (ACC) deaminase enzyme under harsh circumstances such as drought, salinity, and accumulation of toxic metals. Promotion of plant growth by ACC deaminase results in ethylene reduction that enables the plant to withstand pressed abiotic environment factors (Nadeem et al. 2013). These ACC-producing microbes function as a sink of ethylene destruction.

3.8.4 IAA (Indole-3-Acetic Acid) production

Auxins are naturally occurring plant growth hormones formed in the apical areas of plants that induce cell division, root elongation, cell extension, ultimately increasing the absorption of metal ions (Parker et al. 1992). IAA is important for plant-microbe relations, particularly among plants and rhizobacteria, which stimulate the production via expansive root systems and defend the plant from abiotic stress (Spaepen and Vanderleyden 2011). Augmentation of PGPB substantially improves the production of biomass and thus promote phytoremediation (Asghar et al. 2013). PGPB have significant effects on metal transport and their speciation in the rhizosphere (Lors et al. 2004). IAA produced by rhizobacteria can promote root elongation and increase the root surface area, and thus improve the accessibility of nutrients for the plant. This eventually results in the maximization of root exudates, which in turn provide more nutrients to the soil bacteria (Ahemad and Kirret 2014). Table 3.7 reviews the studies on phytoremediation of metals facilitated by PGPR and the growth-regulating compounds associated with them.

3.9 Arbuscular Mycorrhizal Fungi in the Remediation of HM Toxicity

Several microorganisms can live in association with the roots of plants. Arbuscular mycorrhizal fungi (AMF) can establish a symbiotic relationship with some plant species that correspond to heavy metal tolerance. They exist in relationships with roots of higher plants in various forms such as arbuscular mycorrhizas, orchid mycorrhizas, and ectomycorrhizas (Sheng and Xia 2006). AMF is important for plant growth in many forms as it can increase the bioavailability of plant nutrients such as Ca, N, P, S, Zn, K, and Cu in the soil through their extensive hyphal network (Zaidi et al. 2006).

It has also been shown that, in some highly polluted environments (smelting and mining tailings) containing significantly high concentrations of toxic metals, AMF can potentially support host plants by the promotion of water absorption and uptake of minerals and other nutrients by the plant. AMF can enhance plant tolerance under pressed circumstances and can combat plant pathogen infections (Trotta et al. 2006). Results further indicate that the impact of AMF on HM absorption by host plants can be affected by HM speciation, total metal concentrations, physico-chemical characteristics of the substrates, the combined effect of AMF isolates, and the host plants as well as cultivation conditions (Leyval et al. 1997). The endomycorrhizal hyphae of *Glomus intraradices* can transfer inaccessible forms of metals (Cd/Zn) to water-soluble forms making them accessible for the plant (Giasson et al. 2005).

Mycorrhiza-based phytoextraction can be accomplished by a variety of mechanisms: (i) by strengthening growth and biomass production of the plant; (ii) enhanced plant resistance to metals; (iii) increased metal accumulation in plants. The association among *Elsholtzia splendens* and various AMF such as *Gigaspora* spp., *Scutellospora* spp, *Acaulospora* spp., and *Glomus* spp. enhances plant growth and metal accumulation (Wang et al. 2005). Inoculation of fungi and growth-promoting rhizobacteria was observed to improve the metal (Zn and Cd) accumulation and translocation in maize plants (Gharemaleki et al. 2010). The symbiotic effect of fungi and PGPR has been reported to effectively promote the growth of plants and help in the mitigation of metals from polluted soils.

Metal resistance PGPB usually eliminates toxic metals by chelation, precipitation, transformation, bioaccumulation, and absorption. Removal of HMs by algae involves the process of biosorption and bioaccumulation. Biosorption is a reasonably expeditious and reliable technique. The use of hydrophytes, particularly macroalgae, has become increasingly important due to their potential to uptake and accumulate HMs from the surrounding aqueous environment (Mitra et al. 2012). Macroalgae have been reported useful as biomonitors and can be used for the elimination of HMs from the aquatic ecosystem (Gosavi et al. 2004). Several species of green algae, such as *Cladophora* and *Enteromorpha,* have been used to test heavy metal levels in many regions of the world (Al-Ghanayem et al. 2011).

Table 3.7: Phytoremediation of heavy metals facilitated by PGPR

PGPR	Sources	Metal resistances	PGPR-traits	Plant	Functions	Reference
Pseudomonas sp. PsA, *Bacillus* sp. Ba32 (RS)	Cr contaminated soil	Cr	ACC-deaminase, siderophore, IAA, P solubilization, N_2 fixation	Brassica juncea	Increase plant biomass (Phytostabilization)	Wu et al. (2006)
Bradyrhizobium sp. (vigna) RM8 (RS)	Green gram nodules grown in metal-polluted soils	Ni	IAA, siderophore, HCN, NH_3 production	*Vine radiata*	Stimulate growth, nodulation, N_2 content, and seed output (Phytostabilization)	Wani et al. (2007)
20 Promising isolates (RS)	Rhizosphere of saltbush plants grown in tailings; bulk tailings, and mine tailings	Zn, Pb	ACCD, siderophore, IAA	*Atriplex lentiformis*	Increase plant establishment and biomass (Phytostabilization)	Grandlic et al. (2008)
Pseudomonas fluorescence G10	Root of *B. Napus* grown in the heavy metal polluted area	Pb	ACCD, siderophore, IAA	*Brassica napus*	Enhances the root length/ shoot length, and plant biomass; increased Pb uptake in shoots (Phytoextraction)	Sheng et al. (2008)
Pseudomonas sp. PsM6, *P. jessenii* PjM15 (RS)	Serpentine soil, Bragança, Portugal	Ni, Cu, Zn	ACCD, siderophore, IAA	*Ricinus communis*	Increased dry wt of shoot and root, increased Zn translocation and uptake	Rajkumar and Freitas (2008)
Pseudomonas sp. RJ10, *Bacillus* sp. RJ16 (RS)	HM-polluted soil in Nanjing, China	Pb, Cd	ACC deaminase, IAA, siderophore	*Lycopersicon esculentum*	Cd, Pb mobilization increases root/shoot length and shoot tissue, dry weight in the pot experiment (Phytoextraction)	He et al. (2009)

(Contd.)

Table 3.7: (*Contd.*)

PGPR	Sources	Metal resistances	PGPR-traits	Plant	Functions	Reference
Bacillus sp. SLS18	Stem of Mn-hyperaccumulator *Phytolacca acinosa*	Mg, Cd	IAA, siderophores, ACC deaminase	*Sorghum bicolor*, *Phytolacca acinosa* Roxb. and *Solanum nigrum*	Increased plant biomass and uptake of Mn/Cd	Luo et al. (2012)
Rhizobium sp. RL9	Root nodules of lentil	Cd, Ni, Zn, Pb	IAA, siderophores, HCN, NH_3	*Lens culinaris* var. Malka	Increased growth, chlorophyll, nodulation, leghemoglobin, N_2 content, seed protein, and seed yield; decreased uptake of Ni	Wani and Khan (2013)
Rahnella sp. JN27	Roots of *Zea mays*	Cd, Zn, Cu	IAA, siderophores, ACC deaminase, phosphate solubilization	*Solanum nigrum*, *Zea mays*	Increased biomass; increased Cd content (Phytostabilization)	Ming et al. (2014)
Bradyrhizobium sp. YL-6	Root nodules of *Glycine max*	Cd	IAA, ACC deaminase activity, siderophores, phosphate solubilization	*Lolium multiflorum* and *Glycine max*	Increased overall physiology and shoot dry weight, Cd content in roots (Phytostabilization)	Guo and Chi (2014)
Bacillus safensis FO-036b (T), *P. fluorescens* p.f.169 along with SiO_2 and zeolite nano-particles	Rhizosphere of sunflower	Pb, Zn	IAA, siderophore, ACCD	*H. annuus*	Enhanced plant growth, dry weight, and Pb/Zn uptake (Phytoextraction)	Seyed et al. (2018)

3.10 Conclusion and Future Trends in Bioremediation

Heavy metal(loid)s pollution is a significant environmental and health concern. Hence, successful remediation methods are needed. Chemical and physical approaches for cleaning and restoring HM-polluted soils are limited by the high cost of implementation, non-reversible changes in soil features, degradation of indigenous soil microbiota, and the formation of secondary contamination problems. In this situation, bioremediation has gained prominence because it is low-cost and environment friendly than conventional methods. It is evolving as a new *in situ* remediation technology to clean metal(loid)s polluted soil, sediment, and water. Improved understanding of heavy metal absorption by plants would also help to facilitate phytomining, which can be used to extract metals from plants. Bioremediation is projected to be an economically feasible technology for the remediation of toxic metal(loid)s. At present, phytoremediation innovations are in infancy phase and several technological hurdles are needed to be tackled. However, the aid of new advanced agronomic practices such as molecular and genetic engineering tools have accelerated the bioremediation advancement and give rise to specifically designed microorganisms for various metal removal processes. Besides that, there is still a need for a comprehensive understanding of the complexities of plant-heavy metal-microbe associations and the proper disposal of biomass. To establish a healthy soil environment, it is essential to maintain proper vegetation with adequate bioavailability and sustainability of soil microbes. The key objectives of soil remediation include restoration of the microbial population, overall ecological structure of the soil, and the elimination of soil contaminants. In this context, the implementation of plant growth-promoting bacteria as bio inoculants could be a promising approach, considering that they possess the potential to enhance both plant growth and nutrient bioavailability in soil. It could be possible to utilize helpful microorganisms to modify metal bioavailability and enhance the phytoremediation of metal(loid)s on a large scale in the environment.

Acknowledgment

This study was supported by Fundação para a Ciência e Tecnologia (FCT), through the strategic project UID/MAR/04292/2019 granted to Marine and Environmental Sciences Centre (MARE). We thank University Grants Commission, New Delhi for providing funding (Grant sanction F. no. 43-100/2014 SR) for the work. Harsh Kumar is grateful for the fellowship under the project. Authors are thankful to Dr. B. R. Ambedkar of the Department of Botany, St John's College, Agra for providing the necessary facilities and equipment for conducting the research.

References

Adriano, D.C. 1986. Trace Elements in the Environment. Springer-Verlag, New York.

Adriano, D.C. 1992. Biogeochemistry of Trace Metals. Lewis Publishers, CRC Press, Boca Raton, FL, USA.

Agency for Toxic Substances and Disease Registry (ATSDR). 2019. Substance priority list. Atlanta, GA: U.S. Department of Health and Human Services, Public Health Service.

Ahemad, M. and M. Kibret. 2014. Mechanisms and applications of plant growth-promoting rhizobacteria: Current perspective. J. King Saud Univ. Sci. 26: 1-20.

Ahirwar, N.K., G. Gupta, R. Singh and V. Singh. 2016. Isolation, identification, and characterization of heavy metal resistant bacteria from industrial affected soil in central India. Int. J. Pure App. Biosci. 4: 88-93.

Ahmed, E. and S.J. Holmström. 2014. Siderophores in environmental research: Roles and applications. Microb. Biotechnol. 7: 196-208.

Alaraidh, I.A., A.A. Alsahli and E.S.A. Razik. 2018. Alteration of antioxidant gene expression in response to heavy metal stress in *Trigonella foenum-graecum* L. South Afri. J. Bot. 115: 90-93.

Al-Homaidan, A.A., A.A. Al-Ghanayem and A.H. Alkhalifa. 2011. Green Algae as Bioindicators of Heavy Metal Pollution in Wadi Hanifah Stream, Riyadh, Saudi Arabia. Int. J. Water Resour. Arid Environ. 1(1): 10-15.

Alina, M., A. Azrina, M.A.S. Yunus, M.S. Zakiuddin, M. Izuan, H. Effendi and M.R. Rizal. 2012. Heavy metals (mercury, arsenic, cadmium, plumbum) in selected marine fish and shellfish along the Straits of Malacca. Int. Food Res. J. 19: 135-140.

Alloway, B.J. 1995. Soil processes and the behavior of metals. pp. 38-57. *In:* B.J. Alloway [ed.]. Heavy Metals in Soils. Blackie Academic & Professional, London.

Andrianisa, H.A., A. Ito, A. Sasaki, J. Aizawa and T. Umita. 2008. Biotransformation of arsenic species by activated sludge and removal of bio-oxidized arsenate from wastewater by coagulation with ferric chloride. Water Res. 19: 4809-4817.

Arisi, A.C.M., B. Mocquot, A. Lagriffoul, M. Mench and L. Jouanin. 2008. Responses to cadmium in leaves of transformed poplars overexpressing Γ-glutamylcysteine synthetase. Physiol. Plantarum. 109: 143-149.

Asghar, H.N., M.A. Zafar, M.Y. Khan and Z.A. Zahir. 2013. Inoculation with ACC deaminase-containing bacteria to improve plant growth in petroleum-contaminated soil. Rom. Agric. Res. 30: 281-289.

Baes, C.F. and R.E. Mesmer. 1976. The Hydrolysis of Cations. Krieger Publishing Company, Malaber, F.L. pp. 489.

Baryla, A., P. Carrier, F. Franck, C. Coulomb, C. Sahut and M. Havaux. 2001. Leaf chlorosis in oilseed rape plants (*Brassica napus*) grown on cadmium-polluted soil: Causes and consequences for photosynthesis and growth. Planta. 212: 696-709.

Bhattacharyya, P. and D. Jha. 2012. Plant growth-promoting rhizobacteria (PGPR): Emergence in agriculture. World J. Microbiol. Biotechnol. 28: 1327-1350.

Broadley, M.R., P.J. White, J.P. Hammond, I. Zelko and A. Lux. 2007. Zinc in plants. New Phytol. 173: 677-702.

Caregnato, F.F., C.E. Koller, G.R. MacFarlane and J.C.F. Moreira. 2008. The glutathione antioxidant system as a biomarker suite for the assessment of heavy metal exposure and effect in the grey mangrove, *Avicennia marina* (Forsk.) Vierh. Mar. Pollut. Bull. 56: 1119-1127.

Chaney, R.L. 1983. Plant uptake of inorganic waste constitutes. pp. 50-76. *In:* J.F. Parr, P.B. Marsh and J.M. Kla [eds.]. Land Treatment of Hazardous Wastes. Park Ridge, NJ: Noyes Data Corp.

Chaney, R.L. and P.M. Giordano. 1977. Microelements as related to plant deficiencies and toxicities. pp. 234-279. *In:* L.F. Elliotand and F.J. Stevenson [eds.]. Soils for Management of Organic Wastes and Waste Waters. American Society of Agronomy, Madison, WI.

Choudhary, D.K., A. Varma and N. Tuteja. 2017. Plant-microbe interaction. pp. 61-74. *In*: An Approach to Sustainable Agriculture. Springer: Mycorrhiza – Function, Diversity, State of the Art.

Chowdhury, U.K., B.K. Biswas, T.R.Chowdhury, G. Samanta, B.K. Mandal, G.C. Basu, et al. 2000. Groundwater arsenic contamination in Bangladesh and West Bengal, India. Environ. Health Perspect. 108: 393-397.

Chunningham, J.E. and C. Kuiack. 1992. Production of citric and oxalic acids and solubilization of calcium phosphate by *Penecillium bilaii.* Applied and Environ Microbiol. 58: 1451-1458.

Chun-xi, L., F. Shu-li, S. Yun, J. Li-na, Xu-yang and H. Xiao-li. 2007. Effects of arsenic on seed germination and physiological activities of wheat seedlings. J. Environ. Sci. 19: 725-732.

Clarkson, D.T. and J.B. Hanson. 1980. The mineral nutrition of higher plants. pp. 239. *In:* W.R. Briggs, P.W Green and J.L. Jones [eds.]. Annual Review of Plant Physiology. Ann Reviews Inc, Palo Alto, CA.

Clemens, S.M., G. Palmgren and U. Kramer. 2002. A long way ahead: Understanding and engineering plant metal accumulation. Trends Plant Sci. 7: 309-315.

Clemente, R., D.J. Walker, T. Pardo, D. Martínez-Fernandez and M.P. Bernal. 2012. The use of a halophytic plant species and organic amendments for the remediation of trace elements-contaminated soil under semi-arid conditions. J. Hazard. Mater. 15: 223-224.

Compant, S., B. Clément and A. Sessitsch. 2010. Plant growth-promoting bacteria in the rhizo- and endosphere of plants: Their role, colonization, mechanisms involved and prospects for utilization. Soil Biol. Biochem. 42: 669-678.

De Abreu, C.A., M.F. De Abreu and J.C. De Andrade. 1998. Distribution of lead in the soil profile evaluated by DTPA and Mehlich-3 solutions. Bragantia. 57: 185-192.

Dermont, G., M. Bergeron, G. Mercier and M. Richer-Laflèche. 2008. Soil washing for metal removal: A review of physical/chemical technologies and field applications. J. Hazard. Mater. 152: 1-31.

Dixit, R., D. Malaviya, K. Pandiyan, U.B. Singh, A. Sahu, R. Shukla, B.P. Singh, J.P. Rai, P.K. Sharma, H. Lade and D. Paul. 2015. Bioremediation of heavy metals from soil and aquatic environment: An overview of principles and criteria of fundamental processes. Sustainability 7: 2189-2212.

Doelsch, E., G. Moussard and H.S. Macary. 2008. Fractionation of tropical soilborne heavy metals—Comparison of two sequential extraction procedures. Geoderma. 143: 168-179.

Dushenkov, V., P.B.A. Nanda, H. Motto and I. Raskin. 1995. Rhizofiltration: The use of plants to remove heavy metals from aqueous streams. Environ. Sci. Technol. 29: 1239-1245.

Ebbs, S.D. and L.V. Kochian, 1997. Toxicity of Zinc and Copper to *Brassica* species: Implications for phytoremediation. J. Environ. Qual. 26: 776-781.

Eissa, M.A. 2015. Impact of Compost on Metals Phytostabilization Potential of Two Halophytes species. Int. J. Phytorem. 17: 662-668.

Ernst, W. 1980. Biochemical aspects of cadmium in plants. pp. 639-653. *In:* J. Nriagu [ed.]. Cadmium in the Environment. Wiley and Sons, New York.

Evangelou, V.P. 1998. Environmental Soil and Water Chemistry: Principles and Applications. Wiley-Interscience Publications, New York.

Evanko, C.R. and D.A. Dzombak. 1997. Remediation of Metals-Contaminated Soils and Groundwater, Technology, Evaluation Report, Pittsburgh, PA.

Friberg, L., G. Nordberg and V. Vouk. 1986. Handbook on the Toxicology of Metals. 2nd ed. Elsevier, Amsterdam.

Fritsche, W. and M. Hofrichter. 2008. Aerobic degradation by microorganisms. pp. 289-320. *In:* H.J. Rehm and G. Reed [eds.]. Biotechnology Set, Second Edition. Wiley-VCH Verlag GmbH, Weinheim, Germany.

Fu, H., H. Yu and T. Li. 2017. Influence of cadmium stress on root exudates of high cadmium accumulating rice line (*Oryza sativa* L). J. Ecotoxicol. Environ. Saf. 150: 168.

Gajewska, E., M. Wielanek, K. Bergier and M. Skłodowska. 2009. Nickel-induced depression of nitrogen assimilation in wheat roots. Acta Physiol. Plant. 31: 1291-1300.

Garbisu, C. and I. Alkorta. 2003. Basic concepts on heavy metal soil bioremediation. The Eur. J. Mineral Process. Environ Protect. 3: 58-66.

Garbisu, C., I. Alkorta, M.L. Llama and J.L. Serra. 1998. Aerobic chromate reduction by *Bacillus subtilis*. Biodegradation 9: 133-141.

Gharemaleki, T., M.H. Rasouli-Sadaghiani, H. Besharati and A. Tavasoli. 2010. Plant growth-promoting micro-organisms effect on Cd uptake by *Zea mays* in a contaminated soil. Intern Soil Science Congress on "Management of Natural Resources to Sustain Soil Health and Quality", Samsun, Turkey 5: 1135-1140.

Ghosh, M. and S.P. Singh. 2005. A review on phytoremediation of heavy metals and utilization of its byproducts. Appl. Ecol. Environ. Res. 3: 1-18.

Giasson, P., A. Jaouich, S. Gagné and P. Moutoglis. 2005. Arbuscular mycorrhizal fungi involvement in zinc and cadmium speciation change and phytoaccumulation. Remediation J. 15: 75-81.

Glick, B.R. 2003. Phytoremediation: Synergistic use of plants and bacteria to clean up the environment. Biotechnol. Adv. 21: 383-393.

Glick, B.R., D.M. Penrose and J. Li. 1998. A model for lowering of plant ethylene concentrations by plant growth promoting bacteria. J. Theor. Biol. 190: 62-68.

Gosavi, K., J. Sammut and J. Jankowski. 2004. Macro-algal biomonitors of trace metal contamination in acid sulfate soil aquaculture ponds. Sci. Tot. Environ. 324: 25-39.

Grandlic, C.J., M.O. Mendez, J. Chorover, B. Machado and R.M. Maier. 2008. Plant growth-promoting bacteria for phytostabilization of mine tailings. Int. J. Environ. Sci. Technol. 42: 2079-2084.

Greger, M. 1999. Metal availability and bioconcentration in plants. pp. 1-27. *In:* M.N.V. Prasad and J. Hagemeyer [eds.]. Heavy Metal Stress in Plants: From Molecules to Ecosystems. Springer, New York.

Gruenhage, L. and I.I.J. Jagar. 1985. Effect of heavy metals on growth and heavy metals content of *Allium Porrum* and *Pisum sativum*. Angew Bot. 59: 11-28.

Gunes, A., D.J. Pilbeam and A. Inal. 2009. Effect of arsenic-phosphorus interaction on arsenic-induced oxidative stress in *chickpea* plants. Plant Soil. 314: 211-220.

Guo, J. and J. Chi. 2014. Effect of Cd-tolerant plant growth-promoting rhizobium on plant growth and Cd uptake by *Lolium multiflorum* Lam. and *Glycine max* (L.) Merr. in Cd-contaminated soil. Plant Soil 375: 205-214.

Ha, H., J. Olson, L. Bian and P.A. Rogerson. 2014. Analysis of heavy metals sources in soil using kriging interpolation on principal components. Environ. Sci. Technol. 48: 4999-5007.

Hambidge, K.M. and N.F. Krebs. 2007. Zinc deficiency: A special challenge. J. Nutr. 137: 1101-1105.

Hawkes, J.S. 1997. Heavy metals. J. Chem. Edu. 74: 1369-1374.

He, L.Y., Z.J. Chen, G.D. Ren, Y.F. Zhang, M. Qian and X.F. Sheng. 2009. Increased cadmium and lead uptake of a cadmium hyperaccumulator tomato by cadmium-resistant bacteria. Ecotoxicol. Environ. Saf. 72: 1343-1348.

Hinchman, R.R. and N.M. Cristina. 1997. Providing the baseline science and data for real-life phytoremediation applications – Partnering for success. pp. 18-19. *In:* Proceedings of the 2nd Intern. Conference on Phytorem, Seattle, WA.

Huang, C.V., F.A. Bazzaz and L.N. Venderhoef. 1974. The inhibition of soya bean metabolism by Cd and Pb. Plant Physiol. 34: 122-124.

Ishtiyaq, S., H. Kumar, M. Varun, B. Kumar and M.S. Paul. 2018. Heavy metal toxicity and antioxidative response in plants: An overview: Responses, tolerance and remediation. pp.

77-106. *In:* M. Hasanuzzaman, K. Nahar and M. Fujita [eds.]. Plants Under Metal and Metalloid Stress. Springer Intern Publishing, Switzerland.

Ishtiyaq, S., H. Kumar, O.O. Clement, M. Varun and M.S. Paul. 2020. Role of secondary metabolites in salt and heavy metal stress mitigation by halophytic plants: An overview.pp. 307-321. *In:* M. Hasanuzzaman and M.N.V. Prasad [eds.]. Handbook of Bioremediation. Academic Press, Elsevier.

Jiang, Q.Q. and B.R. Singh. 1994. Effect of different forms and sources of arsenic on crop yield and arsenic concentration. Water Air Soil Pollut. 74: 321-343.

Kachout, S.S., A.B. Mansoura, R. Mechergui, J.C. Leclerc, M.N. Rejeb and Z. Ouerghi. 2012. Accumulation of Cu, Pb, Ni and Zn in the halophyte plant *Atriplex* grown on polluted soil. J. Sci. Food Agric. 92: 336-342.

Kaji, M. 2012. Role of experts and public participation in pollution control: The case of *Itai-itai* disease in Japan. Ethics Sci. Environ. Polit. 12: 99-111.

Khan, M.A., I. Ahmad and I. Rahman. 2007. Effect of environmental pollution on heavy metals content of *Withania somnifera*. J. Chin. Chem. Soc. 54: 339-343.

Khan, S., A.E.L. Hesham, M. Qiao, S. Rehman and J.Z. He. 2010. Effects of Cd and Pb on soil microbial community structure and activities. Environ. Sci. Pollut. Res. 17: 288-296.

Kholodova, V.P., K.S. Volkov and V.L.V. Kuznetsov. 2005. Adaptation of the common ice plant to high copper and zinc concentrations and their potential using for phytoremediation. Fiziol Rast. 52: 848-858.

Khurana, N. and C. Chatterjee. 2001. Influence of variable zinc on yield, seed oil content, and physiology of sunflower. Soil Sci. Plant Annals 32: 3023-3030.

Kong, Z., B.R. Glick, J. Duan, S. Ding, J. Tian, B.J. McConkey and G. Wei. 2015. Effects of 1-aminocyclopropane-1-carboxylate (ACC) deaminase-overproducing *Sinorhizobium meliloti* on plant growth and copper tolerance of *Medicago lupulina*. Plant Soil 391: 383-398.

Kumar, D., D.P. Singh, S.C. Barman and N. Kumar. 2016. Heavy metal and their regulation in plant system: An overview. pp. 19-38. *In:* Plant Responses to Xenobiotics, Springer, New York.

Kumar, P., V. Dushenkov, V.H. Motto and I. Raskin. 1995. Phytoextraction – The use of plants to remove heavy-metals from soils. Environ. Sci. Technol. 29: 1232-1238.

Lasat, M.M. 2002. Phytoextraction of toxic metals: A review of biological mechanisms. J. Environ. Qual. 31: 109-120.

Leterrier, M., M. Airaki, J.M. Palma, M. Chaki, J.B. Barroso and F.J. Corpas. 2012. Arsenic triggers the nitric oxide (NO) and S-nitrosoglutathione (GSNO) metabolism in *Arabidopsis*. Environ. Pollut. 166: 136-143.

Leyval, C., K. Turnau and K. Haselwandter. 1997. Effect of heavy metal pollution on mycorrhizal colonization and function: Physiological, ecological and applied aspects. Mycorrhiza 7: 139-153.

Lors, C., C. Tiffreau and A. Laboudigue. 2004. Effects of bacterial activities on the release of heavy metals from contaminated dredged sediments. Chemosphere 56: 619-630.

Luo, S., T. Xu, C. Liang, J. Chen, R. Chan and X. Xiao. 2012. Endophyte-assisted promotion of biomass production and metal-uptake of energy crop sweet sorghum by plant-growth-promoting endophyte *Bacillus* sp. SLS18. Appl. Microbio. Biotechnol. 93: 1745-1753.

Ma, Y., M.N. Prasad, M. Rajkumar and H. Freitas. 2011. Plant growth promoting rhizobacteria and endophytes accelerate phytoremediation of metalliferous soils. Biotechnol. Adv. 29: 248-258.

MacFarlane, G.R. and M.D. Burchett. 2001. Photosynthetic pigments and peroxidase activity as indicators of heavy metal stress in the grey mangrove, *Avicennia marina* (Forsk) *Vierh.* Mar. Pollut. Bull. 42: 233-240.

Maestri, E., M. Marmiroli, G. Visioli and N. Marmiroli. 2010. Metal tolerance and hyperaccumulation: Costs and trade-offs between traits and environment. Environ. Exp. Bot. 68: 1-13.

Marques, A.P.G.C., A.O.S.S. Rangel and P.M.L. Castro. 2009. Remediation of heavy metal contaminated soils: Phytoremediation as a potentially promising clean-up technology. Crit. Rev. Environ. Sci. Technol. 39: 622-654.

Martin, S. and W. Griswold. 2009. Human health effects of heavy metals. Environ. Sci. Technol. Briefs Citizens. 15: 1-6.

McIntyre, T. 2003. Phytoremediation of heavy metals from soils. Adv. Biochem. Engin/Biotechnol. 78: 97-123.

Melato, F.A., N.S. Mokgalaka and R.I. McCrindle. 2016. Adaptation and detoxification mechanisms of Vetiver grass (*Chrysopogon zizanioides*) growing on gold mine tailings. Int. J. Phytorem. 18: 509-520.

Mendez, M.O., E.P. Glenn and R.M. Maier. 2007. Phytostabilization potential of quailbush for mine tailings: Growth, metal accumulation, and microbial community changes. J. Environ. Qual. 36: 245-253.

Ming, Y., H. He, X. Li, T. Zhong, L. Hui and S. Li. 2014. Enhancement of Cd phytoextraction by two *Amaranthus* species with endophytic *Rahnella* sp. JN27. Chemosphere 103: 99-104.

Mitra, N., Z. Rezvan, M. Seyed Ahmad, M. Gharaie and M. Hosein. 2012. Studies of water arsenic and boron pollutants and algae phytoremediation in three springs, Iran. Int. J. Ecosys. 2: 32-37.

Mohanpuria, P., N.K. Rana and S.K. Yadav. 2007. Cadmium-induced oxidative stress influence on glutathione metabolic genes of *Camellia sinensis* (L.) O. Kuntze, Environ. Toxicol. 22: 368-374.

Mohee, R. and A. Mudhoo. 2012. Bioremediation, and Sustainability: Research and Applications. Hoboken, New Jersey: John Wiley & Sons, Inc.; Salem, Massachusetts: Scrivener Publishing LLC.

Mokhtar, H., N. Morad and F.F. Ahmad Fizri. 2011. Hyperaccumulation of copper by two species of aquatic plants. Intern Conference on Envirn. Sci. and Eng. IPCBEE Vol 8. IACSIT Press, Singapore.

Molas, J. 1997. Changes in morphological and anatomical structure of cabbage (*Brassica oleracea* L.) outer leaves and in ultrastructure of their chloroplasts caused by an *in-vitro* excess of nickel. Photosynth. 34: 513-522.

Morel, J.L. 1997. Bioavailability of trace elements to terrestrial plants. pp. 141-176. *In:* J. Tarradellas, G. Bitton and D. Rossel [eds.]. Soil Ecotoxicology. Lewis Publishers: NY, USA.

Morzck, E. and J.N.A. Funicclli. 1982. Effect of Pb and on germination of *Spartina alterniflora* Loisel seeds at various salinities. Environ. Exp. Bot. 22: 23-92.

Mulligan, C.N., R.N. Yong and B.F. Gibbs. 2001. Remediation technologies for metal contaminated soils and groundwater: An evaluation. Eng. Geol. 60: 193-200.

Nadeem, S.M., Z.A. Zahir, M. Naveed and S. Nawaz. 2013. Mitigation of salinity-induced negative impact on the growth and yield of wheat by plant growth-promoting rhizobacteria in naturally saline conditions. Ann. Microbiol. 63: 225-232.

Nakos, G. 1979. Lead pollution: Fate of lead in soil and its effects on *Pinus halepensis*. Plant Soil 50: 159-161.

Narendra, K., G. Ahirwar, V. Singh, R.K. Rawlley and S. Ramana. 2015. Influence on growth, physiology, and fruit yield of tomato (*Lycopersicon esculentum* Mill.) plants by inoculation with *Pseudomonas fluorescence* (SS5): A possible role of plant growth promotion. Int. J. Curr. Microbiol. Appl. Sci. 4: 1024-1029.

Nedjimi, B. and Y. Daoud. 2009. Cadmium accumulation in *Atriplex halimus* subsp. *schweinfurthii* and its influence on growth, proline, root hydraulic conductivity, and nutrient uptake. Flora-Morphology, Distribution, Function. Ecol. Plants 204: 316-324.

Nehra, V. and M. Choudhary. 2015. A review on plant growth-promoting rhizobacteria acting as bioinoculants and their biological approach towards the production of sustainable agriculture. J. Appl. Nat. Sci. 7: 540-556.

NJDEP. 1996. Soil Cleanup Criteria. Department of Environmental Protection. Proposed Cleanup Standards for Contaminated Sites, NJAC 7: 26D, New Jersey.

Nogawa, K., R. Honda, T. Kido, I. Tsuritani and Y. Yamada. 1987. Limits to protect people eating cadmium in rice, based on epidemiological studies. Trace Subst. Environ. Health. 21: 431-439.

NRC. 1997. Challenges of groundwater and soil cleanup. pp. 18-41. *In*: Innovations in Groundwater and Soil Clean-up. Washington, DC, National Academy Press.

Nriagu, J.O., H.K.T. Wong and R.D. Coker. 1982. Deposition and chemistry of pollutant metals in lake around the smeltors of Sunbury, Ontario. Environ. Sci. Technol. 16: 551-560.

Ogunkunle, C.O., A.M. Ahmed El-Imam, E. Bassey, V. Vishwakarma and P.O. Fatoba. 2020. Co-application of indigenous arbuscular mycorrhizal fungi and nano-TiO_2 reduced Cd uptake and oxidative stress in pre-flowering cowpea plants. Environ. Technol. Innov. 20: 101163.

Pajuelo, E., I.D. Rodriguez-Llorente, A. Lafuente and M.A. Caviedes. 2011. Legume-*Rhizobium* symbiosis as a tool for bioremediation of heavy metal polluted soils. pp. 95-123. *In:* M. Saghir Khan, A. Zaidi, R. Goel and J. Musarrat [Eds.]. Bio-management of metal-contaminated soils. Berlin, Germany, Springer Netherlands.

Parker, D.R., J.J. Aguilera and D.N. Thomason. 1992. Zinc-phosphate interactions in two cultivars of tomato has grown in chelator buffered nutrient solution. Plant Soil 143: 163-177.

Peters, W.R. 1999. Chelant extraction of heavy metals from contaminated soil. J. Hazard. Mater. 66: 151-210.

Pilon-Smits, E. and M. Pilon. 2000. Breeding mercury-breathing plants for environmental clean-up. Trends Plant Sci. 5: 235-236.

Poschenrieder, C.H., B. Gunse and J. Barcelo. 1989. Influence of cadmium on water relations, stomatal resistance, and abscisic acid content in expanding bean leaves. Plant Physiol. 90: 1365-1371.

Poskuta, J.W., E. Parys and E. Ramanowska. 1987. The effects of lead on the gaseous exchange and photosynthetic carbon metabolism of pea seedlings. Acta Societatis Botanicorum Poloniae. 56: 127-137.

Prasad, N. 1995. Cadmium toxicity and tolerance in vascular plants. Environ. Exp. Bot. 35: 535-545.

Pulford, I. and C. Watson. 2003. Phytoremediation of heavy metal-contaminated land by trees – A review. Environ. Int. 29: 529-540.

Rajkumar, M. and H. Freitas. 2008. Influence of metal resistant-plant growth-promoting bacteria on the growth of *Ricinus communis* in soil contaminated with heavy metals. Chemosphere 71: 834-842.

Rajkumar, M., N. Ae, M.N.V. Prasad and H. Freitas. 2010. Potential of siderophore-producing bacteria for improving heavy metal phytoextraction. Trends Biotechnol. 28: 142-149.

Rakhshaee, R., M. Giahi and A. Pourahmad. 2009. Studying effect of cell wall's carboxyl-carboxylate ratio change of *Lemna minor* to remove heavy metals from aqueous solution. J. Hazard. Mater. 163: 165-173.

Raskin, I., R.D. Smith and D.E. Salt. 1997. Phytoremediation of metals: Using plants to remove pollutants from the environment. Plant Biotechnol. 8: 221-225.

Riley, R.G., J.M. Zachara and F.J. Wobber. 1992. Chemical contaminants on DOE Lands and Selection of Contaminated Mixtures for Subsurface Science Research, US-DOE Off, Energy Res Suburb Science Programme, Washington DC.
Rodrigues, S., B. Henriques, A. Reis, A. Duarte, E. Pereira and P.F.A.M. Romkens. 2012. Hg transfer from contaminated soils to plants and animals. Environ. Chem. Lett. 10: 61-67.
Sabey, B.R., R.L. Pendleton and B.L. Webb. 1990. Effect of municipal sewage-sludge application on the growth of two reclamation shrub species in copper mine spoils. J. Environ. Qual. 19: 580-586.
Salt, D.E., M. Blaylock, N.P. Kumar, V. Dushenkov, B.D. Ensley, I. Chet and I. Raskin. 1995a. Phytoremediation—A novel strategy for the removal of toxic metals from the environment using plants. Biores. Technol. 134: 68-74.
Salt, D.E., R.C. Prince, I.J. Pickering and I. Raskin. 1995b. Mechanisms of cadmium mobility and accumulation in Indian mustard. Plant Physiol. 109: 1427-1433.
Sánchez-Martin, M.J., M. Garcia-Delgado, L.F. Lorenzo, M.S. Rodriguez-Cruz and M. Arienzo. 2007. Heavy metals in sewage sludge amended soils determined by sequential extractions as a function of incubation time of soils. Geoderma. 142: 262-273.
Sardar, K., A. Shafaqat, H. Samra, H.M. Tauqeer, S. Fatima, M.B. Shakoor, S.A. Bharwana and H.M. Tauqeer. 2013. Heavy metals contamination and what are the impacts on living organisms. Greener J. Environ. Manage. Public Saf. 2: 172-179.
Seneviratne, M., G. Seneviratne, H. Madawala and M. Vithanage. 2017. Role of rhizospheric microbes in heavy metal uptake by plants. pp. 147-163. *In:* J.S. Singh and G. Seneviratne [eds.]. Agro-Environmental Sustainability: Managing Environmental Pollution. Cham: Springer Intern Publishing.
Seregin, I.V. and A.D. Kozhevnikova. 2006. Physiological role of nickel and its toxic effects on higher plants. Fiziol Rast. 53: 285-308.
Seyed, M.M., M. Babak, M.H. Hossein, A. Hoseinali and A.Z. Ali. 2018. Root-induced changes of Zn and Pb dynamics in the rhizosphere of sunflower with different plant growth-promoting treatments in a heavily contaminated soil. Ecotoxicol. Environ. Saf. 147: 206-216.
Sheng, X.F. and J.J. Xia. 2006. Improvement of rape (*Brassica napus*) plant growth and cadmium uptake by cadmium-resistant bacteria. Chemosphere 64: 1036-1042.
Sheng, X.F., J.J. Xia, C.Y. Jiang, L.Y. He and M. Qian. 2008. Characterization of heavy metal-resistant endophytic bacteria from rape (*Brassica napus*) roots and their potential in promoting the growth and lead accumulation of rape. Environ. Pollut. 156: 1164-1170.
Shri, M., S. Kumar, D. Chakrabarty, P.K. Trivedi, S. Mallick, P. Misra, D. Shukla, S. Mishra, S. Srivastava, R.D. Tripathi and R. Tuli. 2009. Effect of arsenic on growth, oxidative stress, and antioxidant system in rice seedlings. Ecotoxicol. Environ. Saf. 72: 1102-1110.
Siedlecka, A., G. Samuelsson, P. Gardenstrom, L.A. Kleczkowski and Z. Krupa. 1998. The activatory model of plant response to moderate cadmium stress-relationship between carbonic anhydrase and Rubisco. pp. 2677-2680. *In:* G. Garab [ed.]. Photosynthesis: Mechanisms and Effects. Kluwer Academic, Dordrecht, Boston, London.
Spaepen, S. and J. Vanderleyden. 2011. Auxin and plant-microbe interactions. Cold Spring Harb. Perspect. Biol. 3: a001438.
Sresty, T.V.S. and R.K.V. Madhava. 1999. Ultrastructural alterations in response to zinc and nickel stress in the root cells of *pigeon pea*. Environ. Exp. Bot. 41: 3-13.
Srivastava, P.C. and U.C. Gupta. 1996. Trace Elements in Crop Production. Science Publishers Inc., Lebanon, New Hampshire, USA.
Stegmann, B. and M. Calmano. 2001. Treatment of Contaminated Soil, Fundamentals, Analysis, Applications. Springer. Berlin, London.
Stiborova, M., M. Pitrichova and A. Brezinova. 1987. Effect of heavy metal ions in growth

and biochemical characteristics of photosynthesis of barley and maize seedling. Biol. Plant. 29: 453-467.

Suciu, I., C. Cosma, M. Todica, S.D. Bolboaca and L. Jantschi. 2008. Analysis of soil heavy metal pollution and pattern in Central Transylvania. Int. J. Mol. Sci. 9: 434-453.

Sun, L., Y. Ma and H. Wang. 2018. Overexpression of PtABCC1 contributes to mercury tolerance and accumulation in *Arabidopsis* and *poplar*. J. Biochem. Biophy. Res. Commun. 497: 997-1002.

Tam, Y.L. and B. Singh. 2004. Heavy metals availability at industrially contaminated soils in NSW, Australia. pp. 97-120. *In*: A.L. Juhaz, G. Magesan and R. Naidu [eds.]. Waste Management. Science Publishers Plymouth.

Tariq, M., M. Ali and Z. Shah. 2006. Characteristics of industrial effluents and their possible impacts on quality of underground water. Soil Environ. 25: 64-69.

Teige, M., B. Huchzermeyer and G. Schultz. 1990. Inhibition of chloroplast ATP synthase/ ATPase is a primary effect of heavy metal toxicity in spinach plants. Biochem. Physiol. 186: 165-168.

Thangavel, P. and C. Subbhuraam. 2004. Phytoextraction: Role of hyperaccumulators in metal contaminated soils. Proc. Ind. Natl. Sci. Acad. 70: 109-130.

Tisdale, S.L., W.L. Nelson and J.D. Beaton. 1985. Soil Fertility and Fertilizers. 4th edn. pp. 358-403. MacMillan Publishing Co. New York.

Trotta, A., P. Falaschi, L. Cornara, V. Minganti, A. Fusconi, G. Drava and G. Berta. 2006. Arbuscular mycorrhizae increase the arsenic translocation factor in the As hyperaccumulating fern *Pteris vittata* L. Chemosphere 65: 74-81.

United States Environmental Protection Agency (USEPA). 2002. Arsenic Treatment Technologies for Soil, Waste, and Water. Report EPA-542-R-02-004. Washington, DC.

Vaclavikova, M., G.P. Gallios, S. Hredzak and S. Jakabsky. 2008. Removal of arsenic from water streams: An overview of available techniques. Clean Technol. Environ. 10: 89-95.

Vamerali, T., M. Bandiera and G. Mosca. 2010. Field crops for phytoremediation of metal-contaminated land: A review. Environ. Chem. Lett. 8: 1-17.

Van Assche, F. and H. Clijsters. 1990. Effects of metals on enzyme activity in plants. Plant Cell Environ. 13: 195-206.

Van der, D., P. Corbisier, L. Diels, A. Gilis, C. Lodewyckx, M. Mergeay, S. Taghavi, N. Spelmans and J. Vangronsveld. 1999. The role of bacteria in the phytoremediation of heavy metals. pp. 265-281. *In:* N. Terry and G. Bañuelos [eds.]. Phytoremediation of Contaminated Soil and Water. Lewis Publishers, Boca Raton, FL.

Vaughan, G.T. 1993. The environmental chemistry and fate of arsenical pesticides in cattle tick dip sites and banana land plantations. pp. 123. CSIRO Division of Coal Industry, Centre for Advanced Analytical Chemistry, NSW, Melboune. Australia.

Wahid, A., A. Ghani and F. Javed. 2008. Effects of cadmium on photosynthesis, nutrition and growth of mung bean *Vigna radiata* L. Agron. Sustain. Devlop. 28: 273-280.

Wang, F.Y., X. Lin and R. Yin. 2005. Heavy metal uptake by arbuscular mycorrhizas of *Elsholtzia splendens* and the potential for phytoremediation of contaminated soil. Plant Soil 269: 225-232.

Wani, P.A. and N.S. Khan. 2013. Nickel detoxification and plant growth promotion by multi metal resistant plant growth-promoting *Rhizobium* species RL9. Bull. Environ. Contamin. Toxicol. 91: 117-124.

Wani, P.A., M.S. Khan and A. Zaidi. 2007. Effect of metal tolerant plant growth promoting *Bradyrhizobium* sp. (*vigna*) on growth, symbiosis, seed yield, and metal uptake by green gram plants. Chemosphere 70: 36-45.

Watanabe, M.E. 1997. Phytoremediation on the brink of commercialization. Environ. Sci. Technol. 31: 182-186.

Watmough, S.A., T.C. Hutchinson and P.J. Dillon. 2004. Lead dynamics in the forest floor and mineral soil in south-central Ontario. Biogeochem. 71: 43-68.

Watt, N.R. 2007. Testing Amendments for increasing soil availability of radionuclides. Phytorem. Methods Rev. 23: 131-137.

Wei, C., C. Wang and L. Yang. 2008. Characterizing spatial distribution and sources of heavy metals in the soils from mining-smelting activities in Shuikoushan Hunan Province, China. J. Environ. Sci. 21: 1230-1236.

Wenzel, W.W. 2009. Rhizosphere processes and management in plant assisted bioremediation (phytoremediation) of soils. Plant Soil 321: 385-408.

World Health Organization (WHO). 1991. Nickel: Environmental Health Criteria, 108: World Health Organization, Geneva.

Wu, S.C., K.C. Cheung, Y.M. Luo and M.H. Wong. 2006. Effects of inoculation of plant growth-promoting rhizobacteria on metal uptake by *Brassica juncea*. Environ. Pollut. 140: 124-135.

Wuana, R.A. and F.E. Okieimen. 2011. Heavy metals in contaminated soils: A review of sources, chemistry, risks and best available strategies for remediation. ISRN Ecol. 1-20.

Yusof, A. M. and N.A.N. Malek. 2009. Removal of Cr (VI) and As (V) from aqueous solutions by HDTMA-modified zeolite Y. J. Hazard. Mater. 162: 1019-1024.

Zaidi, S.S., B.R. Usmani and M.J. Singh. 2006. Significance of *Bacillus subtilis* strain SJ–101 as a bioinoculant for concurrent plant growth promotion and nickel accumulation in *Brassica juncea*. Chemosphere 64: 991-997.

Zaier, H., T. Ghnaya, A. Lakhdar, R. Baioui, R. Ghabriche, M. Mnasri et al. 2010. Comparative study of Pb-phytoextraction potential in *Sesuvium portulacastrum* and *Brassica juncea*: Tolerance and accumulation. J. Hazard. Mater. 183: 609-615.

Zayed, A.M. and N. Terry. 2003. Chromium in the environment: Factors affecting biological remediation. Plant and Soil 249: 139-156.

Zhang, C. 2006. Using multivariate analyses and GIS to identify pollutants and their spatial patterns in urban soils in Galway, Ireland. Environ. Pollut. 142: 501-511.

Zhao, F.J., E. Lombi and S.P. McGrath. 2003. Assessing the potential for zinc and cadmium phytoremediation with the hyperaccumulator *Thlaspi caerulescens*. Plant Soil 249: 37-43.

Zhu, Y.L., A.M. Zayed, J.H. Quian, M. D'souza and N. Terry. 1999. Phytoaccumulation of trace elements by wetland plants: II. J. Environ. Qual. 28: 339-344.

CHAPTER

4

Phytoremediation Potential of Non-Edible Biofuel Plants: Physic Nut (*Jatropha curcas*), Castor Bean (*Ricinus communis*) and Water Hyacinth (*Eichhornia crassipes*)

Reckson Kamusoko* and Raphael M. Jingura
Chinhoyi University of Technology, Chinhoyi, Zimbabwe

4.1 Introduction

There has always been a concern about the pollution of environmental resources such as soil and water due to urbanization and industrialization (Padmavathiamma and Li 2007, Alaboudi et al. 2018, Ansari et al. 2020). Potent environmental contaminants include heavy metals, organic and inorganic substances. Environmental contaminants originate mainly from anthropogenic sources, including energy supply, transport, agriculture, mining, and metal and chemical production (Alaboudi et al. 2018, Grzegorska et al. 2020). Other contaminants are found naturally in the environment due to geogenic activities. Emerging contaminants include hormones, nanomaterials, industrial and household chemicals, disinfectants and their transformation products, and pharmaceutical and medical care products (Kamusoko and Jingura 2017).

Accumulation of contaminants in the environment poses serious health threats and leads to loss of biodiversity. Many treatment technologies are available to ameliorate contaminated water and soil systems. Traditional methods include physical, thermal, and chemical measures (Ali et al. 2013, Kamusoko and Jingura 2017, Alaboudi et al. 2018). These methods have several limitations: high cost, labor-intensive, change in soil structure and texture, the release of secondary pollutants, and perturbation of natural soil biodiversity (Ali et al. 2013, Alaboudi et al. 2018, Grzegorska et al. 2020). Phytoremediation can provide an innovative, simple, efficient, eco-friendly, and cheap alternative to most conventional technologies of environmental purification (Ali et al. 2013, Alaboudi et al. 2018). The technology

Corresponding author: rkamsoko.kamusoko@gmail.com

uses both terrestrial and aquatic plants as phyto pumps to extract and concentrate particulate pollutants (Kamusoko and Jingura 2017).

Phytoremediation is a green technology that utilizes vegetation and its associated microorganisms, soil stabilizers, and agro-methods to render contaminated environments safe (Buono et al. 2020). Plant mechanisms such as rhizofiltration, phytostabilization, phytoextraction, phytodegradation, and phytovolatilization are exploited in phytoremediation (Kamusoko and Jingura 2017, Alaboudi et al. 2018, Ansari et al. 2020). Phytoremediation depends on the plant's physiological processes such as transpiration, photosynthesis, translocation, evaporation, and respiration. A gamut of plant species varying from aquatic microphytes to deep tap-rooted terrestrial macrophytes are utilized as hyperaccumulators (Kamusoko and Jingura 2017). The database of potential phytoremediators comprises more than 400 known plant species (Ndimele et al. 2014). Examples consist of *Pterocarpus indicus*, *Jatropha curcas*, *Amaranthus* spp., and duckweed (Kamusoko and Jingura 2017), sunflower (*Helianthus annuus*) (Alaboudi et al. 2018), aquatic species (*Azolla*, *Eichhornia*, *Lemna*, *Potamogeton*, *Spirodela*, *Wolfia* and *Wolfialla*) (Ansari et al. 2020), *Pongamiapinnata*, *Panicumvirgatum*, *Ricinuscommunis*, *Miscanthusgiganteus*, and *Azadirachtaindica* (Tripathi et al. 2016). These plants are potential remediators of various metal ions, radionuclides, organic pollutants, pesticides and herbicides, explosives, chlorinated solvents, chlorinated hydrocarbons, inorganic compounds, and several industrial organic wastes (Favas et al. 2014).

Although phytoremediation is a promising technology, it is often a slow process. This limits the utility of contaminated sites for various other purposes (Shrestha et al. 2019). However, plant biomass produced is harvested for multiple uses. For example, it can be a feedstock for a biorefinery. Kiran and Prasad (2017) suggest that coupling hyperaccumulators with phytoremediation, biofuels, and other value-added byproducts can potentially promote the global economy in a sustainable way (Fig. 4.1). In this model, phytoremediation can be accompanied by producing valuable phytoproducts, including bioethanol, biodiesel, fiber, wood, charcoal, alkaloids, and bioplastics.

Amongst phytoproducts, the interest in bioenergy is increasingly escalating as it can be regarded as an eco-friendly and flexible supply of energy (Tripathi et al. 2016, Abdelsalam et al. 2019). For example, *H. annus* can remediate persistent organic pollutants (POPs) and heavy metals, and be used for bioenergy, bioethanol and charcoal production. *Miscanthus sinensis* is suitable for the removal of nutrients and heavy metals, and it is also used as a feedstock for bioethanol, and biogas production (Tripathi et al. 2016). *Acacia nilotica* can concurrently extract heavy metals, and provide biomass and charcoal (Tripathi et al. 2016).

This chapter discusses the potential application of three non-edible hyperaccumulators as potential energy crops. These are physic nut (*J. curcas*), castor bean (*R. communis*), and water hyacinth (*E. crassipes*). Mechanisms utilized by plants during phytoremediation of contaminated sites are also discussed. The potency of phytoremediation as a green solution to the environment is also discussed in this chapter. The different types of contaminants, their sources, and toxicological effects are also explained. Lastly, research and development agenda to enhance the phytoremediation potential of bioenergy plants are outlined.

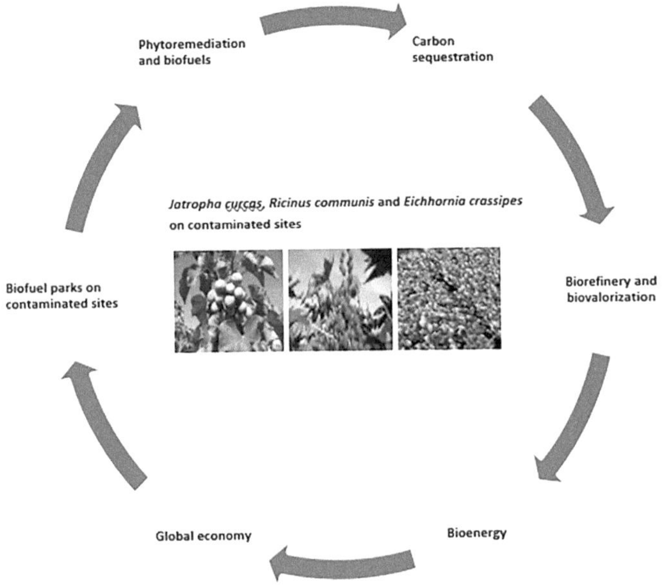

Figure 4.1: Coupling *Jatropha curcas, Ricinus communis,* and *Eichhornia crassipes* with phytoremediation, biofuels, and other valorized phytoproducts can promote the global economy (Adapted from Kiran and Prasad 2017).

4.2 Contaminants: Definition, Sources, and Toxicity

Pollutants and contaminants are two diverse ecological substances that are found throughout the world. However, contaminants can be pollutants; therefore these two terms are often used interchangeably. A contaminant is defined as a substance, material, or agent present at an unacceptable level in an environment (Masindi and Muedi 2018). A pollutant is a contaminant present in the environment or can enter the environment and pose detrimental effects to the natural ecosystem (Masindi and Muedi 2018). Contaminants can be either natural substances or synthetic byproducts of industry and agro-based activities. These can be grouped into four main categories: organic, inorganic, biological, and radiological contaminants (Sharma and Bhattacharya 2017, Masindi and Muedi 2018). Besides, some contaminants are not extensively researched but were found in nature over a long period. These are known as "emerging contaminants" (Kamusoko and Jingura 2017).

4.2.1 Organic contaminants

Organic contaminants are biodegradable substances that are naturally present in the environment or originate from anthropogenic activities. Examples of organic contaminants include human waste, food waste, polychlorinated biphenyls (PBCs), polybrominated diphenyl ethers (PBDEs), polycyclic aromatic hydrocarbons (PAHs), pesticides, petroleum, volatile organic chemicals (VOCs), dyes, and organo-chlorine pesticides (OCPs) (Sharma and Bhattacharya 2017, Masindi and Muedi 2018). Some organic contaminants are more persistent due to their high lipid solubility, stability, lipophilicity, and hydrophobicity (Sand-Jensen 2013). Bioaccumulation of POPs may have several adverse effects on organisms (Masindi and Muedi 2018). POPs can cause deoxygenation, and liberate ammonia and other mineral nutrients that can pose a severe threat to aquatic health (Sand-Jensen 2013). In humans, they can cause chronic health challenges such as cancers, hormonal disruptions, and nervous system disorders (Sharma and Bhattacharya 2017).

4.2.2 Inorganic contaminants

Inorganic contaminants are elements or compounds of mineral origin available in the form of dissolved anions or cations. They include toxic metals, and various types of nutrients and salts (Goldscheider 2010). Inorganic contaminants originate from either geogenic activities or anthropogenic activities such as mining, industry, and agriculture (Sharma and Bhattacharya 2017). The toxicity of inorganic contaminants is a serious global challenge. The toxicological effects of heavy metals at elevated concentrations are well researched. Lead (Pb), cadmium (Cd), mercury (Hg), and arsenic (As) are more likely to cause severe damage to man and his natural ecosystem (Kamusoko and Jingura 2017). Adverse effects of these metals on the environment are also notable even at low concentrations (Raikova et al. 2019).

4.2.3 Biological contaminants

Biological contaminants are generally living organisms and their derivatives that are present in the environment. They are known as microbes or microbiological contaminants. Examples include pathogenic microbes such as bacteria, viruses, fungi, algae, and parasites (microscopic protozoa and worms). Microbiological contaminants can be transmitted through human and animal wastes (Sharma and Bhattacharya 2017). Pathogenic microbes are the most prevalent agents of acute diseases and death in the developing world. Fatal diseases such as cholera, dysentery, and typhoid are caused by pathogenic contaminants (Behnam et al. 2013).

4.2.4 Radiological contaminants

Radiological contaminants can be described as chemicals that contain unstable atoms that can emit ionizing radiation. Radioactive elements come from soils, rocks, or industrial waste. Examples of radioactive contaminants are cesium, plutonium, and uranium (Kamusoko and Jingura 2017, Sharma and Bhattacharya 2017). Radioactive materials are usually not destructive as they can be naturally found in the environment at low levels. However, human-made activities such as nuclear weapon testing, nuclear waste disposal, and nuclear power plant emissions and spillage release radionuclides that may damage the environment (Chakravarty and Kumar 2019). Most radioactive materials are carcinogenic (Sharma and Bhattacharya 2017).

4.2.5 Emerging contaminants

Emerging contaminants can be defined as synthetic or naturally occurring substances or any microorganisms that are not regularly monitored in the ecosystem but are likely to enter the environment and pose some unfavorable ecological conditions (Rosenfeld and Feng 2011). These include medicines, disinfectants, detergents, contrast media, surfactants, herbicides, pesticides, dyes, paints, food additives, preservatives, personal care products, and nanomaterials. They also comprise endocrine-disrupting compounds, analgesics, antibiotics, hormones and pharmaceutical compounds such as anti-inflammatory, anti-diabetic, and anti-epileptic drugs. The principal sources of emerging contaminants are many products from household activities, agriculture, chemical, and pharmaceutical industries (Rosenfeld and Feng 2011, Boxal 2012, Kamusoko and Jingura 2017). The main concern of emerging contaminants is that these compounds' fate in the environment is not well-known. Furthermore, most of these substances cannot be monitored, and they degrade in the environment giving rise to new chemicals with unknown properties (Rosenfeld and Feng 2011).

4.3 The Concept of Phytoremediation

Phytoremediation is derived from a Greek word "*phyton*" meaning "plant," and Latin-derived suffix "*remedium*" meaning "to correct or clean or restore" (Padmavathiamma and Li 2007, Favas et al. 2014, Grzegorska et al. 2020). It can be defined as a cleanup technology that utilizes plant-based mechanisms and associated microorganisms to remove, detoxify or immobilize pollutants in water and soil systems (Kamusoko and Jingura 2017, Kiran and Prasad 2017, Ansari et al. 2020). Phytoremediation is an environmentally friendly technology that embraces many advantages over other technologies. Phytoremediation is an efficient technology to eradicate organic and inorganic substances from the environment (Kamusoko and Jingura 2017). However, the method is suitable for sites where there is low contamination.

Plant species that are used as phyto pumps range from aquatic microphytes to terrestrial macrophytes. Transgenic plants are also promising hyperaccumulators of environmental contaminants (Kamusoko and Jingura 2017). Phytoremediation involves absorption of contaminants by roots, accumulation in plant tissues, degradation, and transformation of contaminants to less detrimental forms (Kamusoko and Jingura 2017, Ansari et al. 2020). The selection of plants with a greater ability to absorb, and accumulate soluble nutrients, metals and other pollutants are a pre-requisite to phytoremediation. These plants should have a fast growth rate, and easy to handle and harvest. The extent of contamination is also an important criterion in choosing plant species for phytoremediation (Ansari et al. 2020).

4.3.1 Advantages and disadvantages of phytoremediation

Phytoremediation offers many advantages over traditional methods that determine its suitability as a cleanup technology (Rungwa et al. 2013, Favas et al. 2014, Kamusoko and Jingura 2017, Buono et al. 2020). The advantages are both technical and cost-driven. Despite several benefits, phytoremediation can be associated with many drawbacks. It is considered a slow process that is still at its infant stage of development. The advantages and limitations of phytoremediation over traditional remedial technologies are summarized in Table 4.1.

Table 4.1: Advantages and disadvantages of phytoremediation

Advantages	Disadvantages
• Economically viable technology that uses solar energy	• Highly dependent on agro-climatic conditions of the site
• *In-situ* and passive method	• Time-consuming and slower than non-biological methods
• Eco-sustainable technology	• The final fate of contaminants might be unknown
• Disposal sites are not required	• Effects on the natural ecosystem are not fully understood
• High acceptance by the public	• Highly variable results
• Minimizes leaching and dispersal of contaminants	• Contaminant solubility is increased, leading to environmental pollution and leaching
• Can treat more than one type of contaminant at a given site	• Most useful for slightly hydrophobic pollutants
• Harvested biomass have multiple applications	• Pollutants are mobilized into groundwater
• Easy to implement and maintain	• Toxicity and high metal concentration destroy the plants as they are selective to metal remediation
• Various mechanisms are available for phytoremediation	• Limited tolerance of plants to pollutants
• Conserves natural resources	• Contaminants are found in woody tissues utilized in biofuel parks
• More aesthetically pleasing than conventional methods	• Agricultural equipment and knowledge of farming, genetics, reproduction, and diseases of the plants are required for large scale operations
• Does not require to wait for emerging plant species to re-establish the degraded site	• Possibility of re-incorporating contaminants in senescing tissues into the environment
• Harvesting of the plant biomass is easily achieved due to the availability of technology	• Applicable to shallow soils (<5 m)
• Reduces surface run-off	• The technology is not easily recognized by many regulatory agencies as it is still naive

4.3.2 Phytoremediation techniques

There are several types of technologies and applications that are used for phytoremediation of contaminated sites (Padmavathiamma and Li 2007, Favas et al. 2014, Ansari et al. 2020, Buono et al. 2020, Grzegorska et al. 2020). Their taxonomy can be based on the mechanism triggered by plant species to eliminate contaminants from soil or water ecosystems (Buono et al. 2020). Phytoremediation techniques vary in the process by which contaminants can be eliminated, immobilized, or metabolized by plant species (Bolan et al. 2011). There are six fundamental categories of phytotechnologies. These include phytostimulation, phytovolatilization, phytoextraction, phytostabilization, phytodegradation, and phytofiltration (Fig. 4.2). A single plant species may concurrently utilize more than one phytoremediation technology depending on the type and depth of contamination (Favas et al. 2014, Buono et al. 2020). Table 4.2 is a description of the different mechanisms of phytoremediation.

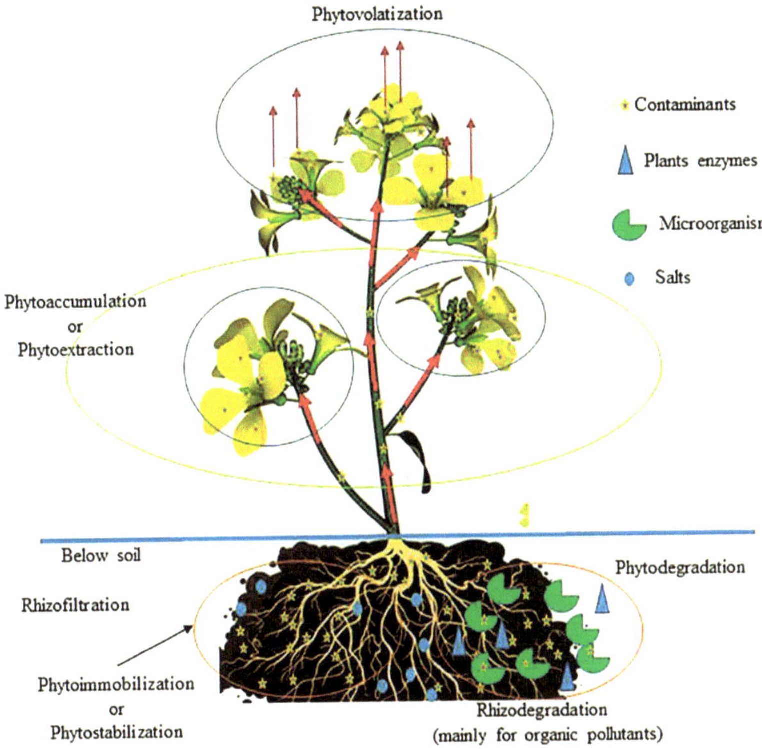

Figure 4.2: Schematic diagram of phytotechnologies (Adapted from Ifon et al. 2019).

4.4 Phytoremediation Potential of Non-Edible Plants Combined with Biofuel Production

4.4.1 Physic nut (*Jatropha curcas*)

Jatropha curcas L. (physic nut) is a fast-growing perennial shrub with multiple applications. The plant is used for biodiesel, organic fertilizer, medicine, and animal feed production (Aggangan et al. 2016). *J. curcas* is native to Mexico and Central America, but it is now naturalized in tropical and sub-tropical regions such as South East Asia and Africa. It is mainly grown as hedge-row to protect crops from domestic animals. The plant belongs to the family, Euphorbiaceae, and grows to a height of

Table 4.2: Various mechanisms of phytoremediation

Phytotechnology	Description	Mechanism	Contaminants
Phytostimulation (rhizodegradation)	Various organic compounds are exuded in the rhizosphere to stimulate microbial communities that participate in the biodegradation of contaminants. The exudates and metabolites can be used as a source of carbon and energy for microbes.	Rhizosphere stimulation	Organic compounds
Phytovolatilization	Based on the ability of plants to absorb some metals/metalloids and release them into the atmosphere.	Volatilization by leaves	Organic and inorganic compounds
Phytoextraction (phytoaccumulation, phytoabsorption, or phytosequestration)	Contaminants can be absorbed by roots followed by translocation and hyperaccumulation in various harvestable tissues of the plant.	Hyperaccumulation	Organic, inorganic, and radiological compounds
Phytostabilization (phytoimmobilization)	Involves the incorporation of contaminants into the lignin of the root cell wall or the humus. Mobilization and diffusion of contaminants into the soil can be minimized.	Complexation	Organic and inorganic compounds
Phytodegradation (phytotransformation)	Contaminants are metabolized inside plant cells by specific enzyme systems, e.g. nitroreductases, dehalogenases, oxygenases, and laccases.	Metabolism	Organic compounds
Phytofiltration	Utilizes plants to absorb, concentrate, and precipitate contaminants from an aqueous solution through their roots or other submerged parts.	Rhizosphere accumulation	Inorganic, organic, and radiological compounds

Source: Padmavathiamma and Li 2007, Bolan et al. 2011, Favas et al. 2014, Grzegorska et al. 2020

about 6 m. *J. curcas* produces crimson color flower and seeds that contain up to 30 percent oil (Mekala et al. 2014, González-Chávez et al. 2017, Primandari et al. 2018). It can survive on a wide range of marginal soils and agro-climatic conditions, often drought-tolerant, and cannot compete for land with food crops (Herrera et al. 2019, Jonas et al. 2020). For these reasons, it is a truism that *J. curcas* is one of the most suitable candidates for phytoremediation. Phytoremediation is mainly designated to edible crops with low biomass and contaminant extraction ability, shorter lifespan, and tolerance of many biotic and abiotic conditions (Bauddh et al. 2015).

The utilization of *J. curcas* for phytoremediation of contaminated sites is increasingly growing as it is a non-edible crop. *J. curcas* is used to restore soils mainly contaminated with POPs, and heavy metals combined with biodiesel, biofertilizer, and charcoal production (Tripathi et al. 2016). Multiple energy carriers and industrial chemical products are also derived from *J. curcas* press-cake after oil extraction in a biorefinery. The oil can be used for biodiesel production. The most promising energy carriers from press-cake include bioethanol, briquettes, pyrolytic products, biogas, and syngas. Various physical, chemical, thermochemical, and biological technologies are available to convert *J. curcas* biomass into valuable products (Jingura and Kamusoko 2017).

Literature shows that *J. curcas* production helps reclaim wasteland, and degraded mine polluted sites (González-Chávez et al. 2017). *J. curcas* has the potential to facilitate the sequestration, uptake, translocation and detoxification of different contaminants (Kamusoko and Jingura 2017). The utility of *J. curcas* for phytoremediation was mainly studied at a small-scale under laboratory and greenhouse conditions. Valuable information was generated regarding the use of the plant in phytoremediation. For example, a greenhouse study by González-Chávez et al. (2017) demonstrated that *J. curcas* could be useful in the uptake and phytostabilization of metals in mine tailings.

The ability of *J. curcas* to extract heavy metals from mining waste was also evaluated and coupled with biocatalyst production. Results showed that *J. curcas* could absorb very high concentrations of iron (Fe) (>3,000 mg kg^{-1} plant) with minimal effects on As. On the other hand, *J. curcas* has been found to accumulate Hg in roots compared to other tissues of the plant (Marrugo-Negrete et al. 2015). A summary of some experiences with the phytoremediation capacity of *J. curcas* is presented in Table 4.3. Undoubtedly, *J. curcas* is most suitable for the restoration of contaminated soils. It has limited utility for the rehabilitation of an aqueous environment.

4.4.2 Castor bean (*Ricinus communis*)

Castor bean (*Ricinus communis*) is a soft-wooded, perennial shrub found in the tropics and warm temperate areas. The flowering plant belongs to the spurge family, Euphorbiaceae. *R. communis* originated from the southeastern Mediterranean Basin, Eastern Africa, and India. It is now widely distributed in the tropical climates (Kuete 2014). This plant is cultivated commercially for its multipurpose castor oil. *R. communis* contains about 60 percent oil, of which 90 percent is ricinoleic acid (Maheshwari and Kovalchuk 2016). This oil can be utilized in cosmetics, coatings,

Table 4.3: Selected examples of phytoremediation potential of *Jatropha curcas*

Mechanism	Media	Target contaminants	Growth conditions	Main results	Reference
Rhizodegradation	Soil	Lubricating oil from automobiles	Room conditions	High removal of lubricating oil in soil amended with organic waste.	Agamuthu et al. (2010)
Rhizoremediation	Garden soil	Lindane	Glasshouse conditions	Lindane accumulated (up to 20.85 μg g^{-1}) in *Jatropha* with a corresponding reduction of up to 80% in soil.	Abhilash et al. (2013)
Phytoextraction	Coal fly ash	Heavy metals (aluminum (Al), manganese (Mn), Cr, Cu, Fe)	Greenhouse pot experiment	Heavy metal accumulation improved by 117% in the root, 62% in stem, and 86% in leaves with the addition of EDTA to fly ash.	Jamil et al. (2009)
Phytoextraction	A mixture of peat and mining soil	Heavy metals	Pot experiment	*J. curcas* translocated large amounts of metals to above root biomass. Translocation factors of >1 were achieved.	Martin et al. (2020)
Phytostabilization	Hydroponic cultures	Zn, Pb	Pot experiment	Some tolerance mechanisms to the toxicity of Zn and Pb were exhibited by *J. curcas*.	Marques et al. (2017)
Rhizodegradation	Soil	1,4-dichlorobenzene (DCB)	*In-vitro* conditions	Rhizospheric isolates of *J. curcas* degraded more than 97% of DCB.	Pant et al. (2016)
Phytostabilization	Polymetallic acid mine tailings	Al, Cu, Zn, Pb, Cd	Greenhouse pot experiment	High accumulation of metals in roots and a high tolerance index (>90%).	Wu et al. (2011)
Phytoaccumulation	Field soil	Cd, Cr, Pb, Zn, Ni, Cu	Field-scale	Best extraction ability for Cd, Cr, Zn, and Ni.	Kun et al. (2012)

Phytoaccumulation	Gold mining soil	Hg	Field experiment	*J. curcas* reported the highest bioconcentration (0.99) in a range of 0.28 to 0.99 amongst 24 different plant species.	Marrugo-Negrete et al. (2016a)
Phytoaccumulation	Hydroponic cultures	Hg	Pot experiment	Hg accumulation in roots was 7 and 12-times high as compared to leaves and shoots, respectively.	Marrugo-Negrete et al. (2016b)
Phytoaccumulation	Soil	Heavy metals	Field study	*J. curcas* had the best extraction ability for Cd, Cr, Ni, and Zn.	Chang et al. (2014)
Uptake and phytostabilization	Mine tailings	Cu, Cd, Pb, Zn	Greenhouse experiment	*J. curcas* was able to grow in mine residue without phytotoxicity symptoms.	González-Chávez et al. (2016)

industrial, and automotive applications. The conversion of castor oil into biodiesel is a promising technology. Other uses of *R. communis* include societal development, employment creation, carbon sequestration, and the reduction of greenhouse gas emissions (Bauddh et al. 2015, Coppock and Dziwenka 2015, Panda et al. 2020). A 0.5-0.6 ton of oil extracted from castor bean produces 1 ton of press-cake. The cake is a useful fertilizer and soil enhancer, and can produce several other phytoproducts (Coppock and Dziwenka 2015). *R. communis* is highly toxic to humans and animals. This toxicity can be derived from a cytotoxic component of a lectin called recin (Kuete 2014, Maheshwari and Kovalchuk 2016). Generally, *R. communis* is a non-edible plant, although some other countries supplement stock feed with castor bean residual cake (Coppock and Dziwenka 2015).

Despite the manufacture of numerous phytoproducts, *R. communis* is now an emerging non-edible phyto pump. The viability of *R. communis* is vested in the fact that it is an energy cash crop in contemporary farming, which is coupled with phytoremediation of contaminated soils (Huang et al. 2011). The capability to remove several toxic metals such as Cd, Pb, Ni, As, Cu, Zn, etc., and some organic contaminants can be unlocked in *R. communis* (Bauddh et al. 2015). This potential is derived from its rapid growth rate and ability to withstand harsh agro-climatic conditions, often arid, hot, and saline (Huang et al. 2011, Yashim et al. 2016). Several authors have studied the phytoremediation efficacy of *R. communis* to ameliorate contaminated sites (Table 4.4). The general information provided is that *R. communis* was tried on several occasions to extract heavy metals in contaminated soil using pot experiments.

4.4.3 Water hyacinth (*Eichhornia crassipes)*

Water hyacinth (*Eichhornia crassipes*) is a perennial, free-floating, invasive aquatic macrophyte that originated in the tropical areas in the Amazon basin of Brazil and other surrounding South American countries. The monocotyledonous plant is now well established in other tropical continents of the world, excluding the Antarctica. It belongs to the Pontederiaceae family (Stohlgren et al. 2013, Priya and Selvan 2014, Rezania et al. 2015). *E. crassipes* is a decorative plant, but several phytoproducts are derived from the plant. The multiplicity of product streams include animal and fish feed, organic fertilizer or mulch, fiberboard, brequettes, bioethanol, and biogas (Abdel-sabour 2010, Rezania et al. 2015).

Typically, water hyacinth contains 30-50 percent cellulose, 20-40 percent hemicellulose, and 15-30 percent lignin. Its high nitrogen content makes it a suitable substrate for co-digestion with cow dung towards biogas production. A mixture of cow dung and dry water hyacinth in the ratio of 1:3 can produce good methane yields. One tone of dry water hyacinth can produce 0.37 megaliters of biogas (Abdel-sabour 2010). An amalgamation of microbial and dilute acid pretreatment is the most effective way to enhance bioethanol production from water hyacinth. This combination can yield 430.66 mg g^{-1} of reducing sugars, and 39.4 percent cellulose (Zhang et al. 2018). However, the mode of reproduction and rapid propagation of water hyacinth is accountable to eutrophication and biodiversity loss in some surface water bodies (Priya and Selvan 2014).

Table 4.4: Selected examples of phytoremediation potential of *Ricinus communis*

Mechanism	Media	Target contaminants	Growth conditions	Main results	Reference
Phytostabilization	Cu-mine soils and slags	Cu and other heavy metals	-	High tolerance index (TI) in Cu-mine soils. Translocation factor for most metals was >1.	Palanivel et al. (2020)
Phytoextraction	Soil	Heavy metals (Cu, Cr, Fe, Mn, Ni, Pb, Zn)	Pot experiment	Considerable amounts of heavy metals accumulated in plant biomass with a significant reduction in their soil content.	Khan et al. (2019)
Mobilization and uptake	Soil	Cd, DDTs	Pot experiment	Total uptake of DDTs and Cd by 24 *R. communis* genotypes ranged from 83.1 to 267.8 and 66.0 to 155.1 per pot, respectively.	Huang et al. (2011)
Phytoextraction	Soil	Cd, Pb	Pot experiment	Decrease of Cd and Pb in soil.	Zhang et al. (2015)
Phytoextraction	Hydroponic conditions	Cd	Greenhouse experiment	The highest uptake and bioaccumulation was 25 and 10 mg $L^{-1}$of Cd, respectively.	Hadi et al. (2015)
Phytoextraction	Gold mining soil	Pb, Cd, Cu, Zn	Pot experiment	*R. communis* was an excellent accumulator of Pb, Cu, Cd, and Zn.	Fanna et al. (2018)
Phytoextraction	Soil	Cu	Pot experiment	Enhanced phytoremediation efficiency with total Cu removal of 121.3 μg Cu $plant^{-1}$.	Zhou et al. (2020)
Phytostabilization	Fly ash	Metals	Field study	Translocation factors for all metals were less than one.	Pandey (2013)
Phytoaccumulation and phytostabilization	Mine tailings	Cu, Zn, Mn, Pb, Cd	Field study	Phytostabilization of tailings with little effects on the accumulation of Ni, Zn, or Cd.	Olivares et al. (2013)
Phytoextraction	Soil	Cd, Zn	*In-situ* experiment	Increased cumulative concentration of Cd and Zn in *R. communis* by 1.14 and 2.19 times, respectively.	Xiong et al. (2018)
Phytoaccumulation	Fly ash	Heavy metals (Al, Fe, Zn, Mn, Cu, Cr)	*In-situ* experiment	Increase in plant biomass and metal tolerance index under 50% fly ash-amended soil.	Panda et al. (2020)

E. crassipes is resistant and tolerant to high levels of heavy metals, phenols, formaldehyde, formic acids, acetic acids and oxalic acids (Rezania et al. 2015, Ansari et al. 2020). Water hyacinth can be an excellent water and wastewater purifier. This is attributed to its high biomass production, high tolerance to pollution, and heavy metal and nutrient biosorption capacities. Its potential for phytoremediation was tapped in the last few decades (Priya and Selvan 2014). The efficacy of E. crassipes for phytoremediation of different contaminants found in industrial wastewater was well evaluated (Priya and Selvan 2014). It *is efficient for the elimination of approximately 69 percent K in water (Fox et al. 2008). The removal efficiency of water hyacinth varies from 60-80 percent nitrogen (N) in aquatic systems (Zhou et al. 2007). Water hyacinth can also extract Cd and Zn from water. Its optimal growth and physiological performance depend on environmental conditions such as pH, temperature, salinity, and solar emission (Ansari et al. 2020). Some studies on the potential of water hyacinth for phytoremediation are shown in Table 4.5. It is without a doubt that water hyacinth can be applied mainly to treat wastewater.*

4.5 Techno-economic Aspects

Data on the cost and performance of soil and water restoration can be often provided for well-established *in-situ* and *ex-situ* conventional technologies. It is not so for emerging technologies such as phytoremediation and bioremediation. Therefore, information on the cost and performance of these new treatment technologies is patchy (Vangronsveld et al. 2009). Generally, phytoremediation has lower operational costs than conventional remediation technologies (Wan et al. 2016, Ali et al. 2017). The cost of phytoremediation is estimated at 25-100 US\$ t^{-1}, while traditional excavation cost is around 150-350 US\$ t^{-1} (Ali et al. 2017). Phytoremediation costs are mainly influenced by the local situation, viz. the type and depth of contamination, agro-climatic conditions, disposal of biomass, and other factors (Vangronsveld et al. 2009).

From an economic and food security point of view, the use of multipurpose non-edible crops for phytoremediation can be highly recommended. The main drawback to the commercialization of phytoremediation by non-edible crops as a sustainable way to ameliorate contaminated sites is the long duration of the process. The cost of biomass production is countered by combining biomass valorization with other land use functions that are economically valuable (Lewandowski et al. 2006). However, data on the profitability of various phytoremediation mechanisms is limited and highly variable. The cost of phytoextraction of around 60,000 to 100,000 US\$ ha^{-1} is lower than the cost of *ex-situ* remediation technologies of about 1 million US\$ ha^{-1} (Robinson et al. 2003). Combining Ni phytomining and energy generation can be predicted to make a profit of 11,500 AU\$ ha^{-1} of harvested biomass. On the other hand, phytoextraction of Ni-contaminated soils can be expected to generate 16,000 US\$ ha^{-1} of harvested crop (Vangronsveld et al. 2009). The cost of phytomanagement of heavy metals from contaminated soil in China was 37.7 US\$ m^{-3} (Wan et al. 2016). Table 4.6 presents a comparison of cost data for phytoremediation and other remediation technologies. Noteworthy is that phytoremediation is relatively less expensive as compared to bioremediation and *ex-situ* remediation.

Table 4.5: Selected examples of phytoremediation potential of water hyacinth (*Eichhornia crassipes*)

Mechanism	Media	Target contaminants	Growth conditions	Main results	Reference
Uptake, phytodegradation, and microbial degradation	Culture solutions	Phosphorous pesticide ethion	Pot experiment	Uptake and phytodegradation reduced ethion by 69%, while microbial degradation reduced ethion by 12%.	Xia and Ma (2006)
Phytotransformation and uptake	Wastewater	Nitrogen	Pilot plant experiment	Total N was reduced by 63.9%. Denitrification nitrogen 73.8%, while sedimentation and uptake contributed to 16.7% and 9.5 nitrogen removal, respectively.	Mayo and Hanai (2016)
Uptake and phytoextraction	Eutrophicated water	Nitrogen	Pond experiment	Reduction in total N, ammonia, nitrite, and chlorophyll-a concentration with an increase in water transparency.	Fang et al. (2007)
Phytoextraction	Wastewater	COD, BOD, Total Kjeldahl Nitrogen, ammonium, PO_4^{3-}, Cd, Cu, Zn, Cr	Aquarium experiment	Water hyacinth had higher phytoremediation efficiency than water lettuce.	Kouamé et al. (2016)
Phytoextraction and phytoaccumulation	Hydroponic cultures	Fe, Cu	Pot experiment	Fe showed a better accumulation than Cu, while Cu showed better translocation capability than Fe.	Ndimele et al. (2014)
Bioaccumulation	Contaminated water	Crude oil	Screen house pot experiment	*E. crassipes* affected the physicochemical quality of water and improved the degradation of crude oil.	Ochekwu and Madagwa (2013)
Phytoextraction	Domestic sewage	Nutrients (heavy metals, nitrate, sulfate)	Plastic drum experiment	Nutrient concentration reduction ranged from 63 to 100%.	Ajibade et al. (2013)

(*Contd.*)

Table 4.5: (*Contd.*)

Mechanism	Media	Target contaminants	Growth conditions	Main results	Reference
Phytoextraction	Wastewater from Cr-contaminated mines	Cr^{6+} and other water quality parameters	Small and large-scale hydroponic experiment	Removed 99.5% Cr^{6+} and reduced the levels of other water quality parameters.	Saha et al. (2017)
Phytoextraction	Hydroponic cultures	Cu, Cd	Pot experiment	More than 90% of Cu and Cd removed by each of the plants.	Swain et al. (2014)
Phytoextraction	Hydroponic cultures	Heavy metals (K, sodium (Na), calcium (Ca), Zn, Pb, Fe, Cd, Mg, Cu)	Tank experiment	The metal uptake ranged from 0.01 (Pd) to 13.52 ppm (K).	Ukunowo and Ogunkanmi (2010)

Table 4.6: Comparison of cost data for phytoremediation, bioremediation, and conventional methods

Remediation technology	Medium	Contaminant	Cost (range or min value)/ US$ m^{-3} unless stated
Phytoremediation	Soil	Metals	147-483
In-situ bioremediation	Soil	VOCs, heavy metals	432
In-situ bioremediation	Soil	BTEX, PHC	226
Phytoremediation	Soil	Metals	13-131
Soil metal washing	Soil	Metals	39-392
Phytostabilization	Soil	As	54
Excavation	Soil	-	270-460 US$ t^{-1}
Excavation	Soil	-	324-552
Phytoremediation	Groundwater	Metals	4.8-6.9 US$ m^{-2}
In-situ bioremediation	Groundwater	VOCs	381

Source: Vangronsveld et al. (2009)

4.6 Research and Development Initiatives for Integrating Phytoremediation and Biofuel Parks

The utility of non-edible biofuel plants for phytoremediation is still emerging, and their technoeconomics is not yet fully understood. The phytoremediation potential of these plants coupled with biofuel production needs to be supported and enhanced using a multidisciplinary approach. Data provided in the literature is insufficient to persuade all the stakeholders involved in such initiatives. Empirical evidence from available data shows that most research on phytoremediation was demonstrated on a small-scale. The transfer of knowledge from the laboratory to field-scale is needed. The best way to achieve this is to set up research and development (R & D) agenda. This will provide the information required to promote the performance of non-edible plants as phytoremediators of the degraded environment as well as feedstocks for biofuel parks. Several criteria can be set as part of the R & D efforts to support the scaling up of phytoremediation combined with biofuel generation (Tripathi et al. 2016, Kamusoko and Jingura 2017). A full state-of-the-art package includes:

- Cost-benefit analysis and socio-economic impact assessment to evaluate the viability of biofuel production from the contaminated environment
- Selection of the mechanisms that are most viable for commercial application
- Optimization of appropriate agronomic practices to reduce the toxicity of pollutants and maximize growth, and yield of plants in contaminated sites towards biofuel production
- Development of suitable plantation models in order to optimize the technoeconomics of restoration activities
- Selection of the most suitable plant genotypes for biofuel parks

- Safety of phytoproducts derived from such remediation activities
- Certification and marketing of phytoproducts
- Up-scaling of experiments to commercial-scale production

4.7 Conclusions

Our natural environment is heavily affected by contaminants that originate from geogenic activities and anthropogenic sources such as mining, industry, and agriculture. Environmental contaminants are broadly classified into four categories: organic, inorganic, biological and radiological contaminants. The presence of contaminants in the environment at elevated levels causes severe toxicological effects on man, and his environment. Phytoremediation using non-edible plants (*J. curcas*, *R. communis* and *E. crassipes*) offers a promising approach to reclaim contaminated sites in the foreseeable future. The technology is an innovative, simple, efficient, eco-friendly, and cheap alternative to other remediation technologies. However, phytoremediation is considered a slow process that is still at its infant stage of development. This chapter has provided information on six basic mechanisms that can be utilized by non-edible plants *to remove contaminants from soil and water. These comprise* phytostimulation, phytovolatilization, phytoextraction, phytostabilization, phytodegradation, and phytofiltration. The most widely studied mechanism is the phytoextraction of heavy metals.

An integrated approach of coupling phytoremediation and valorization of biomass into multiple product streams in a biorefinery is likely to be sustainable. This needs to be enhanced and promoted. Energy carriers are the most dominant product streams of *J. curcas*, *R. communis*, and *E. crassipes* that are obtained from this multidisciplinary approach. Non-edible crops grown on contaminated sites can produce a wide range of biofuel products such as biodiesel, bioethanol, biogas, briquettes, syngas, charcoal, pyrolytic products and so on. This chapter also presents information on the profitability of different remediation technologies. Although there are little cost and performance data on phytoremediation, this technology seems to be a cheaper alternative to bioremediation and conventional methods. The estimated cost of phytoremediation is 25-100 US$ t^{-1}. The chapter has also shown some studies that have and are still to be done on non-edible energy crops in order to optimize phytoremediation in biofuel parks and commercialize small-scale knowledge. Results appear to be encouraging, but variations are available among disparate phytoremediation mechanisms.

References

Abdel-sabour, M.F. 2010. Water hyacinth: Available and renewable resource. Elec. J. Env. Agricult. Food Chem. 9(11): 1746-1759.

Abdelsalam, I.M., M. Elshobary, M.M. Eladawy and M. Nagah. 2019. Utilization of multi-tasking non-edible plants for phytoremediation and bioenergy source – A review. Phyton-Int. J. Exp. Bot. 88(2): 69-90.

Abhilash, P.C., B. Singh, P. Srivastava, A. Schaeffer and N. Singh. 2013. Remediation of lindane by *Jatropha curcas* L: Utilization of multipurpose species for rhizoremediation. Biomass Bioenergy 51: 189-193.

Agamuthu, P., O.P. Abioye and A.A. Aziz. 2010. Phytoremediation of soil contaminated with used lubricating oil using *Jatropha curcas*. J. Hazard. Mater. 179(1-3): 891-894.

Aggangan, N., N. Cadiz, A. Llamado and A. Raymundo. 2016. *Jatropha curcas* for bioenergy and bioremediation in mine tailing area in Mogpog, Marinduque, Philippines. Energy Procedia. Australia 110: 471-478.

Ajibade, F.O., K.A. Adeniran and C.K. Egbuna. 2013. Phytoremediation efficiencies of water hyacinth in removing heavy metals in domestic sewage (a case study of University of Ilorin, Nigeria). Int. J. Eng. Sci. 2(12): 16-27.

Alaboudi, K.A., B. Ahmed and G. Brodie. 2018. Phytoremediation of Pb and Cd contaminated soils by using sunflower (*Helianthus annuus*) plant. Ann. Agric. Sci. 63: 123-127.

Ali, A., D. Guo, A. Mahar, W. Ping, F. Wahid, F. Shen, R. Li and Z. Zhang. 2017. Phytoextraction and the economic perspective of phytomining of heavy metals. Solid Earth Discuss. https://doi.org/10.5194/se-2017-75

Ali, H., E. Khan and M.A. Sajad. 2013. Phytoremediation of heavy metals – Concepts and applications. Chemosphere 97(1): 869-881.

Ansari, A.A., M. Naeem, S.S. Gill and F.M. AlZuair. 2020. Phytoremediation of contaminated waters: An eco-friendly technology based on aquatic macrophytes application. Egypt. J. Aquat. Res. (in press).

Behnam, H., S. Saeedfar and F.S. Mojaveryazdi. 2013. Biological contaminants of water and its effects. Technology, Education and Science International Conference (TESIC). Available at file:///C:/Users/user/AppData/Local/Temp/PaperforTesic2013.pdf

Bauddh, K., K. Singh, B. Singh and R.P. Singh. 2015. *Ricinus communis*: A robust plant for bio-energy and phytoremediation of toxic metals from contaminated soil. Ecol. Eng. 84: 640-652.

Bolan, N.S., J.H. Park, B. Robinson, R. Naidu and K.Y. Huh. 2011. Phytostabilization: A green approach to contaminant containment. pp. 145-204. *In*: D.L. Sparks [ed.]. Advances in Agronomy, Vol. 112. Elsevier.

Boxal, A.B.A. 2012. New and Emerging Water Pollutants Arising from Agriculture. Organization for Economic Co-operation and Development (OECD).

Buono, D.D., R. Terzano, I. Panfili and M.L. Bartucca. 2020. Phytoremediation and detoxification of xenobiotics in plants: Herbicide-safeners as a tool to improve plant efficiency in the remediation of polluted environments: A mini-review. Int. J. Phytoremediation 22(8): 789-803.

Chakravarty, P. and M. Kumar. 2019. Floral species in pollution remediation and augmentation of micrometeorological conditions and microclimate: An integrated approach. pp. 203-219. *In*: V.C. Pandey and K. Bauddh [eds.]. Phytomanagement of Polluted Sites: Market Opportunities in Sustainable Phytoremediation. Elsevier.

Chang, F., C. Ko, M. Tsai, Y. Wang and C. Chung. 2014. Phytoremediation of heavy metal contaminated soil by *Jatropha curcas*. Ecotoxicology 23: 1969-1978.

Coppock, R.W. and M. Dziwenka. 2015. Potential agents that can cause contamination of animal feedstuffs and terror. pp. 781-790. *In*: R.C. Gupta [ed.]. Handbook of Toxicology of Chemical Warfare Agents, 2nd ed.. Elsevier.

Fang, Y.Y., X.E. Yang, H.Q. Chang, P.M. Pu, X.F. Ding and Z. Rengel. 2007. Phytoremediation of nitrogen-polluted water using water hyacinth. J. Plant Nutr. 30(11): 1753-1765.

Fanna, A.G., G. Yadji, T.D.B. Abdourahmane, O.I. Zakaria and A.J.M. Karimou. 2018. Phytoextraction of Pb, Cd, Cu and Zn by *Ricinus communis*. Environ. Water Sci. Public Health Territ. Intell. J. 2(3): 56-62.

Favas, P.J.C., J. Pratas, M. Varun, R. D'Souza and M.S. Paul. 2014. Phytoremediation of soils contaminated with metals and metalloids at mining areas: Potential of native flora. pp. 485-517. *In*: C. Maria and S. Hernandez [eds.]. Environmental Risk Assessment of Soil Contamination. InTech, Shanghai, China.

Fox, L.J., P.J. Struik, B.L. Appletona and B.L. Rule. 2008. Nitrogen phytoremediation by water hyacinth (*Eichhornia crassipes* (Mart.) Solms). Water Air Soil Pollut. 194: 199-207.

Goldscheider, N. 2010. Delineation of spring protection zones. pp. 305-338. *In*: N. Kresic and Z. Stevanovic [eds.]. Groundwater Hydrology of Springs: Engineering, Theory, Management and Sustainability. Elsevier, Burlington, USA/Oxford, UK.

González-Chávez, Ma. del Carmen A., R. Carrillo-González, M.H. Godínez and S.E. Lozano. 2017. *Jatropha curcas* and assisted phytoremediation of a mine tailing with biochar and a mycorrhizal fungus. Int. J. Phytoremediation. 19(2): 174-182.

Grzegorska, A., P. Rybarczyk, A. Rogala and D. Zabrocki. 2020. Phytoremediation: From environment cleaning to energy generation – current status and future perspectives. Energies 13: 2905.

Hadi, F., M.Z. Ul Arifeen, T. Aziz, S. Nawab and G. Nabi. 2015. Phytoremediation of cadmium by *Ricinus communis* L. in hydroponic condition. American-Eurasian J. Agric. Environ. Sci. 15(6): 1155-1162.

Herrera, J.M., X. Sanchez-Chino, G. Davila-Ortiz and C.J. Martinez. 2019. Comparative extraction of *Jatropha curcas* L. lipids by conventional and enzymatic methods. Food Bioprod. Process 118: 32-39.

Huang, H., N. Yua, L. Wang, D.K. Gupta, Z. He, K. Wang, Z. Zhu, X. Yan, T. Li and X. Yang. 2011. The phytoremediation potential of bioenergy crop *Ricinus communis* for DDTs and cadmium co-contaminated soil. Bioresour. Technol. 102: 11034-11038.

Ifon, B.E., A.C.F. Togbé, L.A.S. Tometin, F. Suanon and A. Yessoufou. 2019. Metal-contaminated soil remediation: Phytoremediation, chemical leaching and electrochemical remediation. pp. 534-554. *In*: Z.A. Begum, I.M.M. Rahman and H. Hasegawa [eds.]. Metals in Soil-Contamination and Remediation. InTech, London, UK.

Jamil, S., P.C. Abhilash, N. Singh and P.N. Sharma. 2009. *Jatropha curcas*: A potential crop for phytoremediation of coal fly ash. J. Hazard. Mater. 172(1): 269-275.

Jingura, R.M. and R. Kamusoko. 2017. Technical options for valorization of *Jatropha*: A review. Waste and Biomass Valor. 9(5): 701-713.

Jonas, M., C. Ketlogetswe and J. Gandure. 2020. Variation of *Jatropha curcas* seed oil content and fatty acid composition with fruit maturity stage. Heliyon. 6: e02385.

Kamusoko, R. and R.M. Jingura. 2017. Utility of *Jatropha* for phytoremediation of emerging contaminants of water resources: A review. CLEAN – Soil Air Water 45(11): 1-8.

Khan, M.J., N. Ahmed, W. Hassan, T. Saba, S. Khan and Q. Khan. 2019. Evaluation of phytoremediation potential of castor cultivars for heavy metals from soil. Planta Daninha. v37: e019180998.

Kiran, B.R. and M.N.V. Prasad. 2017. *Ricinus communis* L. (castor bean), a potential multipurpose environmental crop for improved and integrated phytoremediation. The EuroBiotech J. 1(2): 101-116.

Kouamé, K.V., S. Yapoga, K.N. Kouadio, A.S. Tidou and B.C. Atsé. 2016. Phytoremediation of wastewater toxicity using water hyacinth (*Eichhornia crassipes*) and water lettuce (*Pistia stratiotes*), Int. J. Phytoremediation 18(10): 949-955.

Kuete, V. 2014. Physical, hematological, and histopathological signs of toxicity induced by African medicinal plants. pp. 635-657. *In*: V. Kuete [ed.]. Toxicological Survey of African Medicinal Plants. Elsevier, London, UK.

Kun, K., L. Fang-Yan and S. Yong-Yu. 2012. Research development and utilization status on *Jatropha curcas* in China. pp. 161-171. *In*: A.A. Oteng-Amoako [ed.]. New Advances and Contributions to Forestry Research. InTech, Croatia.

Lewandowski, I., I. Schmidt, M. Londo and A. Faaij. 2006. The economic value of the phytoremediation function – Assessed by the example of cadmium remediation by willow (*Salix* ssp). Agric. Syst. 89(1): 68-89.

Maheshwari, P. and I. Kovalchuk. 2016. Genetic transformation of crops for oil production. pp. 379-412. *In*: T.A. Mckeon, D.G. Hayes, D.F. Hildebrand and R.J. Weselake [eds.]. Industrial Oil Crops. Elsevier.

Marrugo-Negrete, J., J. Durango-Hernández, J. Pinedo-Hernández, J. Olivero-Verbel and S. Díez. 2015. Phytoremediation of mercury-contaminated soils by *Jatropha curcas*. Chemosphere 127: 58-63.

Marrugo-Negrete, J., S. Marrugo-Madrid, J. Pinedo-Hernández, J. Durango-Hernández and S. Díez. 2016a. Screening of native plant species for phytoremediation potential at a Hg-contaminated mining site. Sci Total Environ. 542 (Part A): 809-816.

Marrugo-Negrete, J., J. Durango-Hernández, J. Pinedo-Hernández, G. Enamorado-Montes and S. Díez. 2016b. Mercury uptake and effects on growth in *Jatropha curcas*. J. Environ. Sci. 48: 120-125.

Martin, J.F.G., Ma del Carmen, G. Caro, Ma del Carmen, L. Barrera, M.T. Garcia, D. Barbin and P.A. Mateos. 2020. Metal accumulation *by Jatropha curcas* L. adult plants grown on heavy metal-contaminated soil. Plants 9: 418.

Marques, M.C., C.W.A. Nascimentoa, A.J. da Silva and A. da Silva Gouveia-Neto. 2017. Tolerance of an energy crop (*Jatropha curcas* L.) to zinc and lead assessed by chlorophyll fluorescence and enzyme activity. S. Afr. J. Bot. 112: 275-282.

Masindi, V. and K.L. Muedi. 2018. Environmental contamination by heavy metals. pp. 115-133. *In*: H.E.M. Saleh and R.F. Aglan [eds.]. Heavy Metals. InTech, London, UK.

Mayo, A.W. and E.E. Hanai. 2016. Modeling phytoremediation of nitrogen-polluted water using water hyacinth (*Eichhornia crassipes*). Phys. Chem. Earth. 100: 170-180.

Mekala, N.K., R. Potumarthi, R.R. Baadhe and V.K. Gupta. 2014. Current bioenergy researches: Strengths and future challenges. pp. 1-21. *In*: V.K. Gupta, M.G. Tuohy, C.P. Kubicek, J. Saddler and F. Xu [eds.]. Bioenergy Research: Advances and Applications. Elsevier, Waltham, MA, USA.

Ndimele, P.E., C.A. Kumolu-Johnson, K.S. Chukwuka and O.R. Adaramoye. 2014. Phytoremediation of iron (Fe) and copper (Cu) by water hyacinth (*Eichhornia crassipes)*. Trends Appl. Sci Res. 9(9): 485-493.

Ochekwu, E.B. and B. Madagwa. 2013. Phytoremediation potentials of water hyacinth. *Eichhornia crassipes* (Mart.) Solms in crude oil-polluted water. J. Appl. Sci. Environ. Manage. 17(4): 503-507.

Olivares, A.R., R. Carrillo-González, Ma del Carmen, A. González-Chávez and R.M.S. Hernández. 2013. Potential of castor bean (*Ricinus communis* L.) for phytoremediation of mine tailings and oil production. J. Environ. Manage. 114: 316-323.

Padmavathiamma, P.K. and L.Y. Li. 2007. Phytoremediation technology: Hyper-accumulation metals in plants. Water Air Soil Pollut. 184: 105-126.

Palanivel, T.M., B. Pracejus and R. Victor. 2020. Phytoremediation potential of castor (*Ricinus communis* L.) in the soils of the abandoned copper mine in Northern Oman: Implications for arid regions. Environ. Sci. Pollut. Res. 27: 17359-17369.

Panda, D., L. Mandal, J. Barik, B. Padhan and S.S. Bisoi. 2020. Physiological response of metal tolerance and detoxification in castor (*Ricinus communis* L.) under fly ash-amended soil. Heliyon. 6: e04567.

Pandey, V.C. 2013. Suitability of *Ricinus communis* L. cultivation for phytoremediation of fly ash disposal sites. Ecol. Eng. 57: 336-341.

Pant, R., P. Pandey and R. Kotoky. 2016. Rhizosphere mediated biodegradation of 1,4-dichlorobenzene by plant growth-promoting rhizobacteria of *Jatropha curcas*. Ecol. Eng. 94: 50-56.

Primandari, S.R.P., A.K.M.A. Islam, Z. Yaakob and S. Chakrabarty. 2018. *Jatropha curcas* L. Biomass waste and its utilization. pp. 273-282. *In*: M. Nageswara-Rao and J. Soneji [eds.]. Advances in Biofuels and Bioenergy. Intech, London, UK.

Priya, E.S and P.S. Selvan. 2014. Water hyacinth (*Eichhornia crassipes*) – An efficient and economic adsorbent for textile effluent treatment – A review. Arab. J. Chem. 10: S3548-S3558.

Raikova, S., M. Piccini, M.K. Surman, M.J. Allen and C.J. Chuck. 2019. Making light work of heavy metal contamination: The potential for coupling bioremediation with bioenergy production. J. Chem. Technol. Biotechnol. 94: 3064-3072.

Rezania, S., M. Ponraj, A. Talaiekhozani, S.E. Mohamad, M.F. Md Din, S.M. Taib, F. Sabbagh and F. Md Sairan. 2015. Perspectives of phytoremediation using water hyacinth for removal of heavy metals, organic and inorganic pollutants in wastewater. J. Environ. Manage. 163: 125-133.

Robinson, B., J. Ferńandez, P. Madejon, T. Marañon, J.M. Murillo and S. Green. 2003. Phytoextraction: An assessment of biogeochemical and economic viability. Plant Soil 249: 117-125.

Rosenfeld, P.E. and L.G.H. Feng. 2011. Risks of Hazardous Wastes. Elsevier, Amsterdam.

Rungwa, S., G. Arpa, H. Sakulas, A. Harakuwe and D. Timi. 2013. Phytoremediation – An eco-friendly and sustainable method of heavy metal removal from closed mine environments in Papua New Guinea. Procedia Earth and Planet Sci. 6: 269-277.

Saha, P., O. Shinde and S. Sarkar. 2017. Phytoremediation of industrial mines wastewater using water hyacinth. Int. J. Phytoremediation 19(1): 87-96.

Sand-Jensen, K. 2013. Freshwater ecosystems, human impact on. pp. 570-586. *In*: S.A. Levin [ed.]. Encyclopedia of Biodiversity (2nd ed.). Academic Press. Amsterdam, The Netherlands/Oxford, UK

Sharma, S. and A. Bhattacharya. 2017. Drinking water contamination and treatment techniques. Appl. Water. Sci. 7: 1043-1067.

Shrestha, P., K. Belliturk and J.H. Gorres. 2019. Phytoremediation of heavy metal-contaminated soil by switchgrass: A comparative study utilizing different composts and coir fiber on pollution remediation, plant productivity, and nutrient leaching. Int. J. Environ. Res. Public Health 16: 1261.

Stohlgren, T.J., P. Pysek, J. Kartesz, M. Nishino, A. Pauchard, M. Winter, Pino Joan, Richardson David M., Wilson John, Murray Brad R., Phillips Megan L., Celesti-Grapow Laura and Graham Jim. 2013. Globalization effects on common plant species. pp. 700-706. *In*: S.A. Levin [ed.]. Encyclopedia of Biodiversity (2nd ed.). Academic Press. Waltham, MA.

Swain, G., S. Adhikari and P. Mohanty. 2014. Phytoremediation of copper and cadmium from water using water hyacinth, *Eichhornia crassipes*. Int. J. Agric. Sci. Technol. 2(1): doi: 10.14355/ijast.2014.0301.01

Tripathi, V., S.A. Edrisi and P.C. Abhilash. 2016. Towards the coupling of phytoremediation with bioenergy production. Renew. Sust. Energy Rev. 57: 1386-1389.

Ukunowo, W.O. and L.A. Ogunkanmi. 2010. Phytoremediation potential of some heavy metals by water hyacinth. Int. J. Biol. Chem. Sci. 4(2): 347-353.

Vangronsveld, J., R. Herzig, N. Weyens, J. Boulet, K. Adriaensen, A. Ruttens, T. Thewys, A. Vassilev, E. Meers, E. Nehnevajova, D. van der Lelie and M. Mench. 2009. Phytoremediation of contaminated soils and groundwater: Lessons from the field. Environ. Sci. Pollut. Res. doi: 10.1007/s11356-009-0213-6

Wan, X., M. Lei and T. Chen. 2016. Cost-benefit calculation of phytoremediation technology for heavy-metal-contaminated soil. Sci. Total Environ. 563-564: 796-802.

Wu, Q., S. Wang, P. Thangavel, Q. Li, H. Zheng, J. Bai and R. Qiu. 2011. Phytostabilization potential of *Jatropha curcas* L. in polymetallic acid mine tailings. Int. J. Phytoremediation. 13: 788-804.

Xia, H. and X. Ma. 2006. Phytoremediation of ethion by water hyacinth (*Eichhornia crassipes*) from water. Bioresour. Technol. 97(8): 1050-1054.

Xiong, P., C. He, OH. Kokyo, X. Chen, X. Liang, X. Liu, X. Cheng, C. Wu and Z. Shi. 2018. *Medicago sativa* L. enhances the phytoextraction of cadmium and zinc by *Ricinus communis* L. on contaminated land *in-situ*. Ecol. Eng. 116: 61-66.

Yashim, Z.I., E.B. Agbaji, C.E. Gimba and S.O. Idris. 2016. Phytoremediation potential of *Ricinus communis* L. (castor oil plant) in Northern Nigeria. Int. J. Plant Soil Sci. 10(5): 1-8.

Zhang, H., X. Chen, C. He, X. Liang, K. Oh, X. Liu and Y. Lei. 2015. Use of energy crop (*Ricinus communis* L.) for phytoextraction of heavy metals assisted with citric acid. Int. J. Phytoremediation 17(7): 632-639.

Zhang, Q., Y. Wei, H. Han and C. Weng. 2018. Enhancing bioethanol production from water hyacinth by new combined pretreatment methods. Bioresour. Technol. 251: 358-363.

Zhou, W., D. Zhu, L. Tan, S. Liao, H. Hu and H. David. 2007. Extraction and removal of potassium water hyacinth (*Eichhornia crassipes*). Bioresour. Technol. 98: 226-231.

Zhou, X., S. Wang, Y. Liu, G. Huang, S. Yao and H. Hu. 2020. Coupling phytoremediation efficiency and detoxification to assess the role of P in the Cu tolerant *Ricinus communis* L. Chemosphere 247: 125965.

CHAPTER

5

Water Hyacinth: A Potent Source for Phytoremediation and Biofuel Production

Elizabeth Cherian* and Shanti Joseph
Department of Botany, CMS College, Kottayam, Kerala, India

5.1 Introduction

Eichhornia crassipes (Mart) Solms. is a rhizomatous aquatic plant belonging to the family Pontederiaceae. Its dark green leaves that float on the water surface are thick, glossy, and ovate with thick spongy petioles. It possesses feathery, purple coloured and free-hanging roots. The inflorescence consists of a single spike of beautiful lilac flowers. The fruit consists of a thin-walled capsule in which 200 tiny seeds are arranged (Pieterse 1997).

E. crassipes, commonly known as water hyacinth, is a plant species of wide distribution. The plant floating on the water surface with its attractive lilac flowers is a beauty to the eyes but a menace to the environment. It is a scenic beauty when it blooms over a wide field. However, the plant is also called the 'lilac devil' for its adverse impacts on water bodies. This aquatic weed is native to South America and was initially introduced for its aesthetic values in Africa and Southern Asia. Soon it extended to other parts of the world due to its invasive nature. It was first spotted in India at the beginning of 1890 in West Bengal and started spreading to other areas. Hence, it is also known to be the "Terror of Bengal". Now, it has infested a large area of water resources in the country. An internationally important Ramsar site, Vembanad Lake in Kerala, is now infested with this weed (Fig. 5.1). Agricultural run-off that contains a high amount of fertilizers promotes the excessive growth of *Eichhornia,* especially in lentic water bodies. The rapid propagation of the weed results in forming a thick layer on the water surface that may cause blockage of rivers, depletion of dissolved oxygen, loss of indigenous species of flora and fauna, thus reducing the biodiversity. Its impacts have extended to various sectors such as irrigation, hydroelectric generation, agriculture, fishing, inland navigation, and tourism (Guragain et al. 2011).

Corresponding author: elizabeth.cherian@cmscollege

Figure 5.1: Vembanad Lake infested with *Eichhornia crassipes.*

5.2 Ecological Consequences of *E. crassipes*

The presence of *E. crassipes* weed in a water body indicates environmental pollution due to eutrophication. Excess nitrogen and phosphorus from the fertilizers used in the agricultural fields, which reach water bodies, have promoted the extensive growth of water hyacinth (Coetzee and Martin 2011). The plant has a very short reproductive cycle; the time required for its biomass to get doubled is just 11-15 days in a favorable environmental condition (Penfound and Earle 1948, Minychl et al. 2019). It is one of the highest productivity rates among the higher plants. Though the common mode by which the plant spreads in an area is vegetative propagation, the seeds may also act as infestation agents (Albano et al. 2011). The plant is viable enough to produce several inflorescences per year, and each inflorescence can release up to 3000 seeds (Barrett 1980). Another interesting factor is the viability of its seeds that may go up to 20 years. The longevity of the seeds stands as a hurdle for its eradication (Cacho et al. 2006, Albano et al. 2011). Hence, *E. crassipes* holds all conducive elements to grow and thrive in a moderate to a humid environment.

The plant has the adeptness to tolerate extreme pH, temperature, toxic substances, and water level fluctuations (Ganguly et al. 2012). The invasive weed, with its rapid propagation, forms a thick mat over the water surface and prevents the sunlight from entering the deeper regions. It hinders the life of underwater flora and fauna. There are reports on its inhibitory property against some microalgal species also (Almeida et al. 2006). *E. crassipes* is capable of absorbing a wide range of nutrients and other compounds, including metals; the death and decay of the plant again liberate these chemicals in the excess amount into the water bodies making it more eutrophic and polluted (Gao and Bo 2004). Over a while, it depletes the amount of oxygen in the water, and the most affected group is the fish population.

Indigenous fish species that are mostly with gills need more dissolved oxygen to survive. Thus, the proliferation of *Eichhornia* may gradually lead to the disappearance of native species and the emergence of exotic species. For example, fish species like African catfish and Tilapia that are exotic and invasive may overgrow in such conditions (Chatterji 2019, Vicente and Fonseca-Alves 2013). The former is an accessorial air-breathing fish that can successfully live at a lower rate of oxygen and can even be found in the most confined spaces at great densities. Hardy fish species like Tilapia can also thrive in an aquatic ecosystem with a depleted oxygen level (Mohsen et al. 2015). The disappearance of existing species and the emergence of new invasive species disrupt the food chain and food web in the aquatic ecosystem and ultimately affect the human population.

5.3 Impacts of *E. crassipes* on Livelihood

World over, *E. crassipes* is considered a noxious weed that affects various sectors and affects the livelihood of the people who depend on agriculture and fishing. In Egypt, most of the drainage and irrigation systems were being interrupted due to infestation of this weed that harmfully affects the people who entirely depend on agriculture (Fayad et al. 2001, Kateregga and Sterner 2009). In New Guinea, this weed is reported to have altered the socio-economic frame in the lower flood plains of the Sepik River. Further, it has been detrimental to the maintenance of gardens, hunting and fishing areas, and markets (Harley et al. 1996). The studies in Ghana reveal that this noxious weed has affected the transportation system, which in turn made it difficult for the children to reach their educational institutions. Water bodies filled with this weed also promote the breeding of mosquitoes that may lead to the spread of diseases (Room and Fernando 1992).

In several regions of India, the infestations of *E. crassipes* hampered the irrigation and transportation channel. The people whose livelihood depends on fishing and agriculture are the worst affected community due to the flourishing of this weed. Fishing becomes a significant challenge for the traditional fishermen, especially with gill nets used to trap gilled fishes when the water bodies are filled with lush hyacinths. Another sector affected by this problem is the agriculture sector. The impact of the weed on the agriculture sector is very crucial, especially on rice cultivation. It is a severe hurdle for the farmers in India, who already have several issues related to crop production. It is very difficult for them, particularly during the monsoon, as the weed grows abundantly when the irrigation water is flushed into the paddy fields (Dipanjan 2010). In West Bengal, the place where the plant was introduced in India, a severe loss was suffered by the rice farmers because of the rapid growth of *Eichhornia* that directly crushed the crop plant from germination and finally the harvesting (Patel 2012).

Indian economy is an agrarian economy, and Kerala is not an exception from the rest of India. Kerala has a predominantly agriculture sector; rice is the staple food crop, and almost 95% of the grains are produced within the state. About 300,000 rice growers in Kerala are mostly small and marginal farmers; the land area they hold on an average is below 0.4 ha, which is one-fifth of the national average (Suchitra 2015). It is distressing that *E. crassipes* cover almost all the paddy fields of Kerala. Overgrowth of this plant has harmed paddy production in Kerala, particularly in the Kuttanad region of Alappuzha district, renowned as Kerala's Rice Bowl. Kuttanad is a low-lying area where rice is the major crop cultivated. As it is a region of low altitude, farming is practised around 3 meters below sea level. The kuttanad farming system has been declared as a Globally Important Agricultural Heritage System (GIAHS) by FAO (Food and Agriculture Organization). The majority of the people in Kuttanad depend on paddy cultivation for their livelihood. The invasive nature of the plant has posed a threat to the farmers for cultivating crops.

The prolific growth of water hyacinth makes a severe profit loss to the farmers of this place as they need it to be removed before the cultivation of crops. At times, they get financial assistance from the government, but it is only a negligible amount as it needs more labour. After the harvest, grains are more often carried out in boats. Transportation through the water bodies is another challenge faced by farmers. Experts suggest that the sole method to reduce the spreading of the weeds is by putting them to different uses (Jafari 2010).

Various attempts have been made to-date to eradicate this plant using physical, mechanical, chemical and biological measures. However, nothing has been successful yet, and the problem has been amplifying day by day. The only way for its effective management is through utilizing the plant for various needs. The ability of the plant to absorb nutrients and other impurities can be exploited by using it for wastewater treatment and phytoremediation processes. It may be used in hydroponics and for the production of biofuel and various value-added goods.

5.4 Value-added Products from *E. crassipes*

The government is now focusing on removing this weed by channelizing it to produce value-added products that can further provide job opportunities for people. The scope of the plant to generate value-added products extends from mushroom cultivation, briquetting, composting, animal feed, pulp-based products to water purification, phytoremediation and biofuel production (Weiping 2018, Opeyemi et al. 2020). According to Rezania et al. (2016), the products like fish feed, compost, biogas, ethanol and power plant energy (briquette) can be obtained as a by-product when the plant is used for wastewater treatment.

In addition to this, the plant is reported for its antimicrobial, antioxidant, and anticancer properties and can hence be used to produce bioactive substances (Kumar et al. 2014, Shanab et al. 2014). Phytochemical constituents of *E. crassipes* include compounds of nutritional and medicinal importance such as alkaloids, terpenoids, saponins, flavonoids, tannins, glycosides and phenols (Malik 2007). The plant extracts can be used to inhibit the corrosion of metals (Shanab and Shalaby 2012). It was also found useful for the production of pigments and glycerol (Shanab et al. 2017).

The uses mentioned above reveal that the plant that forms a menace to the environment can be converted into something beneficial for society. It will be an ideal management strategy from the government's side if the weed is removed regularly and utilized for such purposes. It reduces the negative impacts of the weed on the aquatic ecosystem and provides economic support to the nation. It can be a sustainable practice, especially for developing countries.

5.5 *E. crassipes* for Wastewater Treatment

Excess nitrogen and phosphorus released from the agricultural residues form the major reason for eutrophication, an environmental problem that needs urgent attention. The presence of nitrogen in the form of ammonia is seen in both domestic and industrial wastewater. Water hyacinth is a potent plant source for removing nutrients, organic and inorganic impurities and heavy metals (Rezania et al. 2015). From earlier times itself, water hyacinth was known for its ability to purify water, and it can be utilized effectively with newer technologies in the modern era (Lindsey and Hirt 1999, Rezania et al. 2015). Several reports on water hyacinth show that it can be effectively used for wastewater treatment (Mangabeira et al. 2004, Rezania et al. 2015, Mishra and Maiti 2017, Rajendra et al. 2017). The high capacity to absorb pollutants and tolerance to toxins coupled with high production rate makes the plant well suited for the purpose (Liao and Chang 2004, Jayaweera and Kasturiarachchi 2004).

Artificial wetlands can be constructed to cultivate *E. crassipes* for the treatment of wastewater. It has been used to remove Fe and Al from wastewater (Jayaweera et al. 2007, 2008). According to Jayaweera and Kasturiarachchi (2004), a six-week old plant is best suited for this purpose. They recommended a hydraulic retention time (HRT) of 21 days as the optimum time needed for the removal of nitrogen and phosphorus from the wastewater. They identified that the key mechanisms of the removal of nitrogen are assimilation and denitrification, while that of phosphorus is by assimilation and sorption. The potential of the plant in a continuous system will depend on the optimum growth rate of the plant (Rezania et al. 2016).

5.6 Phytoremediation for the Removal of Heavy Metals

The effluents from chemical factories may often contain toxic components including heavy metals that may persist for a long time. Releasing high amounts of heavy metals into public water bodies may affect aquatic life and ultimately become a threat to human life itself (Monisha et al. 2014). Toxins that enter the food chain may imbalance the whole ecosystem. These chemical contaminants may impart either short-term or long-term toxicity. In recent years, bioremediation is the most advocated process for eliminating the pollutants from the contaminated sites either by using microbes (microbial remediation) (Bayat et al. 2015, Lakhan et al. 2020) or by using plants that can absorb or accumulate impurities (phytoremediation) (Monisha et al. 2014).

Phytoremediation has become a sustainable, eco-friendly approach for removing contaminants from a region. However, selecting a suitable plant is a key factor for the process to be successful. Various aquatic plants have been reported for their water purifying ability; *Eichhornia, Wolffia, Wolffiella, Lemna, Azolla, Spirodela,* and *Potamogeton* have been categorized as phytoremediators. These plants were found to accumulate aquatic impurities in their body tissues through bioaccumulation and reduce pollution (Jasrotia et al. 2017, Abid et al. 2020). *Eichhornia* was observed to be the most resistant plant that can tolerate the toxicity of heavy metals. It can also tolerate toxic compounds such as formaldehyde, acetic acid, formic acid, oxalic acid and phenol, even in high concentrations (Abid et al. 2020, Yan et al. 2020). Odjegba and Fasidi (2007) pointed out that the plant can bioconcentrate heavy metals. Thus, it can be applied for the treatment of wastewater containing low levels of toxic metals like Pb, Cd, Zn, Cr, Ag, Cu, Ni and Hg. The accumulation of metals was observed in both the shoot and root of the plant, but the absorption rate of each compound was found to vary in different parts analyzed. Sabale et al. (2010) studied the phytoabsorption of different metals in the plant root and was observed to be in the order Fe>Mn>Cu>Zn>Cr>Pb>Ni>Co>Cd. According to Agunbiade et al. (2009), an aquatic plant that can contribute large biomass and accumulate toxic contaminants can be an ideal source to be applied in the phytoremediation process.

Phytoremediation is classified into six types based on the mechanism by which the plant eliminates the pollutants viz. phytoextraction, rhizofiltration, phytotransformation, rhizodegradation, phytostabilization and phytovolatilization. Phytoextraction/phytoaccumulation is the process by which the plant roots absorb the heavy metals and is concentrated and gets accumulated in body parts (Blaylock 2000, Lombi et al. 2001). Rhizofiltration/phytofiltration is the mechanism where the plant root functions as a sieve through which the pollutants are filtered out or are adsorbed on the roots (Suresh and Ravishanker 2004). Phytodegradation/phytotransformation is the conversion of contaminants to much simpler compounds with the help of enzymes in the plants or the microflora associated with the plant. It may further metabolize in the shoot or root system (Suresh and Ravishanker 2004). Rhizodegradation/Phytostimulation is a process where microorganisms play a major role. Microbes present in the rhizosphere are potential agents that could degrade the contaminants effectively (Jones et al. 2004). Phytostabilization/phytorestoration stabilizes the contaminants, thus preventing their dispersion. It is more effective in finely textured soils. This method can sustain high toxicity, and it never converts contaminants to consumable aerial parts (Marseille et al. 2000, Bouwman et al. 2001, Trivedi and Ansari 2005). In phytovolatilization, the pollutants absorbed by the plant roots are translocated to the aerial portions and finally to the leaves. There, they get converted into volatile forms during metabolic activities. Finally, by transpiration, it is expelled out from the plant body (Rubin and Ramaswami 2001, Neumann et al. 2003, Cristaldi et al. 2017).

Table 5.1: Heavy metals and the type of phytoremediation in *Eichhornia crassipes*

Sl. No.	Heavy metals	Mode of phytoremediation	References
1	Cd	Rhizofiltration, Phytoaccumulation (Adsorption and Absorption)	Wolverton (1975), Gardea-Torresdey et al. (2005), Odjegba and Fasidi (2007), Mahamadi and Nharingo (2010), Saraswat and Rai (2010), Balaji et al. (2014), Li et al. (2016), Pandey (2016), Eid et al. (2019), Chukwuka et al. (2020)
2	Ni	Rhizofiltration	Wolverton (1975), Gardea-Torresdey et al. (2005), Odjegba and Fasidi (2007), Eid et al. (2019)
3	Zn	Phytoaccumulation and Phytodegradation	Gardea-Torresdey et al. (2005), Odjegba and Fasidi (2007), Mishra and Tripathi (2009), Mahamadi and Nharingo (2010), Saraswat and Rai (2010), Singh and Kalamdhad (2013), Balaji et al. (2014), Li et al. (2016), Eid et al. (2019), Chukwuka et al. (2020)
4	Pb	Rhizofiltration (Adsorption and Absorption)	Wolverton and McDonald (1975), Odjegba and Fasidi (2007), Mahamadi and Nharingo (2010), Baruah et al. (2012), Balaji et al. (2014), Malar et al. (2014), Li et al. (2016), Eid et al. (2019)
5	Cr	Phytoaccumulation, Phytodegradation and Rhizofiltration	Lytle et al. (1998), Odjegba and Fasidi (2007), Mishra and Tripathi (2009), Saraswat and Rai (2010), Balaji et al. (2014), Pandey (2016), Eid et al. (2019)
6	As	Phytoaccumulation	Rahman and Hasegawa (2011)
7	Hg	Rhizofiltration	Wolverton and McDonald (1975), Riddle et al. (2002)
8	Cu	Phytoaccumulation, Rhizofiltration (Adsorption and Absorption)	Gardea-Torresdey et al. (2005), Odjegba and Fasidi (2007), Ndimele et al. (2014), Li et al. (2016), Pandey (2016), Eid et al. (2019)
9	Co	Phytoaccumulation, Rhizofiltration	Eid et al. (2019)
10	Mn	Phytoaccumulation	Li et al. (2016), Eid et al. (2019)
11	Fe	Phytoaccumulation, Rhizofiltration	Odjegba and Fasidi (2007), Jayaweera et al. (2008), Ndimele et al. (2014)

According to Harun et al. (2011), the accumulation of metals by *E. crassipes* may vary depending on the age of the plant. Among the previously mentioned mechanisms, phytoaccumulation, rhizofiltration and phytodegradation are reported in *E. crassipes* (Table 5.1).

Pandey (2016) identified that naturally grown *E. crassipes* could be used in fly ash ponds and found that it could accumulate Cu, Cr, and Cd in the pond. Mishra and Maiti (2017) also supported the ability of water hyacinth to accumulate a wide range of contaminants such as radionuclides, dyes, heavy metals and other organic and inorganic impurities present in the water bodies. Water hyacinth is a widely acceptable plant species for the phytoremediation process due to its characteristics of fast growth rate, adaptability to thrive in varied environmental conditions and the capacity to uptake an array of toxic contaminants (Ting et al. 2018).

5.7 *E. crassipes* as a Source for Biofuel Production

Recent studies show biomass with high energy value can be effectively utilized as an alternative energy source. *E. crassipes*, due to its high availability and biomass yield, is preferred to be a potential plant species for the production of bioethanol. Nowadays, alternative energy resources are in growing demand as our fossil fuels are undergoing a drastic depletion due to their overutilization. Moreover, it takes a long time for fossil fuels to be formed on the Earth's crust. It also contributes to environmental pollution due to the high emission of carbon and other pollutants. Hence, it is the need of the hour to find an alternative sustainable energy source. Several attempts have been made to produce biofuel using various biological sources.

5.7.1 Biofuel production process

Biofuels form a renewable energy source primarily derived from biological materials that turn into liquid fuels (Ohno and Fukaya 2009). It is an attractive alternative to gasoline as it can lower the rate of greenhouse gases released from the transportation sector (Demirbas 2003). The production process of biofuel may vary depending on the final product; biodiesel is produced through transesterification process, biogas through anaerobic fermentation and bioethanol using a simple fermentation technique.

Earlier, plants with sugar or starch content were used to produce bioethanol, which was considered the first-generation biofuel. Sugarcanes and sugar beets were the sugar-based ethanol plants, whereas corns and other grains were utilized for starch-based ethanol production. The production of ethanol from grains includes the milling of grains and the starch liquefaction process. The sugar is converted to starch by the saccharification process. Finally, ethanol is produced by the fermentation process using yeast (*Saccharomyces cerevisiae*). The ethanol obtained was separated and purified, and concentrated by distillation and dehydration process. It is concentrated above 99.7% for fuel applications (Lennartsson et al. 2014). However, the 'first-generation' seemed unsustainable as it contributes to potential stress over food commodities and may further result in food shortage over the long run.

Second-generation biofuels are primarily focused on lignocellulosic biomass and other agricultural wastes. It includes used cooking oil, the waste from food crops, and wood chips (Ali 2017). Wheat straw and corn husks are examples of second-generation feedstock (Mahmood 2017). After food crops are harvested, the agricultural residue contributes enormous biomass, which can be a good source for biofuel production. Crop residues can be transformed into biofuels by the thermochemical or biochemical conversion process. The former includes hydrothermal liquefaction, pyrolysis, gasification and combustion (Ali et al. 2019).

The biomass from lignocellulosic wastes contains cellulose, hemicellulose and lignin. The biomass is allowed to undergo a series of steps such as pre-treatment, enzymatic hydrolysis, fermentation, distillation and dehydration. In the pre-treatment step, the surface area of carbohydrate is increased before the enzymatic saccharification. Enzymatic hydrolysis releases glucose, xylose and various other kinds of sugars. The saccharification and fermentation process can either be carried out independently or simultaneously. The fermentation process is carried out mainly with the help of yeast, though occasionally genetically modified microorganisms are also used. By genetic engineering, it is possible to synthesize pentose-fermenting microorganisms (Robak and Balcerek 2018).

It is worth noting that the plants that form the feedstock for second-generation biofuel do not require any additional fertilizer, water, or land to grow. However, the plants that are used as intercrops for the production of biofuel may grow along with the food crops and absorb the nutrients from the soil. Though the second-generation biofuels depend only on non-food biomass, some of the non-edible plant parts are a food source to animals. Moreover, it requires a land area to a certain extent. Hence, it paved the way to the third generation of biofuels, where algae emerged as a feedstock (Fig. 5.2). Algae are highly preferred for the production of biodiesel due to their high lipid content, simple growth environment and short time growth (Nirupa et al. 2018). Moreover, it needs only a lesser land space compared to terrestrial plants. However, the algal source is more focused on the production of biodiesel rather than that of bioethanol. Immature production techniques remain as another hurdle to reach large scale production of microalgae biofuel (Tahani and Temtamy 2013).

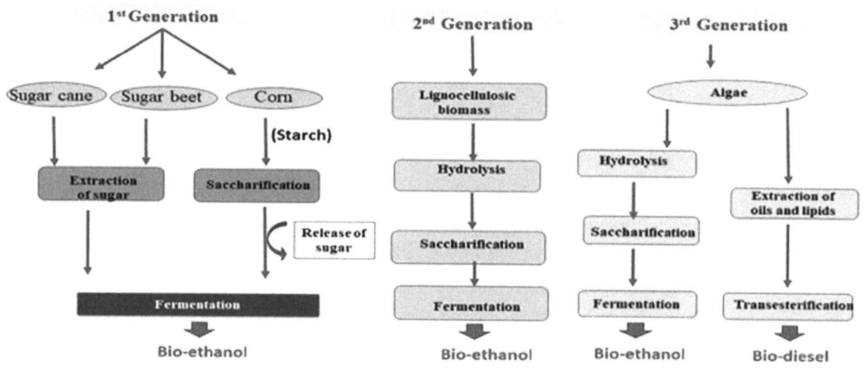

Figure 5.2: Production process of biofuel.

5.7.2 *E. crassipes* as a potent plant species for the production of biofuel

Researchers have been working on different plants for the synthesis of biomass in a shorter time. Biomass is a sustainable alternative energy source as it is economical and easily available (Balat 2011, Lee and Kuan 2015). Several reports are aiming at lignocellulosic biomass like straw or sugarcane bagasse as a feedstock for biofuel production. Recently, the focus has been turned to aquatic plants. Aquatic weeds, which contain cellulose and hemicelluloses content, prove to be beneficial for the production of biofuel (Kumar et al. 2019).

Among the aquatic plants, those with invasive nature can be considered as the best source as it produces a vast amount of biomass in a very short time. Water hyacinth is a powerful resource to be considered for the production of bioethanol as it is easily available and grows abundantly in a range of water bodies all over the world. According to Lu et al. (2007), the plant is capable of producing 140 million offsprings per year. This prolific growth will help to generate greater biomass in a minimum time that can be used to generate biofuel.

Due to the peculiarity in the chemical composition of *E. crassipes,* it can be used to generate bioethanol, biogas and biodiesel. High cellulosic content and low lignin level are the peculiar features of the plant. Hence, the biomass obtained from the plant is preferable for the production of bioethanol and biogas. High lipid content present in the plant favours the production of biodiesel (Poddar et al. 1991, Gressel 2008, Bhattacharya and Kumar 2010).

5.7.3 Bioethanol from *E. crassipes*

For the production of bioethanol, biomass residue with high cellulosic content is the most preferable and cost-effective feedstock. Biomass of *E. crassipes* is rich in cellulose and hemicellulose (58.6%) with a reduced lignin level (Das et al. 2016, Ruan et al. 2016). The bioconversion process is carried out through hydrolysis, fermentation and distillation. According to Shanab et al. (2017), 0.56 kg of ethanol can be obtained from 1 kg of cellulose from *E. crassipes* and hence the plant can be used as a potent source for the production of bioethanol.

5.7.4 Biogas from *E. crassipes*

Biogas is produced by the anaerobic digestion of organic wastes. It is a stepwise process that includes hydrolysis, acidogenesis, acetogenesis, and methanogenesis (Dohanyos and Zabranska 2001, Li and Yang 2016). Mostly, lignocellulosic materials like vegetable waste or cow dung are used as feedstock. According to Bhattacharya and Kumar (2010) and Njogu et al. (2015), *E. crassipes* is a potential feedstock as it contains a high carbon-nitrogen ratio. It can be effectively used in combination with cow dung also. The liquid slurry formed contains a high amount of nitrogen and potassium compared to other feedstocks making it a good fertilizer. Attempts have been made to enhance the biomass yield of *E. crassipes* during the anaerobic fermentation. Njogu et al. (2015) suggested that if the plant is subjected to a preliminary rotting for a period of twenty days, the productivity can be enhanced. Priya et al. (2018) suggested that through additional sun drying, the solid content of the whole plant can be increased to 40%. The yield of biogas methane can be increased (up to 75%) by subsequent digestion. Biogas is comprised of 60% methane and 40% carbon dioxide. However, methane concentration can be increased by lowering the temperature, but it lowers the production yield (Vandiver 1999).

5.7.5 Biodiesel from *E. crassipes*

E. crassipes possess favourable properties for the generation of biodiesel that includes high lipid content and productivity rate (Sagar and Kumari 2013, Bote 2020). Reports show that several lipid classes include glycolipids, non-polar lipids and phospholipids in all parts of *E. crassipes* viz. leaves, petioles, flowers and roots. Leaves contain comparatively more amount of lipid (Arayana et al. 1984, Shanab et al. 2017). Unlike other bioenergy crops, there is no need to sacrifice any land space when using this weed to produce biodiesel, and hence it resolves the problem of food shortage. As it is a hazardous invasive plant, even the crop cultivation phase can be skipped. Compared to the feedstocks of first, second and third generations of biofuel, *E. crassipes* is more economical and acceptable (Lamaisri et al. 2013, Sagar and Kumari 2013). The biodiesel obtained from water hyacinth is a potential substitute to the diesel engine used now and is well accepted for its stability (Shanab 2017, Karthikeyan et al. 2019). He analyzed the property of biodiesel by blending it with petroleum diesel in varied proportions and found that the blend of 20% biodiesel and 80% diesel fuel was comparable to the fossil fuels in terms of smoke emissions and thermal efficiency.

5.8 *E. crassipes* for Phytoremediation and Biofuel Production

Eradication of the noxious weed water hyacinth is an urgent need to be attended to because of the hazardous repercussions it creates in the environment. As experts suggest, utilizing this plant for various purposes is the only solution for its effective removal. Researches over the years have proved that *E. crassipes* has a wide range of possible positive applications, of which utilizing it for the production of biofuel is highly recommended as there is a growing demand for alternative energy resource (Lee and Kuan 2015). Though lignocellulosic biomass forms a major feedstock for the synthesis of bioethanol, degradation of the lignin part remains a stumbling block, thereby making it an expensive procedure (Ali et al. 2019). Biomass of *E. crassipes* with its reduced lignin content could be an ideal source to overcome this problem (Das et al. 2016).

The efficacy to accumulate a wide range of toxic contaminants, including heavy metals, makes *E. crassipes* a suitable plant for phytoremediation. However, after the phytoaccumulation process, the disposal and management of the plants contaminated with toxins in their body is a matter of concern (Yan et al. 2020, Jasrotia et al. 2017). These toxins, if they enter into the food chain, may result in a slew of problems. Therefore, it would be more sensible and sustainable to use such plants for the production of biofuel. *E. crassipes* is a plant that can be effectively applied for both tasks (Luet al. 2007, Ting et al. 2018). As it is an invasive plant, it can generate enormous biomass in a shorter period of time, making it more apt for the purpose. Regular removal of this weed will be beneficial to the aquatic ecosystem and various sectors like agriculture, transportation and tourism. It is highly desirable that the plant that causes unpleasant changes in the environment can be put into productive use.

5.9 Conclusion

Reclamation of water bodies contaminated with heavy metals and other toxic components is a challenging issue where bioremediation is the most acceptable technology. Several aquatic plant species have been attempted, and *E. crassipes* was proved to be potent in biosorption properties. If the same plant can be used for the production of biofuels, it can address several issues such as the safe disposal of toxin accumulated plants, the need of alternative energy resources, cleaning up of water resources and effective utilization of a noxious weed. It also reduces food shortage as no land area is used for cultivating energy crops.

Acknowledgments

I would like to express my special thanks of gratitude to my colleagues Mrs. Ani Merly Paul and Dr. Aleena Manoharan, for reviewing the pre-final draft of this chapter.

References

Abid, A.A., M. Naeem, S.G. Singh and M.A. Fahad. 2020. Phytoremediation of contaminated waters: An eco-friendly technology based on aquatic macrophytes application. The Egyptian Journal of Aquatic Research 46(4): 371-376.

Agunbiade, F.O., B.I. Olu-Owolabi and K.O. Adebowale. 2009. Phytoremediation potential of *Eichhornia crassipes* in metal-contaminated coastal water. Bioresour. Technol. 100(19): 4521-4526.

Albano Pérez, E., J.A. Coetzee, T. Ruiz Téllez and M.P. Hill. 2011. A first report of water hyacinth (*Eichhornia crassipes*) soil seed banks in South Africa. South African Journal of Botany 77: 795-800.

Ali, M., M. Saleem, Z. Khan and I.A. Watson. 2019. The use of crop residues for biofuel production. Biomass, Biopolymer-Based Materials, and Bioenergy: Construction, Biomedical, and other Industrial Applications. Elsevier Ltd.

Ali, M., R. Sultana, S. Tahir, I.A. Watson and M. Saleem. 2017. Prospects of microalgal biodiesel production in Pakistan – A review. Renew Sustain Energy Rev. 80(Suppl. C): 1588-1596.

Almeida, A.S., G. Ana, J. Pereira and J. Gonçalves. 2006. The impact of *Eichhornia crassipes* on green algae and cladocerans. Fresenius Environ. Bulletin 15(12): 1531-1538.

Arayana, G.L., K.S. Rao, A.J. Pantulu and G. Thyagarajan. 1984. Composition of lipids in roots, stalks, leaves and flowers of *Eichhornia crassipes* (Mart.) Solms. Aquat Bot. 20: 219-227.

Balaji, A.R., S. Sakthivel and S. Sheela. 2014. Efficacy of accumulation on heavy metals from aqueous solution using water hyacinth (*Eichhornia crassipes*). Asian Journal of Microbiology Biotechnology and Environmental Sciences 16(1): 115-120.

Balat, M. 2011. Production of bioethanol from lignocellulosic materials via the biochemical pathway: A review. Energy Conversion and Management 52(2): 858-875.

Barrett, S.C.H. 1980. Sexual reproduction in *Eichhornia crassipes*. Fertility of clones from diverse regions. Journal of Applied Ecology 17: 101-112.

Baruah, S., K.K. Hazarika and K.P. Sarma. 2012. Uptake and localization of Lead in *Eichhornia crassipes* grown within a hydroponic system. Advances in Applied Science Research 3(1): 51-59.

Bayat, Z., M. Hassanshahian and S. Capello. 2015. Immobilization of microbes for bioremediation of crude oil polluted environments: A mini review. Open Microbiol. J. 9: 48-54.

Bhattacharya, A. and P. Kumar. 2010. Water hyacinth as a potential biofuel crop. Electronic Journal of Environmental, Agricultural and Food Chemistry 9: 112-122.

Blaylock, M.J. 2000. Phytoextraction of Metals. Phytoremediation of Toxic Metals: Using plants to clean up the environment. Wiley Inter-Science Publication, New York.

Bote, M.A., V.R. Naik and K.B. Jagadeeshgouda. 2020. Review on water hyacinth weed as a potential biofuel crop to meet collective energy needs. Materials Science for Energy Technologies 3: 397-406.

Bouwman, L.A., J. Bloem, P.F.A.M. Romkens, G.T. Boon and J. Vangronsveld. 2001. Beneficial effects of the growth of metal tolerant grass on biological and chemical parameters in copper- and zinc-contaminated sandy soils. Minerva Biotechnologica 13: 19-26.

Cacho, J.O., D. Spring, P. Pheloung and S. Hester. 2006. Evaluating the feasibility of eradicating an invasion. Biological Invasions 8: 903-917.

Chatterjee, P. 2019. African catfish a boon or bane for environment: A critical analysis. Indian Journal of Law and Justice 10(1): 80-90.

Chukwuka, K.S., U.S. Akpabio and U.N. Uka. 2020. Bioaccumulation of zinc and cadmium by two aquatic plants: *Eichhornia crassipes* (Mart.) Solms and *Pistia stratiotes* L. under nursery conditions. Tropical Plant Research 7(1): 229-237.

Coetzee, J. and P. Martin. 2011. The role of eutrophication in the biological control of water hyacinth. *Eichhornia crassipes* in South Africa. Bio Control 57: 247-261.

Cristaldi, A., G.O. Conti, E.H. Jho, P. Zuccarello, A. Grasso, C. Copat and M. Ferrante. 2017. Phytoremediation of contaminated soil by heavy metals and PAHs: A brief review. Environmental Technology and Innovation 8: 309-326.

Das, A., P. Ghosh, T. Paul, U. Ghosh, B.R. Pati and K.C. Mondal. 2016. Production of bioethanol as useful biofuel through the bioconversion of water hyacinth (*Eichhornia crassipes*) 3 Biotech. 6(1): 1- 9.

Demirbas, A. 2003. Current advances in alternative motor fuels. Energy Exploration and Exploitation 21: 457-487.

Dipanjan, G. 2010. Water hyacinth befriending the noxious weed. Science Reporter 46-48.

Dohanyos, M. and J. Zabranska. 2001. Anaerobic digestion. pp. 223-241. *In*: Spinosa, L. and Vesilind, P.A. [eds.]. Sludge into Biosolids. IWA Publishing. London, UK.

Eid, E.M., K.H. Shaltout, F.S. Moghanm, M.S.G. Youssef, E. El-Mohsnawy and S.A Haroun. 2019. Bioaccumulation and translocation of nine heavy metals by *Eichhornia crassipes* in Nile delta, Egypt: Perspectives for phytoremediation. International Journal of Phytoremediation 21(8): 821-830.

Fayad, Y.H., A.A. Ibrahim, A.A. El-Zoghby and F.F. Shalaby. 2001. Ongoing activities in the biological control of water hyacinth in Egypt. pp. 43-46. *In*: Julien, M.H., M.P.

Hill, T.D. Center and D. Jianquig [eds.]. Biological and Integrated Control of Water Hyacinth *Eichhornia crassipes*. ACIAR Proc 102.

Ganguly, A., P.K. Chatterjee and A. Dey. 2012. Studies on ethanol production from water hyacinth – A review. Renewable Sustainable Energy Rev. 16: 966-972.

Gao, L. and L. Bo. 2004. The study of a specious invasive plant, water hyacinth (*Eichhornia crassipes*): Achievements and challenges. Acta Phytoecologica-Sinica 28(6): 735-752.

Gardea-Torresdey, J.L., J.R. Peralta-Videa, G. De La Rosa and J.G. Parsons. 2005. Phytoremediation of heavy metals and study of the metal coordination by X-ray absorption spectroscopy. Coordination Chemistry Reviews 249: 1797-1810.

Gressel, J. 2008. Transgenics are imperative for biofuel crops. Plant Sci. 174: 246-263.

Guragain, Y., J.D. Coninck, F. Husson, A. Durand and S.K. Rakshit. 2011. Comparison of some new pretreatment methods for second generation bioethanol production from wheat straw and water hyacinth. Bioresour. Technol. 102: 4416-4424.

Harley, K.L.S., M.H. Julien and A.D. Wright. 1996. Water hyacinth: A tropical worldwide problem and methods for its control. Proceedings of the Second International Weed Control Congress, Copenhagen, Denmark: 639-644.

Harun, M.Y., A.B. Dayang, Z. Zainal and R. Yunus. 2011. Effect of physical pretreatment on dilute acid hydrolysis of water hyacinth (*Eichhornia crassipes*). Bioresour Technol. 102(8): 5193-5199.

Jafari, N. 2010. Ecological and socio-economic utilization of water hyacinth (*Eichhornia crassipes* (Mart) Solms). J. Appl. Sci. Environ. Manage. 14(2): 43-49.

Jasrotia, S., K. Arun, M. Aradhana. 2017. Performance of aquatic plant species for phytoremediation of arsenic-contaminated water. Appl. Water Sci. 7: 889-896.

Jayaweera, M.W. and J.C. Kasturiarachchi. 2004. Removal of nitrogen and phosphorus from industrial wastewaters by phytoremediation using water hyacinth (*Eichhornia crassipes* (Mart.) Solms). Water Science and Technology 50(6): 217-225.

Jayaweera, M.W., J.C. Kasturiarachchi, R.K. Kularatne and S.L. Wijeyekoon. 2008. Contribution of water hyacinth (*Eichhornia crassipes* (Mart.) Solms) grown under different nutrient conditions to Fe-removal mechanisms in constructed wetlands. J. Environ. Manage. 87(3): 450-460.

Jayaweera, M.W., J.C. Kasturiarachchi, R.K.A. Kularatne and S.L.J. Wijeyekoon. 2007. Removal of aluminium by constructed wetlands with water hyacinth (*Eichhornia crassipes* (Mart.) Solms) grown under different nutritional conditions. Journal of Environmental Science and Health – Part A: Toxic/Hazardous Substances and Environmental Engineering 42(2): 185-193.

Jones, R., W. Sun, C.S. Tang and F.M. Robert. 2004. Phytoremediation of petroleum hydrocarbons in tropical coastal soils. Microbial response to plant roots and contaminant. Environ. Sci. Pollut. Res. 11: 340-346.

Karthikeyan, A., V. Harish, J. Jayaprabakar Jayaraman, V. Dhana Raju, S. Lingesan, A. Prabhu and S. Dhanasekar. 2019. Novel water hyacinth biodiesel as a potential alternative fuel for existing unmodified diesel engine: Performance, combustion and emission characteristics. Energy J. 179: 295-305.

Kateregga, E. and T. Sterner. 2009. Lake Victoria fish stocks and the effects of water hyacinth. Journal of Environment and Development 18: 62-78.

Kumar, A., L.P.S. Rajput, N. Sushma and T. Keerti. 2019. Bioethanol Production from Waste Corn Using *Saccharomyces cerevisiae* and *Aspergillus awamori*. Int. J. Curr. Microbiol. App. Sci. 8(08): 2437-2445.

Kumar, S., R. Kumar, A. Dwivedi and A.K. Pandey. 2014. *In vitro* antioxidant, antibacterial, and cytotoxic activity and *in vivo* effect of *Syngonium podophyllum* and *Eichhornia crassipes* leaf extracts on isoniazid induced oxidative stress and hepatic markers. BioMed. Res. Int. 1-11.

Lakhan, K. and B. Navneeta. 2020. Microbial remediation of heavy metals. Microbial bioremediation and Biodegradation 49-72.

Lamaisri, C., V. Punsuvon, S. Chanprame, A. Arunyanark, P. Srinives and P. Liangsakul. 2015. Relationship between fatty acid composition and biodiesel quality for nine commercial palm oils. Songklanakarin. J. Sci. Technol. 37(4): 389-395.

Lee, W.C. and W.C. Kuan. 2015. Miscanthus as cellulosic biomass for bioethanol production. Biotechnology Journal 10(6): 840-854.

Lennartsson, P.R., P. Erlandsson and M.J. Taherzadeh. 2014. Integration of the first and second generation bioethanol processes and the importance of by-products. Bioresource Technology 165(C): 3-8.

Li, Q., J. Zhan, B. Chen, X. Meng and X. Pan. 2016. Removal of Pb, Zn, Cu, and Cd by two types of *Eichhornia crassipes*. Environmental Engineering Science 33(2): 88-97.

Li, S. and X. Yang. 2016. Biofuel production from food wastes. pp. 617-653. *In*: Luque, R., C.S.K. Lin, K. Wilson, J. Clark [eds.]. Handbook of Biofuels Production – Process and Technologies. Woodhead Publishing Limited, Cambridge, U.K.

Liao, S. and W. Chang. 2004. Heavy metal phytoremediation by water hyacinth at constructed wetlands in Taiwan. J. Aqua. Plant Manage. 42: 60-68.

Lindsey, K. and H.M. Hirt. 1999. Use water hyacinth: A practical handbook of uses for the water hyacinth from across the world. Anamed: Winnen-den 114.

Lombi, E., F. Zhao, S. McGrath, S. Young and G. Sacchi. 2001. Physiological evidence for a high-affinity cadmium transporter highly expressed in a *Thlaspi caerulescens* ecotype. New Phytol. 149: 53-60.

Lu, J., J. Wu, Z. Fu and L. Zhu. 2007. Water hyacinth in China: A sustainability science-based management framework. Environmental Management 40: 823-830.

Lytle, C.M.E.L., F.W. Lytle, N. Yang and J. Qian. 1998. Reduction of Cr (VI) to Cr (III) by wetland plants: Potential for in situ heavy metal detoxification. Environ. Sci. Technol. 32(20): 3087-3093.

Mahamadi, C. and T. Nharingo. 2010. Competitive adsorption of Pb^{2+} Cd^{2+} and Zn^{2+} ions onto *Eichhornia crassipes* in binary and ternary systems. Bioresour. Technol. 101: 859-864.

Mahmood, H., M. Moniruzzaman, T. Iqbal and S. Yusup. 2017. Effect of ionic liquids pretreatment on thermal degradation kinetics of agroindustrial waste reinforced thermoplastic starch composites. J. Mol. Liq. 247: 164-170.

Malar, S., S. Shivendra, F. Paulo and P. Venkatachalam. 2014. Lead heavy metal toxicity induced changes on growth and antioxidative enzymes level in water hyacinths (*Eichhornia crassipes* (Mart.)). Botanical Studies 55: 54.

Malik, A. 2007. Environmental challenge vis a vis opportunity: The case of water hyacinth. Environment International 33: 122-138.

Mangabeira, P.A.O., L. Labejof, A. Lamperti, A.A.F. deAlmeida, A.H. Oliveira, F. Escaig, M.I.G Severo, D. Sijlva, M. Saloes, M.S. Mielke, E.R Lucena, M.C. Martinis, K.B. Santana, K.L. Gavrilov, P. Galle and R. Levi-Setti. 2004. Accumulation of chromium in root tissues of *Eichhornia crassipes* (Mart.) Solms. in Cachoeira river, Brazil. Applied Surface Science 232: 497-501.

Marseille, F., C. Tiffreau, A. Laboudigue and P. Lecomte. 2000. Impact of vegetation on the mobility and bioavailability of trace elements in a dredged sediment deposit: A greenhouse study. Agronomie 20: 547-556.

Minychl, G.D., A.M. Melesse, S.A.Tilahun, M. Abate and D.C. Dagnew. 2019. Potential of water hyacinth infestation on lake Tana. Ethiopia: A prediction using a GIS-based multi-criteria technique. Water 11(9): 1921.

Mishra, S. and A. Maiti. 2017. The efficiency of *Eichhornia crassipes* in the removal of organic and inorganic pollutants from wastewater: A review. Environmental Science and Pollution Research 24(9): 7921-7937.

Mishra, V.K. and B.D. Tripathi. 2009. Accumulation of chromium and zinc from aqueous solutions using water hyacinth (*Eichhornia crassipes*). Journal of Hazardous Materials 164(2–3): 1059-1063.

Mohsen, A., A.E. Hagras, H.A.M. Elbaghdady and M.N. Monier. 2015. Effects of dissolved oxygen and fish size on Nile Tilapia, *Oreochromis niloticus* (L): Growth performance, whole-body composition, and innate immunity Aquaculture. International Journal of the European Aquaculture Society 23(5): 1261-1274.

Monisha, J., T. Tenzin, A. Naresh, B.M. Blessy and N.B. Krishnamurthy. 2014. Toxicity, mechanism and health effects of some heavy metals. Interdiscip. Toxicol. 7(2): 60-72.

Ndimele, P.E., C.A. Kumolu-Joh, K.S. Chukwuka, C.C. Ndimele, O.A. Ayorinde and O.R. Adaramoye. 2014. Phytoremediation of Iron (Fe) and Copper (Cu) by water hyacinth (*Eichhornia crassipes* (Mart.) Solms). Trends in Applied Sciences Research 9(9): 485-493.

Neumann, P.M., M.P. De Souza, I.J. Pickering and N. Terry. 2003. Rapid microalgal metabolism of selenate to volatile dimethyl selenide. Plant Cell Environ. 26: 897-905.

Nirupa, P.K., J. Pickova and G. Francesco. 2018. Stressing algae for biofuel production: Biomass and biochemical composition of *Scenedesmus dimorphus* and *Selenastrum minutum* grown in municipal untreated wastewater. Front. Energy Res. 6: 132.

Njogu, P., Robert Kinyua, Purity Muthoni and Yusuyuki Nemoto. 2015. Biogas production using water hyacinth (*Eichhornia crassipes*) for electricity generation in Kenya. Energy and Power Engineering 7(5): 209-206.

Odjegba, V.J. and I.O. Fasidi. 2007. Phytoremediation of heavy metals by *Eichhornia crassipes*. Environmentalist 27(3): 349-355.

Ohno, H. and Fukaya, Y. 2009. Task specific ionic liquids for cellulose. Chem. Lett. 38: 2-7.

Opeyemi, I.A., A. Tolulope and P.A. Femi. 2020. *Eichhornia crassipes* (Mart) Solms: Uses, challenges, threat, and prospects. The Scientific World Journal (2020): 12.

Pandey, V.C. 2016. Phytoremediation efficiency of *Eichhornia crassipes* in fly ash pond. International Journal of Phytoremediation 18(5): 450-452.

Patel, S. 2012. Threats, management and envisaged utilizations of aquatic weed *Eichhornia crassipes*: An overview. Reviews in Environmental Science and Biotechnology 11(3): 249-259.

Penfound, W.M.T. and T.T. Earle. 1948. The biology of the water hyacinth. Ecological Monographs 18: 447-472.

Pieterse, A.H. 1997. *Eichhornia crassipes* (Mart) Solms. Record from Prosea base. *In*: Faridad Hanum, I. and Van der Maesen, L.J.G. [eds]. PROSEA (Plant Resources of South-East Asia) Foundation, Bogor. Indonesia.

Poddar, K., L. Mandal and G.C. Banerjee. 1991. Studies on water hyacinth (*Eichhornia crassipes*) – Chemical composition of the plant and water from different habitats. Indian Veterinary Journal 68: 833-837.

Priya, P., S.O. Nikhitha, C. Anand, R.S. Dipin Nath and B. Krishnakumar. 2018. Biomethanation of water hyacinth biomass. Bioresource Technology 255(January): 288-292.

Rahman M.A. and H. Hasegawa. 2011. Aquatic arsenic: Phytoremediation using floating macrophytes. Chemosphere 83: 633-646.

Rajendra, B., N.K. Magar, and H. Abdulrazzak Afroz. 2017. Waste water treatment using water hyacinth. 32nd Indian Engineering Congress, The Institution of Engineers (India) Chennai.

Rezania, S., M.F.M. Din, S.M. Taib, F.A. Dahalan, A.R. Songip, L. Singh and H. Kamyab. 2016. The efficient role of aquatic plant (water hyacinth) in treating domestic wastewater in continuous system. International Journal of Phytoremediation 18(7): 679-685.

Rezania, S., M. Ponraj, A. Talaiekhozani, S.E. Mohamad, M.F. Md Din, S.M. Taib and F.M. Sairan. 2015. Perspectives of phytoremediation using water hyacinth for removal of heavy metals, organic and inorganic pollutants in wastewater. Journal of Environmental Management 163: 125-133.

Riddle, S.G., H.H. Tran, J.G. Dewitt and J.C. Andrews. 2002. Field, laboratory, and x-ray absorption spectroscopic studies of mercury accumulation by water hyacinths. Environmental Science and Technology 36(9): 1965-1970.

Robak, K. and M. Balcerek. 2018. Review of second generation bioethanol production from residual biomass. Food Technology and Biotechnology 56(2): 174-187.

Room, P.M. and I.V.S. Fernando. 1992. Weed invasions countered by biological control: Salvinia molesta and *Eichhornia crassipes* in Sri Lanka. Aquatic Botany 42: 99-107.

Ruan, T., R. Zeng, X.Y. Yin, S.X. Zhang and Z.H. Yang. 2016. Water hyacinth (*Eichhornia crassipes*) biomass as a biofuel feedstock by enzymatic hydrolysis. Bio. Resources 11(1): 2372-2380.

Rubin, E. and A. Ramaswami. 2001. The potential for phytoremediation of MTBE. Water Res. 35: 1348-1353.

Sabale, S., Vikas Jadhav, Deepali Jadhav, B.S. Mohite and Kesharsingh Patil. 2010. Lake contamination by accumulation of heavy metal ions in *Eichhornia crassipes*: A case study of Rankala Lake Kolhapur (India). Journal of Environmental Science & Engineering 52: 155-156.

Sagar, C.V. and N.A. Kumari. 2013. Sustainable biofuel production from water hyacinth (*Eichhornia crassipes*). International Journal of Engineering Trends and Technology 4(10): 4454-4458.

Saraswat, S. and J.P.N. Rai. 2010. Heavy metal adsorption from aqueous solution using *Eichhornia crassipes* dead biomass. Int. J. Mineral Proc. 94: 203-206.

Shanab, S.M.M. and E.A. Shalaby. 2012. Biological activities and anticorrosion efficiency of water hyacinth (*Eichhornia crassipes*). Journal of Medicinal Plants Research 6(23): 3950-3962.

Shanab, S.M.M., E.A. Hanafy and E.A. Shalaby. 2014. Biodiesel production and antioxidant activity of different Egyptian Date Palm seed cultivars. Asian J. Biochem. 9 (3): 119-130.

Shanab, S.M.M., E.A. Hanafy and E.A. Shalaby. 2017. Water hyacinth as non-edible source for biofuel production. Waste and Biomass Valorization 9(2): 255-264.

Singh, J. and A.S. Kalamdhad. 2013. Reduction of bioavailability and leachability of heavy metals during vermicomposting of water hyacinth (*Eichhornia crassipes*). Environ. Sci. Pollut. Res. 20: 8974-8985.

Suchitra, M. 2015. Rice at risk. Down to earth. https://www.downtoearth.org.in/coverage/rice-at-risk-43367.

Suresh, B. and G. Ravishankar. 2004. Phytoremediation – A novel and promising approach for environmental clean-up. Crit. Rev. Biotech. 24: 97-124.

Tahani, S.G. and S.A. El-Temtamy. 2013. Commercialization potential aspects of microalgae for biofuel production: An overview. Egyptian Journal of Petroleum 22(1): 43-51.

Ting, W.H.T., I.A.W. Tan, S.F. Salleh and N.A. Wahab. 2018. Application of water hyacinth (*Eichhornia crassipes*) for phytoremediation of ammoniacal nitrogen: A review. Journal of Water Process Engineering 22(February): 239-249.

Trivedi, S. and A.A. Ansari. 2005. Molecular mechanism in the phytoremediation of heavy metals from coastal waters. Phytoremediation 219-231.

Vandiver, V.V. 1999. Florida Aquatic Weed Management Guide. Univ. of FL, IFAS, Cooperative Extension Service, Publ. SP-55, 130 pp.

Vicente, I.S.T. and C.E. Fonseca-Alves. 2013. Impact of introduced Nile Tilapia (*Oreochromis niloticus*) on non-native aquatic ecosystems. Pakistan Journal of Biological Sciences 16(3): 121-126.

Weiping, S., Q. Sun, M. Xia, Z. Wen and Z. Yao. 2018. The resource utilization of water hyacinth (*Eichhornia crassipes* (Mart) Solms) and its challenges. Resources 7(3): 46.

Wolverton, B.C. 1975. Water hyacinths for removal of cadmium and nickel from polluted waters. NASA Technical Memorandum TM-X-72721.

Wolverton, B.C. and R.C. McDonald. 1975. Water hyacinths and alligator weeds for removal of lead and mercury from polluted waters. NASA Technical Memorandum TM-X- 72723.

Yan, A., W. Yamin, N.T. Swee, L.M.Y. Mohamed, G. Subhadip and C. Zhong. 2020. Phytoremediation: A promising approach for revegetation of heavy metal polluted land. Frontier Plant Science 11: 359.

CHAPTER
6

Algae: Source of Biofuel and Phytoremediation

N.K. Aliya[1], C.M. Jijeesh[2] and K.C. Jisha[1]*

[1] Department of Botany, MES Asmabi College, P. Vemballur; Kerala 680671, India
[2] Department of Silviculture and Agroforestry, College of Forestry, Kerala Agricultural University, Thrissur, Kerala 680656, India

6.1 Introduction

Algae, the organisms with photosynthetic capacity, are endowed with immense economic importance. They are regarded as the most effective organisms on Earth for utilizing sun's energy for producing organic compounds through photosynthesis (Shalaby 2011). Algae are found predominantly in water, but can also be seen everywhere on Earth including soils, hot springs, snow fields, cold deserts, etc. (Abdel-Raouf et al. 2012). They lack stems or roots or leaves, unlike angiosperms. It is well known that the algae constitute one of the sources of oxygen used on Earth by humans and other terrestrial organisms. Besides, this vital function is also employed for various purposes like as food, fish and animal feed, in medicines, cosmetics etc. (Goswami et al. 2015).

Algae could be used efficiently to treat various kinds of waste waters, such as nutrient-rich household waste water, water-containing animal waste, or industrial waste water. Along with the treatment of waste waters, they effectively utilize the phosphorus and nitrogen present in these waste waters, thus reducing their content in waste water. Moreover, they release different types of biofuels as their metabolic by-products (Bilanovic et al. 1988, Benemann and Oswald 1996, Bigogno et al. 2002). The major reason for using algae for various purposes may be due to its high growth rate when compared to other plant groups and also may be due to its wide adaption to any growing habitat (Goswami et al. 2015).

Nowadays, we are more concerned about our environment and trying to practice more eco-friendly techniques for energy production and also for reducing the environmental pollution. Recently, many research works were carried out in the biofuel production from algae as well as phytoremediation using algae. In this review

Corresponding author: jishakc123@gmail.com

paper, the ecological significance of algae in terms of phytoremediation and biofuel production is discussed.

6.2 Phytoremediation

Environmental pollution severely affects the ecosystem functioning. Several efforts were made to alleviate the harmful effects of environmental pollution. Heavy metals contribute a major share in polluting our environment. There are several physico-chemical methods like soil washing, application of electric field, excavation, etc., which can be practiced for removing the heavy metal content from the ecosystems (Yan et al. 2020). But all these methods have certain limitations such as deterioration of soil quality, secondary pollution etc. (Yan et al. 2020). In addition, the chemical methods employed to minimise the content of heavy metals in an ecosystem require a significant surplus of chemicals which, in turn, increases the cost because of creating the voluminous sludge. The conventional methods used to eliminate the heavy meal contents have definite benefits as well as drawbacks (EPA 1997), but generally none of these is lucrative (Volesky 2001, Rai 2009). Thus, an eco-friendly approach which is cost-effective also is to be opted.

Several research works have been carried out to improve the quality of the environment by natural means to address this problem. Phytoremediation is such an approach, which has seen an increase in demand in the recent years. It is a phyto-based process, which uses the plants to absorb pollutants from the atmosphere and thus reduce pollutants in the ecosystem (Berti and Cunningham 2000). Currently, this approach is a promising novel method since plants are solar-driven, rendering this an economically viable method with immense potential to achieve the goal of sustainable environment.

Phytoremediation includes eliminating pollutants in contaminated soil, water or air by using natural, transgenic or genetically modified plants that can collect, destroy or wipe out heavy metals, various solvents, pesticides, crude oil, etc. (Flathman and Lanza 1998). Some plants are hyper accumulators of heavy metals and they can remove abnormally high amount of heavy metals from ecosystem through their roots and shift them to the shoot system (Baker and Brook 1989, Brown et al. 1994). According to Cobbet and Goldsbrough (2002), every plant species uses a complicated mechanism through which they can effectively manage the heavy metal uptake, accumulation and elimination.

The cost-effective and environmentally sustainable nature is the prime attraction behind the advancement of phytoremediation technologies. Many native plants were tested under field condition for their phytoremediation potential and were shown that they can be effectively employed for the development of tolerant plant population (McGrath and Zhao 2003). These plants used for phytoremediation purposes improve the elimination of pollutants by incorporating a part of them as nutrients (Firz and Wenzel 2002). According to Tong et al. (2004), the criteria for the plants employed for phytoremediation are faster growth, huge biomass production within a short duration, and which can be easily cultivated and harvested, if possible multiple times per year.

Yoshida et al. (2006) classified the phytoremediation of heavy metals into three classes, viz. phytoextraction, rhizofiltration and phytostabilization. Phytoextraction is the heavy metals' absorption and transporting to different parts of their body, while in rhizofiltration, the roots extract and precipitate the metals around them, and in phytostabilization, the plants convert the metals into more stable compounds, thus preventing their circulation again. Recently, three other types of phytoremediation methods were added and they include phytodegradation (conversion of toxic substances to lesser toxic or ones without any toxicity), phytovolatilization (plants convert the hazardous compounds to volatile compounds without toxicity) and phytostimulation (the root system of plants produce exudates and it alleviates the toxicity of pollutants). The different methods of phytoremediation are depicted in Fig. 6.1.

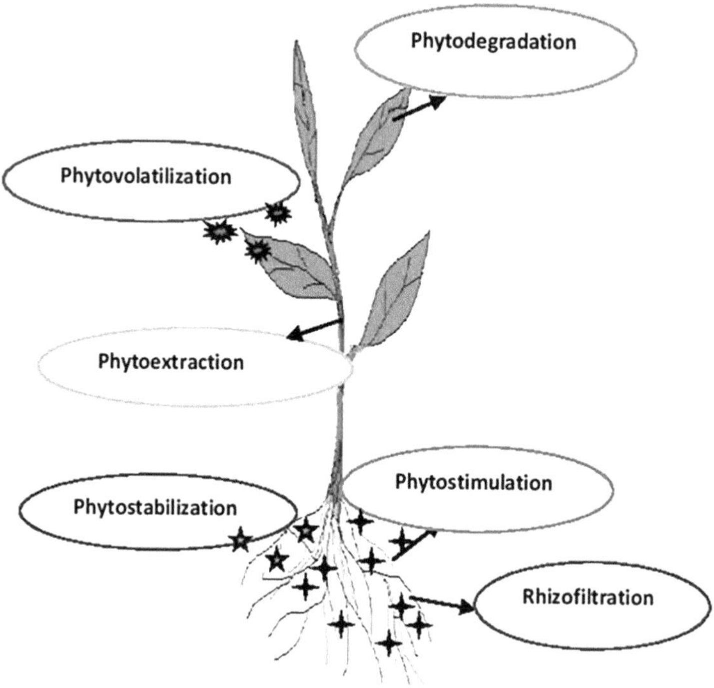

Figure 6.1: Phytoremediation methods in plants.

6.2.1 Phytoremediation by algae

Phytoremediation can be effectively done by using plants having fast growth, high biomass production, and ability to withstand any environmental condition. In this regard, the algae have many advantages. Moreover, algae can be grown everywhere; they can be effectively grown on the soil which is not ideal for other plants, water bodies like seawater, lakes, ponds, waste water etc. (Goswami et al. 2015). These living organisms perform a very significant task in alleviating the heavy metal content in oceans and lakes (Sigg 1985, 1987) and also in their surrounding environment

(Zakeri et al. 2011). The main heavy metal alleviating methods are absorption, sedimentation, precipitation, flocculation, complex formation, microbial activity, cation and anion exchange, oxidation or reduction and uptake. The ability of algae to absorb the metals has been observed since many years. They can directly take up the toxic metals from the surroundings, leading to the increased concentrations of toxic metals in their body when compared to the ambient conditions (Chang et al. 2005, Shamsuddoha et al. 2006). Recently, the usage of macro and micro algae has gained great attention because of their absorption ability towards heavy metals and accepting the toxic heavy metals from the environment and reducing its toxicity (Mitra et al. 2012).

The bioaccumulation studies in algae expose the increased amount of pollutants in the organism through the uptake of water or minerals (Imamul Huq et al. 2007). There are several qualities of algae that make them ideal for heavy metal removal. The high tolerance potential for heavy metals, high surface area/volume ratios, ability to develop both heterotrophically and autotrophically, phytochelatin expression, phototaxy, and capacity for genetic modification are the key characteristics of algae as phytoremediating agents (Cai et al. 1995). Several reports are available on the use of algae as phytoremediating agents. In marine environments around the world, macroalgae have been routinely used to assess heavy metal pollution. It is mainly due to their ability to collect heavy metals within their tissues and hence, in marine systems, they act as metal accessibility biomonitors (Rainbow 1995, Gosavi et al. 2004). The species *Fucus vesiculosu*, a marine brown algae, is employed to absorb benzo[a]pyrene, a polycyclic aromatic hydrocarbon (PAH) and the algal cells were able to absorb 89-99% of hydrocarbons (Kirso and Irha 1998). Nowadays, many green algae like *Cladophora* and *Enteromorpha* are being employed for quantification of heavy metals worldwide (Al-Homaidan et al. 2011).

According to Sjahrul (2012), marine phytoplanktons (*Chaetoceros calcitrans* and *Tetracelmis chuii*) can be effectively used as phytoremediators. Sekabira et al. (2010) reported the capacity of algae to concentrate lead, cadmium, copper and zinc. Thus, algae can be employed for the biomonitoring purposes of heavy metal contamination, especially in the urban water bodies because it can be employed in the quantification of heavy metal pollutants. Thus, for biomonitoring of pollution from heavy metals, especially in water sources in urban areas, algae can be effectively employed for the heavy metal quantification. Heavy metal concentrations in algal biomass can be used as an indicator of the stream water metal load. Therefore, algae can be effectively employed in phytoextraction methods for heavy metal in aquatic urban water bodies and in waste water effluents (Sekabira et al. 2010). Imamul Huq et al. (2007) reported the efficacy of *Navicula* for biosorption of arsenic from the environment. Heavy metal absorption capability of *Chlorella* sp was first identified in the mining related industries and subsequently in many other studies identified as other metal-tolerant algae. For effective Cd^{2+} alleviation, Matsunaga et al. (1999) used microalgae from marine sources. There was an increase in the uptake of cadmium by *Chlamydomonas*, *Chlorella* and *Scenedesmus*, that are green microalgae and the cyanobacteria (*C. paris*) under high pH (Sakaguchi et al. 1979, Les and Walker 1984).

Cyanophycean and chlorophycean members have been found to be hyper-absorbents and hyper-accumulators for metals such as arsenic and boron (Baker 1981), which can absorb and accumulate these metals into their body parts from the surrounding habitat. Thus, the algae belonging to these classes can be used as hyper-phytoremediators and moreover, their occurrence in the water itself decreases the arsenic and boron content of the water (Baker 1981). In a study, *Spirulina maxima* has been reported to have a high ability to extract inorganics from water. It could remove 87 percent of nitrogen and 60 percent of phosphorus from household waste water (Laliberte et al. 1997). Mei et al. (2006) reported the high strontium uptake capacity of *Platymonas subcordiformis*, which is a marine green microalga. Some algal members can be employed for cleaning up of hydrocarbons. The majority of algal species for hydrocarbon utilization include diatoms and among diatoms, Pennates are the predominant ones (Egmond 1995). There should be a sufficient amount of sunshine, carbon dioxide and bio-available nitrogen for algal organisms to function effectively in remediation processes (Scow and Hicks 2005).

Phormidium, the cyanophycean member, can be used as a heavy metal hyper accumulator for Cu, Cd, Ni, Pb and Zn (Wang et al. 1995) and *Caulerpa racemosa* var. *cylindracea* is being utilized as a cost effective biomaterial to eliminate boron from water bodies (Bursali et al. 2009). According to Nielsen et al. (2005), brown algae like *Focus* spp. often grow luxuriantly in the heavy metal contaminated habitats. A green microalga, *Dunaliella salina*, accumulated more amounts of zinc, which was followed by copper and cobalt, while it showed less accumulation towards cadmium (Liu et al. 2002). List of algae employed for phytoremediation is represented in Table 6.1.

6.2.2 Heavy metal tolerance mechanisms in algae

Mejare and Bulow (2001) reported that different organisms react differently to the contamination of heavy metals and the main methods include heavy metal compartmentalization, exclusion, chelation, complex formations, vacuole sequestration, the production of phytochelatins (PCs) and metallothioneins (MTs), which are special binding proteins. Being effective binding ligands for heavy metals, organic acids such as citric and malic acid and amino acids such as histidine (His) and nicotianamine (NA) and phosphate derivatives (phytate) are very important in heavy metal tolerance and detoxification (Mehta and Gaur 1999, Payne 2000, Ahner et al. 2002, Sharma and Dietz 2006, Liu et al. 2011). Darnall et al. (1986) stated that, in general, there are three different biological strategies for the removal of heavy metals in a solution and they are biosorption or adsorption of heavy metals on the body surface of the organism, transport and cellular absorption of metal ions, and microorganisms' chemical alteration of metal ions. Biosorption has been regarded as the most effective form of metal sorption from waste water due to its rapidity in action.

With regard to waste water treatments, the algae employed for treatments showed enhanced heavy metal adsorption and attraction because of highly and negatively charged cell wall components in their body (Sekabira et al. 2010). The diatoms are frequently occurring in aquatic hydrocarbon mitigating areas (Boyd

Table 6.1: List of algae used for phytoremediation

Sl. No	Name of algal species	Type of pollutant removed	References
1	*Anabaena cylindrica*	Lead (Pb)	Swift and Forciniti (1997)
2	*Ascophyllum nodosum*	Gold (Au)	Kuyucak and Volesky (1988)
		Cobalt (Co)	Kuyucak and Volesky (1988)
		Nickel (Ni)	Holan and Volesky (1994)
		Lead (Pb)	Holan and Volesky (1994)
3	*Caulerpa racemosa*	Boron (B)	Bursali et al. (2009)
4	*Chaetoceros calcitrans*	Phosphorus (P)	Sekabira et al. (2010)
		Cadmium (Cd)	Sekabira et al. (2010)
		Copper (Cu)	Sekabira et al (2010)
		Zinc (Zn)	Sekabira et al. (2010)
5	*Chlorella vulgaris*	Chromium (Cr)	Mehta and Gaur (1999)
		Copper (Cu)	Mehta and Gaur (1999)
6	Cyanobacteria	Polyphosphate Bodies (PDB)	Jensen et al. (1982)
7	*Daphnia magna*	Arsenic (As)	Irgolic et al. (1977)
8	*Dunaliella salina*	Zinc (Zn)	Liu et al. (2002)
		Copper (Cu)	Liu et al. (2002)
		Cobalt (Co)	Liu et al. (2002)
		Cadmium (Cd)	Liu et al. (2002)
9	*Enteromorpha* sps	Copper (Cu)	Brinza et al. (2005)
10	*Fucus vesiculosus*	Nickel (Ni)	Holan and Volesky (1994)
		Benzo a Pyrene (BAH)	Kirso and Irha (1998)
		Zinc (Zn)	Fourest and Volesky (1997)
		Polycyclic aromatic hydrocarbon	Kirso and Irha (1998)
11	*Laminaria japonica*	Zinc (Zn)	Fourest and Volesky (1997)
12	Microalgae	Cadmium (Cd)	Matsunaga et al. (1999)
13	*Micrasterias denticulata*	Cadmium (Cd)	Volland et al. (2012)
14	*Navicela*	Arsenic (Ar)	Imamul Haq et al. (2007)
15	*Phormedium* sps.	Cadmium (Cd)	Wang et al. (1995)
		Zinc (Zn)	Wang et al. (1995)
		Lead (Pb)	Wang et al. (1995)
		Nickel (Ni)	Wang et al. (1995)
		Copper (Cu)	Wang et al. (1995)
16	*Phormedium ambiguum*	Mercury (Hg)	Shanab et al. (2012)

(*Contd.*)

Table 6.1: (*Contd.*)

Sl. No	Name of algal species	Type of pollutant removed	References
17	*Phormedium bohner*	Chromium (Cr)	Dwivedi et al. (2010)
18	*Platymonas subcordiformis*	Strontium (Sr)	Mei et al. (2006)
19	*Pseudochlorococcum typicum*	Lead (Pb)	Shanab et al. (2012)
20	*Sargassum filipendula*	Copper (Cu)	Davis et al. (2000)
21	*Sargassum fluitans*	Copper (Cu)	Davis et al. (2000)
		Iron (Fe)	Figueira et al. (1997)
		Zinc (Fe)	Fourest and Volesky (1997)
		Nickel (Ni)	Holan and Volesky (1994)
22	*Sargassum natans*	Lead (Pb)	Holan and Volesky (1994)
23	*Sargassum vulgare*	Lead (Pb)	Holan and Volesky (1994)
24	*Scenedesmus quadricauda* var *qaudrispina*	Cadmium (Cd)	Shanab et al. (2012)
25	*Spirulina maxima*	Nitrogen (N)	Laliberta et al. (1997)
		Phosphorous (P)	Laliberta et al. (1997)
26	*Spirulina platensis*	Cadmium (Cd)	Perez-Rama et al. (2002)
27	*Spirogyra hyalina*	Cadmium (Cd)	Kumar and Oommen (2012)
		Mercury (Hg)	Kumar and Oommen (2012)
		Lead (Pb)	Kumar and Oommen (2012)
		Arsenic (As)	Kumar and Oommen (2012)
		Cobalt (Co)	Kumar and Oommen (2012)
.28	*Tetracelmis chuii*	Lead (Pb)	Sekabira et al. (2010)
		Cadmium (Cd)	Sekabira et al. (2010)
		Copper (Cu)	Sekabira et al. (2010)
		Zinc (Zn)	Sekabira et al. (2010)
29	*Tetraselmis chuil*	Arsenic (As)	Irgolic et al. (1977)

et al. 2001, Ziervogel et al. 2004); however, the role in hydrocarbon removal from the water bodies was not well tracked. According to Kirso and Irha (1998), the algae like *Enteromorpha* sp. and *Cladophora glomerata*, have the ability to oxidize benzo[a] pyrene with the enzymes like cytochrome P450,o-diphenol oxidase, and perioxidase. The high rate of biological degradation takes place in layers of the surface of the organic mats and includes processes such as photo-oxidation, phytoremediation, volatilization and microbiological remediation. In between the hydrocarbon layers, the cyanobacteria present inside the organic mat grow and generate exopolysaccharides that in turn create a matrix to settle remediating bacteria. In this way, a sandwich effect was produced in which the hydrocarbons targeted were placed between two layers of remediation activities, oxic above and anoxic below (Matbiopol 1999).

Different algae show different mechanisms of heavy metal tolerance. *Thalassiosira weissflogii* and *Thalassiosira pseudonana* show the heavy metal tolerance by producing large amount of phytochelatins by the increased enzymatic activity of phytochelatin synthase (Ahner et al. 2002). According to Ren et al. (1998), the Cd^{2+} resistance and its absorption in cyanobacteria and algae were primarily mediated by genes present in plasmids. Polysaccharides, lipids and proteins present in the cell surface of cyanobacteria act as the heavy metal binding site. Within the cell wall, these compounds supply the carboxylic, amino, phosphate, sulfydryl, and thiol groups which can bind heavy metals (Ting et al. 1991). The microalgae remove the heavy metal content from contaminated water by using two different methods of which the first includes metabolism of the algae and uses pollutants in low amounts, while the second method includes biosorption (Matagi et al. 1998). In *Vaucheria* sp., amino group in the cell wall was found as the binding agent with metallic ions. Moreover, both covalent and ionic bonding was reported to have significant role in heavy metal absorption (Crist et al. 1981). Most algal members can transform mercuric or phenylmercuric ions into metallic mercury that volatilizes from the cell and ultimately from the solution (De Filippis 1981).

In the case of metallic cation sequestration, complex formation happens between a metal ion and functional groups on the surface or inside the porous structure of the biological material. In this complexation process, the carboxyl groups present in the alginate provide a significant contribution. The adsorption capacity of different algal species may vary. Because of the high levels of alginates and sulfated polysaccharides in their cell walls and on which metals show a strong attraction, the brown algae are more competent accumulators of metals (Davis et al. 2003). The precipitation of lead phosphate in *Anabaena cylindrica* on the cell wall and inside the cell was recorded by Swift and Forciniti (1997). In their work, it was confirmed that the fast uptake of lead occurred on the cell wall, while the absorption of lead and its subsequent precipitation inside the cell occurred very slowly. The pH can directly affect the metal uptake by algae. When the proton increases or a decrease in pH occurs, the metal ions compete with the protons for the binding site. Therefore, due to lower competition with protons, most cation adsorption occurs at a high pH (Schiewer and Volesky 1995). Due to its rapid sorption rate and its high sorption capacity, *Spirulina platensis* was found to be well suited for the removal of cadmium from waste water. It can also extract cadmium at a wide range of pH and temperature levels.

The accumulation of Cd^{2+} occurs in algae internally (Perez-Rama et al. 2002) in a two-step process of uptake. The first step is the sudden physicochemical adsorption of Cd^{2+} onto the binding sites of the cell wall, involving proteins and polysaccharides. This first phase was followed by a lag period of steady intracellular uptake. This second phase was energy dependent and included a number of transportation systems. The cadmium specifically attacks the photosynthetic structures or organelles of many algae, such as *Chlamydomonas*, *Dunaliella*, and *Nostociella*, according to many workers (Visviki and Rachlin 1994, Fernandez-Pinas et al. 1995). The photosynthetic activity of *Nostoc* has been shown to inhibit cadmium (Fernandez-Pinas et al. 1995).

The presence of high amount of polyphosphate bodies (PPB) in cyanobacteria was reported due to toxicity of heavy metal and these bodies are believed to act as the sites of metal sorption (Jensen et al. 1982). It was known that the cyanophycean granules act as the cell nitrogen storage (Thomas 1993) and they accrue in response to various stresses like nutrient limitation and presence of xenobiotics. Thus, these bodies are suggested to act as the part of the internal cell detoxification system (Fernandez-Pinas et al. 1995).

Mehta and Gaur (1999) reported a higher content of proline in response to Cu and Cr in *Chlorella vulgaris*, the green algae, which revealed the close association of metal uptake, metal toxicity and proline accumulation in these algae. Thus, the heavy metal protective effect of proline was confirmed in *Chlorella* along with the inhibition of lipid peroxidation. The different heavy metal tolerance mechanisms in algae are summarized in Fig. 6.2.

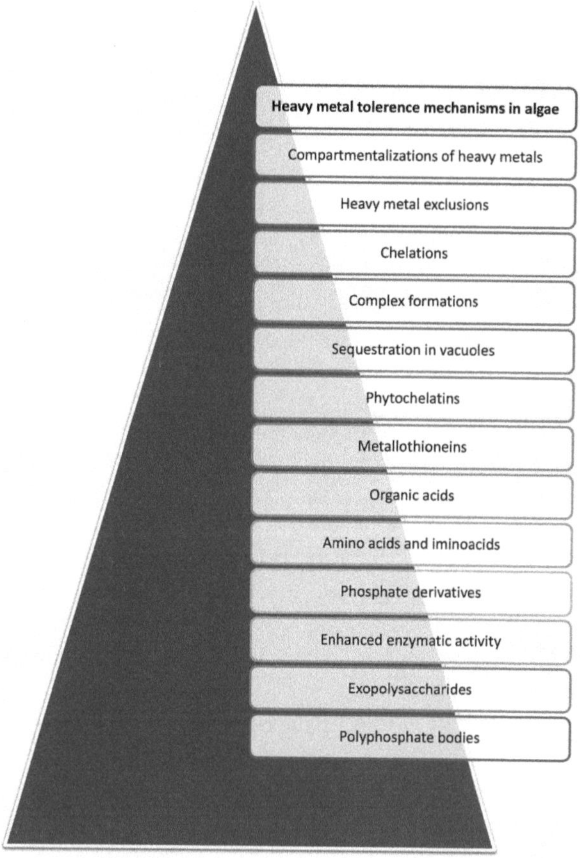

Figure 6.2: Heavy metal tolerance mechanisms in algae.

6.3 Biofuels

Due to continuous use of non-renewable fossil fuels, alternative renewable energy sources were more popularized and the enhanced global warming effects and increasing fuel cost also demanded the search of new renewable energy resources (Mathimani et al. 2015, Chiappe et al. 2016, Subsamran et al. 2018). The biofuels contribute to an alternative to the world energy demand (Bohutskyi et al. 2016). Biofuels are made from many feed stocks and are renewable, non-toxic and eco-friendly fuels (Chi et al. 2018, Sharma et al. 2018). It includes any kind of liquid, solid or gaseous fuel that can be produced from renewable raw substances. The type and quantity of biomass, type of energy and financial benefit from the product are the key concerns when biofuels are generated (Sharma and Sharma 2017).

Various classes of organisms sourced from agricultural, aquatic and forestry ecosystem can be used as the feedstocks for many types of biofuel production like biodiesel (Yanqun et al. 2008), biohydrogen (Marques et al. 2011), bio-oil (Shuping et al. 2010), bioethanol (Behera et al. 2014), and biogas (Behera et al. 2014). Because of their use as food or feed, agricultural crops are regarded as first generation biofuels (Sharma and Sharma 2017). The major classes of biofuels include bioethanol, biodiesel, biohydrogen, biomethanol, bioethers (biodimethyl ether, biomethyltetrabutyl ether (bio-MTBE)) and bioethyltetrabutyl ether (bio-ETBE) (Judge et al. 2003). Agrofuel, a first generation biofuel, employs crops like palm, sugarcane, sugarbeet, soybean, and sweet sorghum as feedstocks for producing biofuels. It is formed by fermentation by yeast; thereby, carbohydrate in plants is converted to bio-ethanol and the biodiesel is produced (Lu et al. 2011). Non-food crop plants such as grass, *Jatropha*, silver grass, switch grass, and non-edible parts of food crops are used in the production of second generation biofuels (Brown and Brown 2013). But uncontrolled use of different existing resources without proper recompense leads to reduction in the biomass availability, which in turn resulted in environmental issues like loss of forests and biodiversity (Goldemberg 2007, Li et al. 2008, Saqib et al. 2013). Algal biofuels are biofuels of the third generation that do not rely on land for growth, so we can reduce the use of land, water and unnecessary pesticides (Saqib et al. 2013), and the algal biofuel production is mainly focused on metabolic engineering of the microalgal genome to increase the biofuel yields, at the same time minimizing the cost (Lu et al. 2011, Dutta et al. 2014). According to Sharma and Sharma (2017), the algal feedstock is better than agricultural feedstock because algal cultivation of algae does not interfere with agriculture since algae do not occur in the agricultural land. The different types of biofuels are represented in Fig. 6.3.

6.3.1 Biofuel production from algae

Compared with other terrestrial biomass sources for making biofuel, algae have a reduced environmental impact. Quick growth rate, potential for rapid growth resulting in high biomass without competition for arable land etc. make algae good source for the sustainable fuel production. Due to the great algal species diversity, this group of organisms offer researchers several options to identify and select the

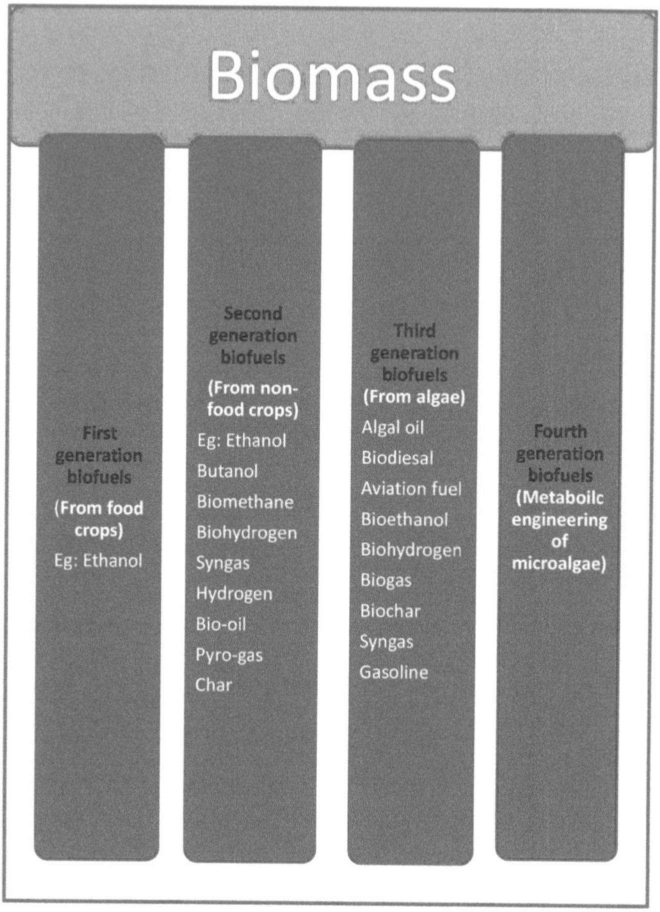

Figure 6.3: Different types of biofuels.

most suitable strains and also offers basis for genetic constitution, which can be used for enhancing the production of biofuels from the selected algal strains (Hannon et al. 2010). In addition, the advantage of algae for biofuel production is that the waste water can be remediated while culturing the algae in these waste waters. The main types of waste waters that can be cleaned by using algae are the municipal waste waters, which are rich in phosphates and nitrates (Fierro et al. 2008, Douskova et al. 2009). Because of these qualities, the algae are endowed with high potential for cost-effective production of biofuels.

Algae effectively use CO_2 and are the organisms which fix more than 40% of the global carbon on the Earth; among the algae, the major share of productivity is from marine microalgae (Falkowski et al. 1998, Parker et al. 2008). Multiplication of algae is a rapid process, and many species of algae become twofold within a short period like 6 h, while some other species have two doublings in a day (Sheehan et al. 1998, Huesemann et al. 2009). All algae can produce energy-rich oils, but many of microalgae can store much higher amount of oils naturally in their body (Rodolfi et al. 2009). It was reported that *Botryococcus* spp. can accumulate long-chain hydrocarbons equivalent to 50% in their dry mass (Kojima and Zhang 1999).

Microalgae are different group of unicellular organisms with capacity to provide many solutions to the depletion of non-renewable energy resources in the developing world. The ancestral relation in microalgae is very diverse, so the genetic engineering is easy in this group (Deschamps and Moreira 2009). Moreover, because of the unicellular nature of microalgae, they can rapidly multiply by cell division and the evolution of new strains can be fast. Algal culture requires nutrients, sunlight, water and CO_2 for good growth. They can effectively store various nutrients such as sulphur, phosphorus, nitrogen etc. and they possess several behavioural, morphological, and chemical attributes for protection from pathogens and diseases. Algae are rich sources of different types of fatty acids, tannin, bromophenol, different polysaccharides, peptides, alcohols, halogenated compounds, amides, alkaloids, etc. (Pesando 1990, Bhadurg and Wright 2004). Because of the presence of various defence methods, the algae can effectively thrive in any type of cultural conditions and can prepare themselves for the biofuel production.

Various types of biofuels like, biogas, biodiesel, biomethane, bioethanol, etc. can be produced from algae. The carbohydrate of algae can be utilized for extracting bioethanol, while the oils from algae are effectively used for producing biodiesel from algae. The residue after the biofuel processing can be used for the production of a variety of protein supplements, nutraceuticals, biocontrol agents, therapeutics, fertilizers, eicosapentaenoic acid (EPA), animal feed and docosahexaenoic acid (DHA) (Hannon et al. 2010). Different species of algae employed for biofuel production are represented in Table 6.2.

6.3.2 Biofuel production methods in algae

Algae with high lipid content are preferred for biofuel production. The algal cultivation for desired biofuel production can be either in open ponds or advanced photobioreactors. We can harvest about 0.06-0.231 g\l\day of algal biomass in open pond systems (Blanco et al. 2007, Moheimani and Borowitzka 2007) and about 3 g\l\day algal biomass in bioreactors (Doucha and Livensky 2006). Through the application of genetic engineering, this production can be enhanced. Both the open and closed methods of algal cultivation have their own merits and demerits. In closed systems, contamination will be less but it requires initial high expense, while the open system of algal cultivation requires only less expense but is always under a threat of contamination by other microorganisms (Lebeau and Robert 2003, Matsudo et al. 2008). For lowering the contamination levels in algal cultures, microalgae are the most promising ones which can thrive under various environmental conditions highly competing with various contaminating microorganisms.

Biodiesel is an example for biodegradable fuel which can cause the reduction of sulphur (Hossain et al. 2008). Generation of biomethane or biogas comes from the anaerobic digestion of organic matter. The anaerobic digestion of organic matter produces biogas or biomethane, which consists mainly of methane and carbon dioxide (Ward et al. 2014). The hydrocarbons present in microalgae can be transformed to

Table 6.2: List of algae used for biofuel production

Sl. No.	Name of algal species	Biofuel type	References
1	*Chlorella* sp.	Biodiesel	Ehimen et al. (2010)
2	*Chlorella pyrenoidosa*	Biodiesel	Huang (2015)
3	*Chlorella protothecoides*	Biodiesel	Gulyurt et al. (2016)
4	*Chlorella vulgaris*	Biodiesel	Mathimani et al. (2017)
5	*Dunaliella tertiolecta*	Biodiesel	Tang et al. (2011)
6	*Dictyochloropsias splendida*	Biodiesel	Carvalho et al. (2011)
7	*Desmodesmus quadricaudatus*	Biodiesel	Shafik et al. (2015)
8	*Nannochloropsis oculata*	Biodiesel	Carvalho et al. (2011)
9	*Nitzschia* sp.	Biodiesel	Minowa et al. (1995)
10	*Neochloris aquatica*	Biodiesel	Jaiswar et al. (2016)
11	*Phormidium* sp.	Biodiesel	Minowa et al. (1995)
12	*Pseudokirchneriella* sp.	Biodiesel	Minowa et al. (1995)
13	*Sprulina* sp.	Biodiesel	El-Shimi et al. (2013)
14	*Schizochytrium limacinum*	Biodiesel	Johnson and Wen (2009)
15	*Oscillatoria* sp.	Biodiesel	Saad and Shafik (2017)
16	*Ascophyllum nodusum*	Bioethanol	Obata et al. (2016)
17	*Chlamydomponas reinhardtii*	Bioethanol	Choi et al. (2010)
18	*Chloroccocum littorale*	Bioethanol	Ueno et al. (1998)
19	*Laminaria digitata*	Bioethanol	Obata et al. (2016)
20	*Ulva lactua*	Bioethanol	El-Sayed et al. (2016)
21	*Anabaena cylindrical*	Biohydrogen	Nayak et al. (2014)
22	*Bacillarophyta* sp.	Biohydrogen	Huang et al. (2016)
23	*Mastiogocladus laminosus*	Biohydrogen	Miyamoto et al. (1979)
24	*Chlamydomonas reinhardtii*	Biohydrogen	Oncel et al. (2015)
25	*Chlorella* sp.	Biohydrogen	Barreiro et al. (2013)
26	*Chlorella vulgaris*	Biohydrogen	Biller et al. (2012)
27	*Chlorogleopsis fritschii*	Biohydrogen	Biller et al. (2012)
28	*Cyanobacteria* sp.	Biohydrogen	Huang et al. (2016)
29	*Dunaliella* sp.	Biohydrogen	Minowa et al. (1995)
30	*Nannochloropsis oculata*	Biohydrogen	Biller et al. (2012)
31	*Nannachloropsis gaditana*	Biohydrogen	Biller et al. (2012)
32	*Phaeodactylum tricornutum*	Biohydrogen	Barreiro et al. (2013)
33	*Porphyridium cruentum*	Biohydrogen	Biller et al. (2012)
34	*Spirulina platensis*	Biohydrogen	Biller et al. (2012)
35	*Scenedesmus dimorphus*	Biohydrogen	Biller et al. (2012)
36	*Tetraselmis* sp.	Biohydrogen	Eboibi et al. (2014)

37	*Emiliania huxleyi*	Biogas	Wu et al. (1991)
38	*Chlorella vulgaris*	Ethanol	Hirano et al. (1997)
39	*Arthrospira maxima*	Methane	Inglesby and Fisher (2012)
40	*Euglena gracilis*	Methane	Nguyen et al. (2015)
41	*Spirulina* sp.	Methane	Zamalloa et al. (2012)
42	*Scenedesmus obliquus*	Methane	Zamalloa et al. (2012)
43	*Chlorella protothecoides*	Oil	Rizzo et al. (2013)
44	*Microcystis aeruginosa*	Oil	Miao et al. (2004)
45	*Chlorella* sp.	Oil/Gas	Babich et al. (2011)
46	*Chlorella vulgaris*	Oil/Gas/Char	Wang et al. (2015)
47	*Duniella tertiolecta*	Oil/Gas/Char	Grierson et al. (2009)
48	*Nannochloropsis* sp.	Oil/Gas/Char	Pan et al. (2010)
49	*Synechococcus* sp.	Oil/Gas/Char	Grierson et al. (2009)
50	*Tetraselmis chuii*	Oil/Gas/Char	Grierson et al. (2009)
51	*Chlorella vulgaris*	Syngas	Onwudili et al. (2013)
52	*Nannochloropsis* sp.	Syngas	Khoo et al. (2013)
53	*Nannochloropsis oculata*	Syngas	Duman et al. (2014)
54	*Nannochloropsis gaditana*	Syngas	Sanchez et al. (2013)
55	*Spirulina platensis*	Syngas	Stucki et al. (2009)
56	*Saccharina latissimi*	Syngas	Onwudili et al. (2013)
57	*Tetraselmis* sp.	Syngas	Alghurabie et al. (2013)

diesel, kerosene, gasoline. *Botryococcus braunii*, a microalga, yield hydrocarbons from which good oil content can be obtained and the extraction of these oils are very easy since these are produced outside the cell (RangaRao and Ravishankar 2007). Through biomass gasification process in oxygenic atmosphere, the formation of bio-syngas occurs which in turn yield water, hydrogen, CO, methane, and ashes (Buxy et al. 2013). High temperature and water content lower than 20% are required for the gasification process to occur (Ghasemi et al. 2012). But in the anoxic condition also, microalgae can definitely yield hydrogen, thus act as a good supply of clean energy without the danger of greenhouse gases by using sunlight and water (Archana and Anjana 2012). Bioethanol is synthesised by yeast through the fermentation of sugars. Algae are known to be the good source of starch (Harun et al. 2010, Markou et al. 2012, Chen et al. 2013) and the hemicellulose and cellulose of many microalgae are used for the synthesis of sugars and from the sugars, ethanol can be produced. The processing of algae in different ways yield a number of energy products. The major types of algal processing include biochemical, thermochemical (Raheem et al. 2015), biochemical, transesterification, and photosynthetic microbial fuel cell conversion processes etc. (Naik et al. 2010).

The harvesting of algal culture is a dewatering process, requires high energy and is the most expensive step in the algal processing. It is carried out in two different methods. One method is floatation and another one is reverse osmosis. The yielded algal biomass is processed and it involves many stages like thickening, filtering, drying etc. The harvesting process results in the formation of an algal slurry with 5-15% of total suspended solids. By chemical and mechanical methods, oil extraction was performed from the harvested algal biomass through cell disruption. In this stage, the oil content and the water content are to be removed. The oil content of algae can be modified by altering the cultivation methods of algae (Hannon et al. 2010). Finally, the obtained lipids are converted into biodiesel and glycerine. This process, which involves the conversion of fatty acids to biodiesel and glycerine, is known as transesterification. From the obtained crude biodiesel, the desired biofuels are produced through various refining processes. The entire steps of algal biofuel production are summarized in Fig. 6.4.

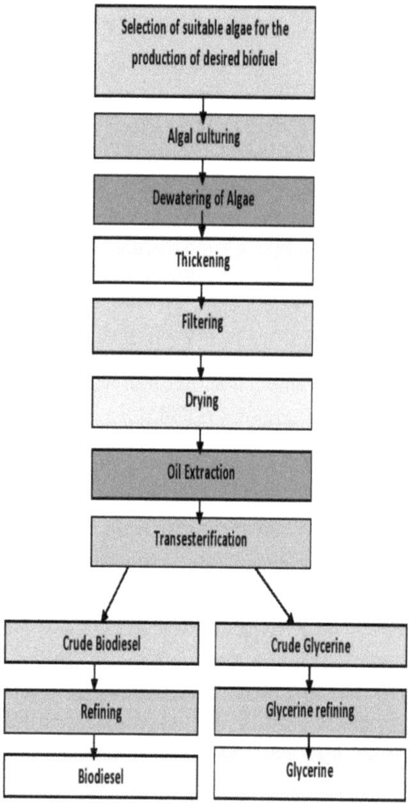

Figure 6.4: Stages of biofuel production from algae.

Even though the algae show a high potentiality for the production of biofuels, the biofuel production techniques still show many bottlenecks before they can become a major source of fuel contributing to the sustainable source of energy. The major threats for the development of these biofuel production techniques from algae include fresh water and fertilizers like nitrogen, phosphorus and other nutrients (Bohutskyi et al. 2016).

6.4 Conclusion

Algae are continuously producing a variety of compounds of high economic value and they are acting as the source of food, animal feed, cosmetics, medicines, and oil for biofuel. They can grow well in diverse habitat and produce huge biomass, thus making these organisms more attractive. Huge quantity of energy in the form of oils and polysaccharides is already stored in algae and it becomes higher with the rapid biomass increase. Because of this high oil content and the rapid biomass production, algae are known as the potential source for the production of various biofuels. In spite of this major sustainable application of algae, they are also routinely employed for the cleaning of polluted water because of their phytoremediation properties. Many algae can be cultivated in waste waters like sewage, industrial effluents and animal wastes and they utilize the nutrients present in the water, thereby purifying it and also producing huge biomass which can be further utilized for biofuel production. Thus, algae form an interesting group of organisms, which possess great importance for the sustainable development.

References

Abdel-Raouf, N., A.A. Al-Homaidan and I.B.M. Ibraheem. 2012. Microalgae and wastewater treatment. Saudi. J. Biol. Sci. 19: 257-275.

Ahner, B.A., L. Wei, J.R. Oleson and N. Ogura. 2002. Glutathione and other low molecular weight thiols in marine phytoplankton under metal stress. Mar. Ecol. Prog. Ser. 232: 93-103.

Alghurabie, I.K., B.O. Hasan, B. Jackson, A.A. Kosminski and P.J. Shman. 2013. Fluidized bed gasification of Kingston coal and marine microalgae in a spouted bed reactor. Chem. Eng. Res. Des. 91: 1614-1624.

Al-Homaidan, A.A., A.A. Al-Ghanayem and A.H. Alkhalifa. 2011. Green algae as bioindicators of heavy metal pollution in Wadi Hanifah Stream, Riyadh, Saudi Arabia. Inter. J. Wat. Resour. Arid Environ. 1: 10-15.

Archana, T. and P. Anjana. 2012. Cyanobacterial hydrogen production: A step towards clean environment. Int. J. Hydrogen Energy 37: 139-150.

Babich, I.V., M. Vander Hulst, L. Lefferts, J.A. Moulijn, P. O'Connor and K. Seshan. 2011. Catalytic pyrolysis of microalgae to high-quality liquid bio-fuels. Biomass Bioener. 35: 3199-3207.

Baker, A.J.M. and R.R. Brooks. 1989. Terrestrial higher plants which hyperaccumulate metallic elements – A review of their distribution, ecology and phytochemistry. Bioreco. 1: 81-126.

Baker, A.J.M. 1981. Accumulators and excluders – strategies in the response of plants to heavy metals. J. Plant Nutri. 3: 643-653.

Barreiro, D.L., W. Prins, F. Ronsse and W. Brilman. 2013. Hydrothermal liquefaction (HTL) of microalgae for biofuel production: State of the art review and future prospects. Biomass Bioener. 53: 113-127.

Behera, S., R. Singh, A. Richa, N.K. Sharma, S. Madhulika and K. Kumar. 2014. Scope of algae as third generation biofuels. Front. Bioeng. Biotechnol. 2: 90-102.

Benemann, J.R. and W.J. Oswald. 1996. Systems and economic analysis of microalgae ponds for conversion of carbon dioxide to biomass (Final Report: Grant No. DEFG22

93PC93204), Pittsburgh Energy Technology Centre, Pittsburgh, PA, US Department of Energy.

Berti, W.R. and S.D. Cunningham. 2000. Phytostabilization of metals. pp. 71-88. *In*: Raskin, I. and B.D. Ensley [eds.]. Phytoremediation of Toxic Metals: Using Plants to Clean Up the Environment. Wiley, New York.

Bhadury P. and P.C. Wright. 2004. Exploitation of marine algae: Biogenic compounds for potential antifouling applications. Planta 219: 561-578.

Bigogno, C., I. Khozin-Goldberg, S. Boussiloa, A. Vonshak and Z. Cohen. 2002. Lipid and fatty acid composition of the green oleaginous alga *Parietochoris incise*, the richest plant source of arachidonic acid. Phytochem. 60: 497-503.

Bilanovic, D., G. Shelef and A. Sukenik. 1988. Flocculation of microalgae with cationic polymers-effect of medium salinity. Biomass. 17: 65-76.

Biller, P., A.B. Ross, S.C. Skill, A. Lea-Langton, B. Balasundaram, C. Hall, R. Riley and C.A. Llewellyn. 2012. Nutrient recycling of aqueous phase for microalgae cultivation from the hydrothermal liquefaction process. Algal Res. 1: 70-76.

Blanco, A.M., J. Moreno, J.A. Del Campo, J. Rivas and M.G. Guerrero. 2007. Outdoor cultivation of lutein-rich cells of *Muriellopsis* sp. in open ponds. Appl. Microbiol. Biotechnol. 73: 1259-1266.

Bohutskyi, P., C. Steven, K. Ben, C.F. Shek, D. Yacar, Y. Tang, M. Zivojnovich, M.J. Betenbaugh and E.J. Bouwer. 2016. Phytoremediation of agriculture runoff by filamentous algae polyculture for biomethane production, and nutrient recovery for secondary cultivation of lipid generating microalgae. Bioresour. Technol. 222: 294-308.

Boyd, J.N., J.H. Kucklick, D.K. Scholz, A.H. Walker, R.G. Pond and A. Bostrom. 2001. Effects of oil and chemically dispersed oil in the environment. No. 4693. Washington, DC: American Petroleum Institute.

Brinza, L., M. Dring and M. Gavrilescu. 2005. Biosorption of Cu^{2+} ions from aqueous solution by *Enteromorpha* sp. Environ. Engineer. Manag. J. 4: 29-46.

Brown, S.L., R.L. Chaney, J.S. Angle and A.J.M. Baker. 1994. Phytoremediation potential of *Thlaspi caerulescens* and bladder compion for zinc- and cadmium-contaminated soil. J. Environ. Qual. 23: 1151-1157.

Brown, T.R. and R.C. Brown. 2013. A review of cellulosic biofuel commercial-scale projects in the United States. Biofuels Bioprod. Biorefin. 7: 235-245.

Bursali, E.A., L. Cavas, Y. Seki, S.S. Bozkurt and M. Yurdakoc. (2009). Sorption of boron by invasive marine seaweed: *Caulerpa racemosa* var. *cylindracea*. Chem. Eng. J. 150: 385-390.

Buxy, S., R. Diltz and P. Pullammanappallil. 2013. Biogasification of marine algae *Nannochloropsis Oculata*. *In*: Wicks, G., J. Simon, R. Zidan, R. Brigmon, G. Fischman, S. Arepalli, A. Norris and M. McCluer. [eds.]. Materials Challenges in Alternative and Renewable Energy II: Ceramic Transactions. John Wiley & Sons, Inc. Hoboken, NJ, USA.

Cai, X.H., T. Logan, T. Gustafson, S. Traina and R.T. Sayre. 1995. Applications of eukaryotic algae for the removal of heavy metals from water. Mol. Mar. Biol. Biotechnol. 4: 338-344.

Carvalho, R.M., J.V.C. Vargas, L.P. Ramos, C.E.B. Marino and J.C.L. Torres. 2011. Microalgae biodiesel via in situ methanolysis. J. Chem. Technol. Biotechnol. 86: 1418-1427.

Chang, P., J.Y. Kim and K.Y. Kim. 2005. Concentrations of arsenic and heavy metals in vegetation at two abandoned mine tailings in South Korea. Environ. Geochem. Health. 27: 109-119.

Chen, C.Y., X.Q. Zhao, H.W. Yen, S.H. Ho, C.L. Cheng, D.J. Lee, F.W. Bai and J.S. Chang. 2013. Microalgae-based carbohydrates for biofuel production. Biochem. Eng. J. 78: 1-10.

Chi, N.T.L., P.A. Duc, T. Mathimani and A. Pugazhendhi. 2018. Evaluating the potential of green alga *Chlorella* sp. for high biomass and lipid production in biodiesel viewpoint. Biocatal. Agric. Biotechnol. 17: 184-188.

Chiappe, C., A. Mezzetta, C.S. Pomelli, G. Iaquaniello, A. Gentile and B. Masciocchi. 2016. Development of cost-effective biodiesel from microalgae using protic ionic liquids. Green. Chem. 18: 4982-4989.

Choi, S.P., M.T. Nguyen and S.J. Sim. 2010. Enzymatic pretreatment of *Chlamydomonas reinhardtii* biomass for ethanol production. Bioresour. Technol. 101: 5330-5336.

Cobbett, C. and P. Goldsbrough. 2002. Phytochelatins and Metallothioneins: Roles in heavy metal detoxification and homeostasis. Ann. Rev. Plant Biol. 53: 159-182.

Crist, R.H., K. Oberholser, N. Shank and M. Nguyen. 1981. Nature of binding between metallic ions and algal cell walls. Environ. Sci. Technol. 15: 1212-1217.

Cunningham, S.C. and W.R. Berti. 2000. Phytoextraction and Phytostabilization: Technical, economic, and regulatory considerations of the soil-lead issue. pp. 359-376. *In*: Terry, N. and G. Banuelos [eds.]. Phytoremediation of Contaminated Soil and Water. Boca Raton, Florida, USA: Lewis Publishers.

Darnall, D.W., B. Greene, M.T. Henzi, J.M. Hosea, R.A. McPherson, J. Sneddon and M.D. Alexander. 1986. Selective recovery of gold and other metal ions from an algal biomass. Environ. Sci. Technol. 20: 206-208.

Davis, T., B. Volesky and A. Muccib. 2003. A review of the biochemistry of heavy metal biosorption by brown algae. Water Res. 37: 4311-4330.

Davis, T.A., B. Volesky and R.H.S.F. Vieira. 2000. Sargassum seaweed as biosorbent for heavy metals. Water Res. 34: 4270.

De Filippis, L.F., R. Hampp and H. Ziegler. 1981. The effects of sub-lethal concentrations of zinc, cadmium and mercury on Euglena-Growth and pigments. Z. Pflanzenphysiol. 101: 37-47.

Deschamps, P. and D. Moreira. 2009. Signal conflicts in the phylogeny of the primary photosynthetic eukaryotes. Mol. Biol. Evol. 26: 2745-2753.

Doucha, J. and K. Livansky. 2006. Productivity, CO/O exchange and hydraulics in outdoor open high density microalgal (*Chlorella* sp.) photobioreactors operated in a middle and southern European climate. J. Appl. Phycol. 18: 811-826.

Douskova, I., J. Doucha, K. Livansky, J. Machat, P. Novak, D. Umysova, V. Zachleder and M. Vitova. 2009. Simultaneous flue gas bioremediation and reduction of microalgal biomass production costs. Appl. Microbiol. Biotechnol. 82: 179-185.

Duman, G., M.A. Uddin and J. Yanik. 2014. Hydrogen production from algal biomass via steam gasification. Bioresour. Technol. 166: 24-30.

Dutta, K., A. Daverey and J. Lin. 2014. Evolution retrospective for alternative fuels: First to fourth generation. Renew. Energy 69: 114-122.

Dwivedi, S., S. Srivastava, S. Mishra, A. Kumar, R.D. Tripathi, U.N. Rai, R. Dave, P. Tripathi, D. Charkrabarty and P.K. Trivedi. 2010. Characterization of native microalgal strains for their chromium bioaccumulation potential: Phytoplankton response in polluted habitats. J. Hazard. Mater. 173: 95-101.

Eboibi, B.E., D.M. Lewis, P.J. Ashman and S. Chinnasamy. 2014. Effect of operating conditions on yield and quality of bio crude during hydrothermal liquefaction of halophytic microalga *Tetraselmis* sp. Bioresour. Technol. 170: 20-29.

Egmond, W.V. 1995. Diatoms [online]. Egmond. Retrieved July 1, 2007, from http://www.microscopy-uk.org.uk/mag/indexmag.html? http://www.microscopy-uk.org.uk/mag/wimsmall/diadr.htm

Ehimen, E.A., Z.F. Sun and C.G. Carrington. 2010. Variables affecting the in situ trans-esterification of microalgae lipids. Fuel 89: 677-684.

El-Sayed, W.M.M., H.A. Ibrahim, U.M. Abdul-Raouf and M. El-Nagar. 2016. Evaluation of bioethanol production from *Ulva lactuca* by *Saccharomyces cerevisiae*. J. Biotechnol. Biomater. 6: 2-10.

El-Shimi, H.I., N.K. Attia, S.T. El-Sheltawy and G.I. El-Diwani. 2013. Biodiesel production from *Spirulina platensis* microalgae by in-situ transesterification process. J. Sustain. Bioene. Syst. 3: 224-233.

EPA (Environmental Protection Agency). 1997. Electrokinetic laboratory and field processes applicable to radioactive and hazardous mixed waste in soil and ground water. EPA 402/R-97/006. Washington, DC.

Falkowski, P.G., R.T. Barber and V.V. Smetacek. 1998. Biogeochemical controls and feedbacks on ocean primary production. Science 281: 200-207.

Fernandez-Pinas, F., P. Mateo and I. Bonilla. 1995. Ultrastructural changes induced by selected cadmium concentrations in the cyanobacterium Nostoc UAM 208. J. Plant Physiol. 147: 452-456.

Fierro, S., M. del Pilar Sánchez-Saavedra and C. Copalcúa. 2008. Nitrate and phosphate removal by chitosan immobilized *Scenedesmus*. Bioresour. Technol. 99: 1274-1279.

Figueira, M.M., B. Volesky and V.S.T. Ciminelli. 1997. Assessment of interference in biosorption of a heavy metals. Biotechnol. Bioeng. 54: 344-350.

Fitz, W.J. and W.W. Wenzel. 2002. Arsenic transformations in the soil-rhizosphere-plant system: Fundamentals and potential application to phytoremediation. J. Biotechnol. 99: 259-278.

Flathman, P.E. and G.R. Lanza. 1998. Phytoremediation – Current views on an emerging green technology. J. Soil Contamination 7: 415-432.

Fourest, E. and B. Volesky. 1997. Alginates properties and heavy metal biosorption by marine algae. Appl. Biochem. Biotechnol. 67: 215-226.

Ghasemi, Y., S. Rasoul-Amini, A.T. Naseri, N. Montazeri-Najafabady and M.A. Mobasher. 2012. Microalgae biofuel potentials. Appl. Biochem. Microbiol. 48: 126-144.

Goldemberg, J. 2007. Ethanol for a sustainable energy future. Science 315: 808-810.

Gosavi, K., J. Sammut and J. Jankowski. 2004. Macroalgal biomonitors of trace metal contamination in acid sulfate soil aquaculture ponds. Sci. Total Environ. 324: 25-39.

Goswami, G., V. Bang and S. Agarwal. 2015. Diverse application of algae. Inter. J. Advance Res. Sci. Eng. 4: 1102-1109.

Grierson, S., V. Strezov, G. Ellem, R. Mcgregor and J. Herbertson. 2009. Thermal characterisation of microalgae under slow pyrolysis conditions. J. Anal. Appl. Pyrolysis 85: 118-123.

Gülyurt, M., D. Özçimen and I. Benan. 2016. Biodiesel production from *Chlorella protothecoides* oil by microwave-assisted transesterification. Int. J. Mol. Sci. 17: 579.

Hannon, M., J. Gimpel, M. Tran, B. Rasala and S. Mayfield. 2010. Biofuels from algae: Challenges and potential. Biofuels 1: 763-784.

Harun, R., M.K. Danquah and G.M. Forde. 2010. Microalgal biomass as a fermentation feedstock for bioethanol production. J. Chem. Technol. Biotechnol. 85: 199-203.

Hirano, A., R. Ueda, S. Hirayama and Y. Ogushi. 1997. CO_2 fixation and ethanol production with microalgal photosynthesis and intra cellular anaerobic fermentation. Energy 22: 137-142.

Holan, Z.R. and B. Volesky. 1994. Biosorption of lead and nickel by biomass of marine algae. Biotech. Bioeng. 43: 1001-1009.

Hossain, A., A. Salleh, A. Boyce, P. Chowdhury and M. Naqiuddin. 2008. Biodiesel fuel production from microalgae as renewable energy. Am. J. Biochem. Biotechnol. 4: 250-254.

Huang, J., J. Xia, W. Jiang, Y. Li and J. Li. 2015. Biodiesel production from microalgae oil catalyzed by a recombinant lipase. Bioresour. Technol. 180: 47-53.

Huang, Y., Y. Chen, J. Xie, H. Liu, X. Yin and C. Wu. 2016. Bio-oil production from hydrothermal liquefaction of high-protein high-ash microalgae including wild *Cyanobacteria* sp. and cultivated *Bacillariophyta* sp. Fuel. 183: 9-19.

Huesemann, M.H., T.S. Hausmann, R. Bartha, M. Aksoy, J.C. Weissman and J.R. Benemann. 2009. Biomass productivities in wild type and pigment mutant of *Cyclotella* sp. (diatom). Appl. Biochem. Biotechnol. 157: 507-526.

Imamul Huq, S.M., M.B. Abdullah and J.C. Joardar. 2007. Bioremediation of arsenic toxicity by algae in rice culture. Land Contam. Reclam. 15: 327-333.

Inglesby, A.E. and A.C. Fisher. 2012. Enhanced methane yields from anaerobic digestion of *Arthrospira maxima* biomass in an advanced flow-through reactor with an integrated recirculation loop microbial fuel cell. Energy Environ. Sci. 5: 7996-8006.

Irgolic, K.J., E.A. Woolso, R.A. Stockton, R.D. Newman, N.R. Bottino, R.A. Zingaro, P.C. Kearney, R.A. Pyles, S. Maeda, W.J. Mcshane and E.R. Cox. 1977. Characterization of arsenic compounds formed by *Daphnia magna* and *Tetraselmis chuii* from inorganic arsenate. Environ. Health Perpect. 19: 61-66.

Jaiswar, S., N.B. Balar, R. Kumar, M. Patel and C. Prakram. 2016. Morphological and molecular characterization of newly isolated microalgal strain *Neochloris aquatica* SJ-1 and its high lipid productivity. Biocatal. Agricul. Biotechnol. 9: 108-112.

Jensen, T.E., M. Baxter, J.W. Rachlin and V. Jani. 1982. Uptake of heavy metals by *Plectonema boryanum* (Cyanophyceae) into cellular components, especially polyphosphate bodies: An X-ray energy dispersive study. Environ. Pollut. 27: 119-127.

Johnson, M.B. and Z. Wen. 2009. Production of biodiesel fuel from the microalga *Schizochytrium limacinum* by direct transesterification of algal biomass. Ener. Fuels 23: 5179-5183.

Judge, D. and D. Earnshaw. 2003. The European Parliament. Palgrave: Basingstoke, UK.

Khoo, H.H., C.Y. Koh, M.S. Shaik and P.N. Sharratt. 2013. Bioenergy co-products derived from microalgae biomass via thermo-chemical conversion – Life cycle energy balances and CO_2 emissions. Bioresour. Technol. 143: 298-307.

Kirso, U. and N. Irha. 1998. Role of algae in fate of carcinogenic polycyclic aromatic hydrocarbons in the aquatic environment. Environ. Toxicol. Environ. Saf. 41: 83-89.

Kojima, E. and K. Zhang. 1999. Growth and hydrocarbon production of microalga *Botryococcus braunii* in bubble column photobioreactors. J. Biosci. Bioeng. 87: 811-815.

Kumar, J.I. and C. Oommen. 2012. Removal of heavy metals by biosorption using freshwater alga *Spirogyra hyaline*. J. Environ. Biol. 33: 27-32.

Kuyucak, N. and B. Volesky. 1988. Biosorbents for recovery of metals from industrial solutions. Biotechnol. Lett. 10: 137-142.

Laliberté, G., E.J. Olguin and J. de la Noüe. 1997. Mass cultivation and wastewater treatment using spirulina. pp. 159-166. *In*: Vonshak, A. [ed.]. *Spirulina platensis* (Arthrospira): Physiology, Cell-biology and Biotechnology. Taylor & Francis Ltd., London, U.K.

Lebeau, T. and J.M. Robert. 2003. Diatom cultivation and biotechnologically relevant products. Part I. Cultivation at various scales. Appl. Microbiol. Biotechnol. 60: 612-623.

Les, A. and R.W. Walker. 1984. Toxicity and binding of copper, zinc and cadmium by the blue-green alga, *Chroococcus paris*. Water Air Soil Pollut. 23: 129-139.

Li. Y., M. Horsman, N. Wu, C.Q. Lan and N. Dubois-Calero. 2008. Biofuels from microalgae. Biotechnol. Progr. 24: 815-820.

Liu, Y., S. Yang, S. Tan, Y. Lin and J. Tay. 2002. Aerobic granules: A novel zinc biosorbent. Appl. Microbiol. 35: 548-551.

Liu, G.Y., Y.X. Zhang and T.Y. Chai. 2011. Phytochelatin synthase of *Thlaspi caerulescens* enhanced tolerance and accumulation of heavy metals when expressed in yeast and tobacco. Plant Cell Rep. 30: 1067-1076.

Lu, J., C. Sheahan and P. Fu. 2011. Metabolic engineering of algae for fourth generation biofuels production. Energ. Environ. Sci. 4: 2451-2466.

Mathimani, T., L. Uma and D. Prabaharan. 2017. Optimization of direct solvent lipid extraction kinetics on marine trebouxiophycean algae by central composite design – Bioenergy perspective. Energy Convers. Manag. 142: 334-346.

Markou, G., I. Angelidaki and D. Georgakakis. 2012. Microalgal carbohydrates: An overview of the factors influencing carbohydrates production, and of main bioconversion technologies for production of biofuels. Appl. Microbiol. Biotechnol. 96: 631-645.

Marques, A.E., A.T. Barbosa, J. Jotta, M.C. Coelho, P. Tamagnini and L. Gouveia. 2011. Biohydrogen production by *Anabaena* sp. PCC 7120 wild-type and mutants under different conditions: Light, nickel, propane, carbon dioxide and nitrogen. Biomass Bioenergy 35: 4426-4434.

Matagi, S.V., D. Swai and R. Mugabe. 1998. A review of heavy metal removal mechanisms in wetlands. Afr. J. Trop. Hydrobiol. Fish. 8: 23-35.

Matbiopol. 1999. Role of microbial mats in bioremediation of hydrocarbon polluted coastal zones – Final integrative report [online]. Universite de Pau et des Pays de L'Adour. Retrieved December 1, 2007, from http://web.univ-pau.fr/RECHERCHE/MATBIOPOL/

Mathimani, T., L. Uma and D. Prabaharan. 2015. Homogeneous acid catalysed trans-esterification of marine microalga *Chlorella* sp. BDUG 91771 lipid – An efficient biodiesel yield and its characterization. Renew. Energy 81: 523-533.

Matsudo, M.C., R.P. Bezerra, S. Sato, P. Perego, A. Converti and J.C.M. Carvalho. 2008. Repeat fed-batch cultivation of *Arthrospira (spirulina) patensis* using urea as nitrogen source. Biochem. Engineer. J. 43: 52-57.

Matsunaga, T., H. Takeyama, T. Nakao and A. Yamazawa. 1999. Screening of marine microalgae for bioremediation of cadmium-polluted seawater. J. Biotechnol. 70: 33-38.

McGrath, S.P. and F.J. Zhao. 2003. Phytoextraction of metals and metalloids from contaminated soils. Curr. Opin. Biotechnol. 14: 277-282.

Mehta, S.K. and J.P. Gaur. 1999. Heavy-metal-induced proline accumulation and its role in ameliorating metal toxicity in *Chlorella vulgaris.* New Phytol. 143: 253-259.

Mei, L., X. Xie, R. Xue and Z. Liu. 2006. Effects of strontium-induced stress on marine microalgae Platymonas subcordiformis (Chlorophyta: Volvocales). Chinese J. Oceanol. Limnol. 24: 154-160.

Mejare, M. and L. Bülow. 2001. Metal-binding proteins and peptides in bioremediation and phytoremediation of heavy metals. Trends Biotechnol. 19: 67-73.

Miao, X., Q. Wu and C. Yang. 2004. Fast pyrolysis of microalgae to produce renewable fuels. J. Anal. Appl. Pyrolysis 71: 855-863.

Minowa, T., S. Yokoyama, M. Kishimoto and T. Okakura. 1995. Oil production from algal cells of *Dunaliella tertiolecta* by direct thermochemical liquefaction. Fuel 74: 1735-1738.

Mitra, N., Z. Rezvan, M.S. Ahmad and M.G.M. Hosein. 2012. Studies of water arsenic and boron pollutants and algae phytoremediation in three springs, Iran. Inter. J. Ecosyst. 2: 32-37.

Miyamoto, K., P.C. Hallenbeck and J.R. Benemann. 1979. Hydrogen production by the thermophilic alga *Mastigocladus laminosus*: Effects of nitrogen, temperature, and inhibition of photosynthesis. Appl. Environ. Microbiol. 38: 440-446.

Moheimani, N.R. and M.A. Borwitzka. 2007. Limits to productivity of the alga *Pleurochrysis carterae* (haptophyta) grown in outdoor raceway ponds. Biotechnol Bioeng. 96: 27-36.

Naik, S.N., V.V. Goud, P.K. Rout and A.K. Dalai. 2010. Production of first and second generation biofuels: A comprehensive review. Renew. Sustain. Energy Rev. 14: 578-597.

Nayak, B.K., S. Roy and D. Das. 2014. Biohydrogen production from algal biomass (*Anabaena* sp. PCC7120) cultivated in airlift photobioreactor. Int. J. Hydrogen Energy 39: 7553-7560.

Nguyen, T., F.A. Roddick and L. Fan. 2015. Impact of green algae on the measurement of *Microcystis aeruginosa* populations in lagoon-treated waste water with an algae online analyser. Environ.Technol. 36: 556-565.

Nielsen, H.D., T.R. Burridge, C. Brownlee and M.T. Brown. 2005. Prior exposure to contamination influences the outcome of toxicological testing of *Fucus serratus* embryos. Mar. Pollut. Bull. 50: 1675-1680.

Obata, O., J. Akunna, H. Bockhorn and G. Walker. 2016. Ethanol production from brown seaweed using non-conventional yeasts. J. Bioethanol. 2: 134-145.

Oncel, S.S., A. Kose, C. Faraloni, E. Imamoglu, M. Elibol, G. Torzillo and F. Vardar Sukan. 2015. Biohydrogen production from model microalgae *Chlamydomonas reinhardtii*: A simulation of environmental conditions for outdoor experiments. Int. J. Hydrogen Energy 40: 7502-7510.

Onwudili, J.A., A.R. Lea-Langton, A.B. Ross and P.T. Williams. 2013. Catalytic hydrothermal gasification of algae for hydrogen production: Composition of reaction products and potential for nutrient recycling. Bioresour. Technol. 127: 72-80.

Pan, P., C. Hu, W. Yang, Y. Li, L. Dong, L. Zhu, D. Tong, R. Qing and Y. Fan. 2010. The direct pyrolysis and catalytic pyrolysis of *Nannochloropsis* sp. residue for renewable bio-oils. Bioresour. Technol. 101: 4593-4599.

Parker, M.S., T. Mock and E.V. Armbrust. 2008. Genomic insights into marine microalgae. Annu. Rev. Genet. 42: 619-645.

Payne, R. 2000. *Spirulina* as bioremediation agent: Interaction with metals and involvement of carbonic anhydrase. M.Sc. Thesis, Rhodes University.

Pérez-Rama, M., J. Abalde, C. Herrero and E.T. Vaamonde. 2002. Cadmium removal by living cells of the marine microalga *Tetraselmis suecica*. Bioresour. Technol. 84: 265-270.

Pesando, D. 1990. Antibacterial and antifungal activities of marine algae. Intro. Applied. Phycol. 3-26.

Raheem, A., W.W. Azlina, Y.T. Yap, M.K. Danquah and R. Harun. 2015. Thermochemical conversion of microalgal biomass for biofuel production. Renew. Sustain. Energy Rev. 49: 990-999.

Raheem, A., P. Prinsen, A.K. Vuppaladadiyam, M. Zhao and R. Luque. 2018. A review on sustainable microalgae based biofuel and bioenergy production: Recent developments. J. Clean. Prod. 181: 42-59.

Rai, P.K. 2009. Heavy metal phytoremediation from aquatic ecosystems with special reference to macrophytes. Critical Rev. Environ. Sci. Technol. 39: 697-753.

Rainbow, P.S. 1995. Biomonitoring of heavy metal availability in the marine environment. Mar. Pollut. Bull. 31: 183-192.

RangaRao, A. and G.A. Ravishankar. 2007. Influence of CO_2 on growth and hydrocarbon production in *Botryococcus braunii*. J. Microbiol. Biotechnol. 17: 414-419.

Ren, L., D. Shi, J. Dai and B. Ru. 1998. Expression of the mouse metallothionein-I gene conferring cadmium resistance in a transgenic cyanobacterium. FEMS Microbiol. Lett. 158: 127-132.

Rizzo, A.M., M. Prussi, L. Bettucci, I.M. Libelli and D. Chiaramon 2013. Characterization of microalga *Chlorella* as a fuel and its thermogravimetric behavior. Appl. Energy 102: 24-31.

Rodolfi, L., C.G. Zittelli, N. Bassi, G. Padowani, N. Biondi, G. Bonini and M.M. Tredici. 2009. Microalgae for oil: Strain selection, induction of lipid synthesis and outdoor mass cultivation in a low-cost photobioreactor. Biotechnol. Bioeng. 102: 100-112.

Saad, M.G. and H.M. Shafik. 2017. The challenges of biodiesel production from *Oscillatoria* sp. J. Int. J. Adv. Res. 5: 1316-1322.

Sakaguchi, T., T. Tsuji, A. Nakajima and T. Horikoshi. 1979. Accumulation of cadmium by green microalgae. Eur. J. Appl. Microbiol. Biotechnol. 8: 207-215.

Sakakibara, M., Y. Ohmori, N.T.H. Ha, S. Sano and K. Sera. 2011. Phytoremediation of heavy metal contaminated water and sediment by *Eleocharis acicularis*. Clean: Soil, Air, Water 39: 735-741.

Sanchez-Silva, L., D. López-González, A.M. Garcia Minguillan and J.L. Valverde. 2013. Pyrolysis, combustion and gasification characteristics of *Nannochloropsis gaditana* microalgae. Bioresour. Technol. 130: 321-323.

Saqib, A., M.R. Tabbssum, U. Rashid, M. Ibrahim, S.S. Gill and M.A. Mehmood. 2013. Marine macroalgae *Ulva*: A potential feed-stock for bioethanol and biogas production. Asian J. Agri. Biol. 1: 155-163.

Schiewer, S. and B. Volesky. 1995. Modeling of the proton-metal ion exchange in biosorption. Environ. Sci. Technol. 29: 3049-3058.

Scow, K.M. and K.A. Hicks. 2005. Natural attenuation and enhanced bioremediation of organic contaminants in groundwater. Curr. Opin. Biotechnol. 16: 246-253.

Sekabira, K., O.H. Origa, T.A. Basamba, G. Mutumba and E. Kakudidi. 2010. Application of algae in biomonitoring and phytoextraction of heavy metals contamination in urban stream water. Int. J. Environ. Sci. Technol. 8: 115-128.

Shafik, H.M., M.G. Saad and H.A. El-Serehy. 2015. Impact of nitrogen regime on fatty acid profiles of *Desmodesmus quadricaudatus* and *Chlorella* sp. and ability to produce biofuel. Acta Bot. Hung. 57: 205-218.

Shalaby, E.A. 2011. Algae as promising organisms for environment and health. Plant Signal. Behav. 9: 1338-1350.

Shanab, S., E. Ashraf and E. Shalaby. 2012. Bioremoval capacity of three heavy metals by some microalgae species (Egyptian Isolates). Plant. Signal. Behav. 7: 1-8.

Shamsuddoha, A.S.M., A. Bulbull and S.M. Imamul Huq. 2006. Accumulation of arsenic in green algae and its subsequent transfer to the soil-plant system. Bangladesh J. Microbiol. 22: 148-151.

Sharma, N. and P. Sharma. 2017. Industrial and biotechnological applications of algae: A Review. J. Advances. Plant Biol. 1: 1-25.

Sharma, J., S.S. Kumar, N.R. Bishnoi and A. Pugazhendhi. 2018. Enhancement of lipid production from algal biomass through various growth parameters. J. Mol. Liq. 269: 712-720.

Sharma, S.S. and K.J. Dietz. 2006. The significance of amino acids and amino acid-derived molecules in plant responses and adaptation to heavy metal stress. J. Exp. Bot. 57: 711-726.

Sheehan, J., T. Dunahay, J. Benemann and P. Roessler. 1998. A Look Back at the US Department of Energy's Aquatic Species Program – Biodiesel from Algae. Vol. 328. National Renewable Energy Laboratory; CO, USA.

Shuping, Z., W. Yulong, Y. Mingde, I. Kaleem, L. Chun and J. Tong. 2010. Production and characterization of bio-oil from hydrothermal liquefaction of microalgae *Dunaliella tertiolecta* cake. Energy 35: 5406-5411.

Sigg, L. 1985. Metal transfer mechanisms in lakes; role of settling particles. pp. 283-310. *In*: Stumm, W. [ed.]. Chemical Processes in Lakes. Wiley Interscience, New York.

Sigg, L. 1987. Surface chemical aspects of the distribution and fate of metal ions in lakes. pp. 319-348. *In*: Stumm, W. [ed.]. Aquatic Surface Chemistry. Wiley Interscience, New York.

Sjahrul, M. 2012. Phytoremediation of Cd^{2+} by Marine Phytoplanktons, *Tetracelmis chuii* and *Chaetoceros calcitrans*. Inter. J. Chem. 4: 69-74.

Stucki, S., F. Vogel, C. Ludwig, A.G. Haiduc and M. Brandenberger. 2009. Catalytic gasification of algae in supercritical water for biofuel production and carbon capture. Energy Environ. Sci. 2: 535-541.

Subsamran, K., P. Mahakhan, K. Vichitphan, S. Vichitphan and J. Sawaengkaew. 2018. Potential use of vetiver grass for cellulolytic enzyme production and bioethanol production. Biocatal. Agric. Biotechnol. 17: 261-268.

Swift, D.T. and D. Forciniti. 1997. Accumulation of lead by *Anabaena cylindrica*:

Mathematical modeling and an energy dispersive X-ray study. Biotechnol. Bioeng. 55: 408-419.
Tang, H., N. Abunasser, M.E.D. Garcia, M. Chen, K.Y. S. Ng and S.O. Salley. 2011. Potential of microalgae oil from *Dunaliella tertiolecta* as a feed stock for biodiesel. Appl. Energy 88: 3324-3330.
Thomas, E.J. 1993. Ultrastructure of microalgae. pp. 8-51. *In*: Tarmar, B. [ed.]. Cyanobacterial Ultrastructure. CRC Press. Boca Raton. Florida.
Ting, Y.P., F. Lawson and I.G. Prince. 1991. Uptake of cadmium and zinc by the alga *Chlorella vulgaris*, Part II: Multi-ion situation. Biotechnol. Bioeng. 37: 445-455.
Tong, Y.P., R. Kneer and Y.G. Zhu. 2004. Vacuolar compartmentalization: A second generation approach to engineering plants for phytoremediation. Trends. Plant Sci. 9: 7-9.
Ueno, Y., N. Kurano and S. Miyachi. 1998. Ethanol production by dark fermentation in the marine green alga, *Chlorococcum littorale*. J. Ferment. Bioengine. 86: 38-43.
Visviki, J. and J.W. Rachlin. 1994. Acute and chronic exposure of *Dunaliella salina* and *Chlamydomonas bullosa* to copper and cadmium: Effects on ultrastructure. Arch. Contam. and Toxicol. 26: 154-162.
Volesky, B. 2001. Detoxification of metal-bearing effluents: Biosorption for the next century. Hydrometallurgy 59: 203-216.
Volland, S., D. Schaumlöffel, D. Dobritzsch, G.J. Krauss and U. Lütz-Meindl. 2012. Identification of phytochelatins in the cadmium-stressed conjugating green alga *Micrasterias denticulata*. Chemosphere 91: 448-454.
Wang, T.C., J.S. Weissman, G. Ramesh, R. Varadarajan and J.R. Benemann. 1995. Bioremoval of toxic elements with aquatic plants and algae. p. 33. *In*: Hinchee, R.E., D.B. Anderson, and R.E. Hoeppel. [eds.]. Bioremediation of Inorganics. Battelle Press, Columbus, OH.
Wang, Y., Y. Yang, F. Ma, L. Xuan, Y. Xu, H. Huo, D. Zhou and S. Dong. 2015. Optimization of *Chlorella vulgaris* and bioflocculant-producing bacteria co-culture: Enhancing microalgae harvesting and lipid content. Lett. Appl. Microbiol. 60: 497-503.
Ward, A.J., D.M. Lewis and F.B. Green. 2014. Anaerobic digestion of algae biomass: A review. Algal Res. 5: 204-214.
Wu, Q., Y. Shiraiwa, H. Takeda, G. Sheng and J. Fu. 1991. Liquid-saturated hydrocarbons resulting from pyrolysis of the marine cocco-lithophores *Emiliania huxleyi* and *Gephyrocapsa oceanica*. Mar. Biotechnol. 1: 346-352.
Yan, A, Y. Wang, S.N. Tan, M.L. Mohd Yusof, S. Ghosh and Z. Chen. 2020. Phytoremediation: A promising approach for revegetation of heavy metal-polluted land. Front. Plant Sci. 11: 359.
Yanqun, L., H. Mark, W. Nan, Q.L. Christopher and D.C. Nathalie. 2008. Biofuels from microalgae. Biotechnol. Progress. 24: 815-820.
Yoshida, N., R. Ikeda and T. Okuno. 2006. Identification and characterization of heavy metal-resistant unicellular alga isolated from soil and its potential for phytoremediation. Bioresour. Technol. 97: 1843-1849.
Zakeri, R., M. Noori, M. Zakeri, S.A. Mazaheri and M.H. Gharaiie. 2011. Studies of the physico-chemical effects of water especially heavy metals concentration on algae density in three springs of Sang-E-Noghreh area (fariman) in winter, 11th National Conference of Psychology of Iran, Shahid Beheshti University, 14-15 September 2011, Tehran-Iran.
Zamalloa, C., N. Boon and W. Verstraete. 2012. Anaerobic digestibility of *Scenedesmus obliquus* and *Phaeodactylum tricornutum* under mesophilic and thermophilic conditions. Appl. Energy 92: 733-738.
Ziervogel, H., A.J. Nelson, E. Murdock, J. Selann and B. Adeney. 2004. Diatom, cyanobacterial and microbial mats as indicators of hydrocarbon contaminated arctic streams and waters. Presented at the Remediation Technology Symposium 2004, Banff, AB, Canada.

CHAPTER
7

Biomass Production and Energetic Valorization in Constructed Wetlands

S.I.A. Pereira[1], P.M.L. Castro[1] and C.S.C. Calheiros[2]*

[1] Universidade Católica Portuguesa, CBQF - Centro de Biotecnologia e Química Fina – Laboratório Associado, Escola Superior de Biotecnologia, Porto, Portugal

[2] Interdisciplinary Centre of Marine and Environmental Research (CIIMAR/CIMAR), University of Porto, Matosinhos, Portugal

7.1 Introduction

The global energy demand will increase in the coming decades, keeping in consideration the expected rising of world population associated with the depletion of fossil fuel reserves and the high industrialization, especially in developed countries (Gaurav et al. 2016, IEA 2019, OECD/FAO 2019). Currently, fossil fuels still constitute the primary energy source worldwide. However, the depletion of natural reserves and the growing concern on their impacts on the environment and human health have encouraged alternative strategies by policymakers (Dale and Holtzapple 2015). In 2018, the European Union (EU) prioritized targets for the reduction of greenhouse gas emissions (GHG) and increase the use of renewable energies in 2030, thus aiming to reduce GHG at least by 40% and increase the renewable energy consumption by 32% across all sectors (Giuntoli 2018). Boosting biofuel production was considered one of the most promising options to achieve these targets. In this alignment, the production of biofuels in the EU has increased from 29.2 petajoules, in 2000, to approximately 649.8 petajoules, in 2019 (Sönnichsen 2020). The recent European Green Deal aims at turning environmental challenges into opportunities constituting a "roadmap for making the EU's economy sustainable" with actions to "boost the efficient use of resources by moving to a clean, circular economy and restore biodiversity and cut pollution", and indeed valorization of biomass derived from constructed wetlands (CWs) embodies such principles (EC 2019).

Biofuels can be classified into primary and secondary. Primary biofuels comprise those in which raw materials are not processed and are used in their natural form, while in secondary biofuels, feedstocks are processed, resulting in solid, liquid, or gaseous forms (Dragone et al. 2010, FAO 2008). Moreover, based on feedstock

Corresponding author: cristina@calheiros.org

used and processing technology, biofuels can also be classified as i) first-generation biofuels: *bioethanol or biobutanol* – derived from the fermentation of food crops rich in starch and sugars, including wheat, barley, corn, potato, sugarcane, and sugar beet; *biodiesel* – produced by transesterification of oil crops such as sunflower, palm, soybean, and rapeseed; ii) second-generation biofuels: *bioethanol and biodiesel* – their production relies on the hydrolysis of lignocellulosic materials followed by the fermentation of starch and sugars of non-food crops, namely cassava, jatropha, miscanthus, as well as, of agricultural/forest by-products; and iii) third-generation biofuels: *biodiesel, bioethanol, and biogas* - derived from marine resources, like macro- and microalgae and cyanobacteria (Dragone et al. 2010, Chauhan and Skakla 2011, Gaurav et al. 2016, Rodionova et al. 2017). A fourth-generation biofuel based on the conversion of vegoil (vegetable oil-based) and biodiesel into biogasoline using the most advanced technology was reported by Chauhan and Skakla (2011).

The sustainability of biofuel production options to counteract the dependence on fossil fuels has been questioned due to the low GHG abatement observed in certain biofuels, the increasing competition for natural resources, including land and water, and the consequent impairment of food supply (FAO 2008, Barnabè et al. 2013). The use of non-food crops in the production of second-generation biofuels alleviates the negative impact of first-generation biofuels on food supply. However, the growing need for land for biomass production may result in land-use changes, which often involve converting former agricultural fields into biofuel feedstock producing areas. These changes jeopardize the available area for agricultural production worldwide, threatening the food supply and security (Barnabè et al. 2013, Popp et al. 2014). Furthermore, the problem of water scarcity in the agricultural sector may be intensified by the large-scale biofuel production that requires high amounts of water, either in the irrigation of crops in the field or in the processing phase (Fingerman et al. 2010).

The use of marginal and contaminated lands for feedstock production to the energy sector appears as a suitable alternative, minimizing the conflict between food and biofuel production (FAO/UNEP 2011, Evangelou et al. 2014, Tripathi et al. 2016). Indeed, several energy crops (e.g. willow, common reed, miscanthus, jatropha, poplar) are known to maintain their fast growth on degraded and polluted soils where food crop production is impossible. This strategy may offer significant advantages, as it accomplishes the current bioenergy needs while remediating degraded soils (Pandey et al. 2016). Biomass harvested from natural resources constitutes another source of raw materials for biofuel production and may include forest, woodland, grassland and, aquatic plants grown in natural and artificial wetlands (FAO/UNEP 2011). Artificial or CWs are one of the most productive ecosystems where biomass production has great relevance, varying according to the plant species, climate, and implementation site (WWAP 2018). Although they are designed primarily for wastewater treatment, the co-benefit of producing biomass that can be harvested for further valorization is pivotal for resource-oriented management programs as sustainable bioenergy sources, not compromising the production of other food or energy crops (Gizińska-Górna et al. 2016, Avellán et al. 2017, Masi et al. 2018, Avellán and Gremillion 2019).

Considering CWs as a bioresource for wastewater treatment and biomass valorization for energy production, 2 out of the 17 sustainable development goals (SDGs) are directly addressed (UN 2015): Goal 6: "Ensure availability and sustainable management of water and sanitation for all" and Goal 7: "Ensure access to affordable, reliable, sustainable and modern energy for all". Further related are SDG 11: "Make cities and human settlements inclusive, safe, resilient and sustainable", in the way that CWs contributes to effective water treatment for small communities, and SDG 13: "Take urgent action to combat climate change and its impacts", in the way that CWs provide a range of ecosystem services and can create GHG sinks. CWs as a nature-based solution (NBS) act as tool for climate adaptation and mitigation (Avellán et al. 2017). Implementation of CWs is in alignment with what has been raised in the progress report of SDG that states that there is a need for higher levels of ambition regarding renewable energy use and more efficient use and management of water, in order to answer to its growing demand, to threats to its security, and the severity and increasing occurrence of droughts and floods as a result of climate change (UN 2019). Furthermore, this approach is also in compliance with the issues stated in the resolution of the "International Decade (2018–2028) for Action – Water for Sustainable Development" by the United Nations General Assembly that foresees "a greater focus on the sustainable development and integrated management of water resources for the achievement of social, economic and environmental objectives" and "highlights the importance of promoting efficient water usage at all levels, taking into account the water, food, energy, environment nexus, including the implementation of national development programmes" (UN 2016).

The present chapter provides an overview related to the potential of energetic valorization of the biomass produced in CWs while contributing to the treatment of wastewater efficiently.

7.2 Constructed Wetlands as a Bioresource

Constructed wetlands are human-made systems for water treatment that intend to mimic several physicochemical and biological processes in natural wetlands. Constructed wetlands can be classified based on water flow regime and direction (free water surface or subsurface flow, vertical or horizontal), type of growing vegetation (submerged, emergent, free-floating, and floating/leaved plants), or even a combination of models as hybrid systems. They have been applied for several wastewater treatment purposes, namely domestic, industrial, stormwater, and runoff water (Calheiros et al. 2018, Stefanakis 2019).

Constructed wetlands can be considered integrated resource recovery systems, and versatile in terms of the scale of application, such as from single houses and small communities to mega-cities (Avellán and Gremillion 2019). Especially, for small communities, they constitute an alternative at a reduced cost of implementation and maintenance compared to conventional wastewater treatment systems. Conventional treatment systems are high energy consumers if compared to CWs, which require far less energy to treat wastewater, besides having the potential to be net suppliers of energy (Shao et al. 2013, Avellán et al. 2017). Constructed wetlands are thus considered a cost-effective NBS and multifunctional system that provides several ecosystem services such as provision of raw material, air quality, climate regulation, moderation of extreme events, water purification, biodiversity and habitats, nutrient cycling, GHG sinks, and ornamental, and aesthetic value and integration (Avellán et al. 2017, WWAP 2018, Calheiros et al. 2018, 2020).

Constructed wetlands can be seen as a bioresource when used for: (i) land fertilization by the incorporation in soils of treated sludge, P-enriched substrates, and composted biomass, and (ii) energetic benefits through microbial fuel cells, reuse of treated wastewater for irrigation, and fertilization of energy crops and biomass valorization (Fig. 7.1).

Figure 7.1: Schematic representation of the different uses of CWs resources.

Sludge treatment wetlands, and reed bed-based systems intend to provide a sustainable solution for the sludge produced in the conventional wastewater treatment plants. They operate based on natural dewatering and mineralization processes with the final product meeting the quality requirements for land application or applied as fertilizer in agriculture (Nielsen and Larsen 2016). Other resources can be recovered from CWs for application as fertilizers, including porous materials with high P retention capacity, such as LECA® and Filtralite-P™, commonly used in CW beds to improve P removal. These materials, on saturation, can be removed and directly applied as soil amendments, reducing P chemical fertilization (Jenssen et al. 2005, Ballantine and Tanner 2010). Constructed wetlands can also integrate bioelectrochemical systems/microbial fuel cells, so-called electroactive wetlands, for electricity generation. This technology has shown promising results for the recovery of energy stored in the form of chemicals in the wastewaters, although more research is needed to support the scale-up of the technology (Srivastava et al. 2020). Water reuse is another benefit provided by CWs, being of great importance to protect the primary resource and to promote the circularity of water management. Water treated by CWs can be forwarded for the irrigation of agricultural lands (Avellán and Gremillion 2019), green areas (Milani et al. 2020) and energy crops irrigation and fertilization (Langergraber et al. 2020). For instance, Barbagallo et al. (2014) reported using CWs-treated effluents in the irrigation of energy crops with high biomass yields grown in open fields, such as *Arundo donax*, *Miscanthus* × *giganteus*, and *Vetiver zizanioides*. Similarly, Licata et al. (2019) used treated wastewater by CW for the irrigation of *Cynodon dactylon*, *Paspalum vaginatum,* and

Lycopersicon esculentum. Despite the higher electroconductivity and Na^+ content in treated wastewaters, no symptoms of toxicity or adverse effects were observed on the growth and productivity of *C. dactylon* and *P. vaginatum*. Treated wastewater is a good source of nutrients, which is an advantage over freshwater; however, the contamination by fecal bacteria, if present, is still an issue, limiting the use of this kind of water for the irrigation of vegetable crops consumed raw or cooked, such as tomatoes.

Excellent prospects can be expected concerning what is envisaged under the umbrella of the new circular economy Action Plan (EC 2020), where the European Commission will "develop an Integrated Nutrient Management Plan, with a view to ensuring more sustainable application of nutrients and stimulating the markets for recovered nutrients" and the perspectivation to assess natural means of nutrient removal. Besides that, the new Water Reuse Regulation will "encourage circular approaches to water reuse in agriculture".

7.3 Biomass Production in CW Systems

Biomass valorization recovered from CWs towards energy production is getting increased attention in the last years (Avellán et al. 2017, WWAP 2018, Licata et al. 2019, Langergraber et al. 2020). Macrophytes used in CW systems need to be pruned at least once a year (Kouki et al. 2016), resulting in high amounts of wastes that should to be adequately disposed. However, if it is given the proper attention, these wastes can constitute a valuable resource.

Besides water treatment, CWs can be implemented with the purpose to collect biomass for energy production; this is an additional value when compared to conventional cultivation of energy crops (Langergraber et al. 2020). However, specific key design parameters must be taken into consideration when doing the dimensioning, namely: i) the nutritional content of the wastewater must be adequate for the needs of the target crops, ii) the optimization of CW system flow, iii) the selection of fast-growing plant species with high yields of aboveground material and with high energetic value, and iv) planning the frequency of harvesting in order to enhance biomass production (Langergraber et al. 2020). Their energy yield is measured from the production of dry biomass per unit area (e.g. tonnes per hectare per year) (Avellán et al. 2017).

In CWs, plant biomass is produced as a consequence of nutrient removal from wastewater that mainly occurs by plant uptake. Microorganisms also play a part in transforming nutrients into inorganic compounds available to plants (Ciria et al. 2005). The bioenergy yield associated to plant biomass can be potentiated by taking advantage of the nitrogen available in wastewater, by optimizing the hydrologic flow and by choosing productive plants (Liu et al. 2012).

Avellán et al. (2017) pointed out three technologies appropriate for CW-derived biomass conversion to energy, namely direct combustion, biogas, and bioethanol production. Concerning direct combustion, dried or briquettes biomass, as pellets and woodchips, can be produced from the plants growing in the CWs, e.g. *Phragmites* sp., *Typha* sp., *Phalaris* sp., *Cyperus* sp. and combusted as a cooking or heating fuel source (Langergraber et al. 2020). For instance, by direct combustion of dry biomass harvested from CWs, 10% of the cooking energy needs of a small village can be fulfilled. That has been supported by a study by Ciria et al. (2005) that suggested the use of harvested biomass from CWs instead of the use of conventional fossil fuels in boilers for building heating (in a small community context). Liu et al. (2012) mentioned the possibility of using CWs biomass to produce power, after gasification, for community residences and even to run facilities at CWs sites. The wetland biomass feedstocks can also supply more complex biogas and bioethanol reactors and take energetic supply to another level of sufficiency. These uses can be of more relevance in small-scale communities, such as in villages and in developing countries (Avellán et al. 2017). The biomass produced in CWs, besides the additional ecosystem benefits already mentioned above, can be an alternative source of cellulosic feedstock for second-generation biofuels (Liu et al. 2019). According to the life cycle assessment carried out by Liu and co-workers (2019) in a 1000 m^2 CW at Zhujiajian (China), this system had a positive net energy balance, i.e. CWs can produce more renewable energy than what is necessary for its implementation and operation. Furthermore, CWs can be implemented on marginal lands not competing for fertile soils, which are of utmost importance for food production, nor causing ecosystem disruptions, standing out as a sustainable and competitive solution when compared to conventional biofuel production systems (Liu et al. 2012, 2019).

Another application for the biomass harvested in CWs is its use as a soil amendment after composting. Kouki et al. (2016) valorized the biomass residues of *A. donax* and *Typha latifolia* grown in a CW for rural wastewater treatment by producing compost with a nutritional load enough to be applied in the soil as an organic fertilizer enabling plant growth improvement. Calheiros et al. (2015) implemented a full-scale constructed wetland (area = 40.5 m^2) for wastewater treatment from a tourism unit. The CW operates with horizontal subsurface flow (H-SSF) and is vegetated with a polyculture of ornamental plants (*Canna flaccida, Zantedeschia aethiopica, Canna indica, Agapanthus africanus*, and *Watsonia borbonica*). This polyculture was aesthetically integrated in the landscape, with a pleasant appearance during the seasons due to different flowering times. Further on, the flowers were used for decorations inside the guest house. Plant biomass also had an added value since it was harvested for on-site composting. The harvesting was done by the local farmworkers that had a brief training previously to procced with the best management techniques (Fig. 7.2).

Figure 7.2: Maintenance operations through plant harvesting for further composting being carried out in a constructed wetland in a tourism unit in Portugal - Paço de Calheiros (photo: Cristina Calheiros).

In the second year of the CW operation, only one prune was made from which ca 800 kg of plant biomass was sent to the farm composting unit. In the third year of operation, two prunes occurred, one in wintertime (ca 500 kg of plant biomass retrieved) and the other in summertime (ca 600 kg of plant biomass retrieved). Each harvest comprised four hours of work of three people, including cutting and disposal of biomass at the composting site.

Another system that, besides water treatment, produces a significant amount of biomass and allows for nutrients recycling via the plant biomass produced, which can be used for energy purposes, is the willow system. A further use that can be given to willow is to strengthen banks of riverbeds. The willow system is a type of CW planted with *Salix* sp., with no outflow, where the design is conceived to treat the wastewater through evapotranspiration (Gregersen and Brix 2001, Langergraber et al. 2020). The harvesting of the willows stems is undertaken regularly to remove nutrients and stimulate plant growth (Fig. 7.3).

Figure 7.3: Willow system in Denmark (photo: Cristina Calheiros).

It is essential that these systems must be designed with projections on how the biomass will be processed, what the combustion process will imply, and who will be the end-user of the produced heat. They are appropriate for areas where wastewater discharge standards are strict, and soil infiltration is not allowed or possible (Gregersen and Brix 2001, Langergraber et al. 2020).

7.4 Constructed Wetlands as a Strategy of Biomass Production and Energetic Valorization

The selection of plants for CWs is an issue of extreme importance since they play pivotal roles not only in promoting water treatment and ecosystem dynamics, but also in biomass production (Brisson and Chazarenc 2009, Calheiros et al. 2020). Factors as local climate, plant tolerance to the wastewater composition, and plant growth and biomass yields must be considered for plant selection (Langergraber et al. 2020). According to Vymazal (2013), more than 150 macrophyte species have been considered in CWs, with *Phragmites*, *Typha*, *Scirpus*, *Juncus*, and *Eleocharis* included in the most commonly used. Liu et al. (2012) considered that *A. donax*, *Phragmites australis*, and *Typha angustifolia* are among the plant species with higher energetic value, which was also already reported by Laurent et al. (2015) for *A. donax* and *Typha* sp. Avellán and Gremillon (2019) analyzed data retrieved from 76 publications on biomass density, energy content, and water balance of 4 macrophytes, *Phragmites* spp., *Typha* spp., *A. donax*, and *C. papyrus,* and reported biomass yields between 1500 g/m^2 for *Typha* spp. and up to 6000 g/m^2 for *A. donax*. Moreover, a recent work conducted by Lin et al. (2020) showed that *P. australis* is among the wetland species containing higher levels of cellulose (39%) and holocellulose (55%) followed by *Thalia dealbata* and *Juncus effusus*, being considered good candidates for the production of second-generation biofuels for this reason. Indeed, the results obtained for bioethanol production using the biomass of these three species are comparable to those obtained for other lignocellulosic feedstocks.

Therefore, the use of plant species that combine the ability to remove pollutants efficiently from wastewater and the high biomass production with the potential of energetic valorization, including the production of second-generation biofuels is a keystone strategy wholly aligned with the concept of a circular economy. A non-exhaustive list of plant species often used in CW beds, with good biomass yields and energetic potential, is presented in Table 7.1.

Biomass production in CWs seems highly dependent on several aspects, including nutrient availability, type of flow system, and frequency of harvesting. Besides a higher number of *Phragmites karka* shoots present at the inlet zones of a CW, Angassa et al. (2019) also reported that those plants were thicker, robust, and taller than those found at the outlet, as the higher nutrient load at the entrance of the bed positively influences plant growth. In another study, the addition of 10% of biochar to the substrate (LECA®) in an H-SSF enhanced wastewater treatment efficiency and increased biomass of *T. latifolia* plants. The above- and belowground biomass in beds amended with biochar was 1.9 and 1.5-fold higher than in non-amended ones, which seems to be related to the improved plant nutrition and reduced phytotoxicity provided by the addition of biochar (Kasak et al. 2018). Biomass yields may also greatly vary according to the type of flow implemented in the CW. Barbera et al. (2009) showed that the H-SSF bed favored *P. australis* biomass production (4701 g/m^2) when compared to the vertical subsurface flow system (V-SSF; 3088 g/m^2). An opposite trend was observed for spontaneous vegetation (e.g. *Phalaris* spp., *Chrysanthemum segetum*), which grew much better in the V-SSF bed (1700 g/m^2) than in the H-SSF system (240 g/m^2).

A growing body of evidence shows that cutting aboveground biomass of CW plants is essential to remove pollutants absorbed in plant tissues (Ciria et al. 2005, Gizińska-Górna et al. 2016, Rozema et al. 2016) and some cases may improve wastewater treatment efficiency (Rozema et al. 2016, Yang et al. 2016).

Table 7.1: Biomass production by different plant species colonizing constructed wetlands (CWs) for wastewater treatment

Plant species	Type of wastewater	Wastewater properties	CWs system classification	Productivity (aboveground biomass)	Energetic characteristic of biomass	References
Phragmites karka	Municipal wastewater	COD: 470-480 mg/L TN: 84-88 mg/L TP: 41-44 mg/L	H-SSF pilot scale	18.2-25.6 kg/m^2	Not reported	Angassa et al. (2019)
Phragmites australis Spontaneous vegetation (e.g. *Phalaris, Chrysanthemum segetum, Silybum marianum; Senecio vulgaris*)	Rural wastewater after primary and secondary sedimentation	N-NO_3^-: 2 mg/L	Two-stage pilot scale: H-SSF + H-SSF H-SSF + V-SSF	*P. australis*: 3894-4701 g/m^2 Spontaneous vegetation: 715-1770 g/m^2	Not reported	Barbera et al. (2009)
Typha latifolia	Municipal raw wastewater	COD: 536-744 mg/L NH_4^+-N: 36-43 mg/L NO_3^--N: 16-22 mg/L NO_2^--N: 0.077-0.099 mg/L P: 23-29 mg/L	H-SSF	2.8 kg/m^2	LHV (MJ/kg): 19.63 Ash content: 10.2%	Ciria et al. (2005)
Arundo donax *P. australis*	Untreated recirculating aquaculture system wastewater	BOD: 28.3 mg/L TN: 35.1 mg/L TP: 14.3 mg/L	H-SSF	*A. donax*: 125 t/ha/yr *P. australis*: 77 t/ha/yr	Not reported	Idris et al. (2012)

P. australis (bed 1) *Salix viminalis* (bed 2) *Helianthus tuberosus* (bed 3) *Miscanthus* x *giganteus* (bed 4)	Domestic wastewater	Not reported	Plant beds in series: V-SSF (bed 1)→ H-SSF (bed 2)→ H-SSF (bed 3)→ V-SSF (bed 4)	*P. australis*: 13.6 Mg/ha *S. viminalis*: 8.7 Mg/ha *H. tuberosus*: 5.9 Mg/ha *M.* x *giganteus*: 9.6 Mg/ha	HHV (MJ/kg): *P. australis*: 17.9 *S. viminalis*: 19.2 *H. tuberosus*: 18.9 *M.* x *giganteus*: 18.8 LHV (MJ/kg): *P. australis*: 16.8 *S. viminalis*: 18.0 *H. tuberosus*: 17.7 *M.* x *giganteus*: 17.6 Ash content (%): *P. australis*: 10.4 *S. viminalis*: 2.2 *H. tuberosus*: 2.4 *M.* x *giganteus*: 2.4	Gizińska-Górna et al. (2016)
Cyperus papyrus	Domestic wastewater	COD: 64.5-92.4 mg/L NH_3: 10.7-21.5 mg/L *o*-PO_4^{3-}: 3.03-5.75 mg/L	Free water surface wetland	2341-3115 g/m^2	Not reported	Perbangkhem and Polprasert (2010)
A. donax	Urban wastewater	COD: 63.6 mg/L TN: 18.2 mg/L TP: 3.62 mg/L	H-SSF	3.88 kg/m^2	HCV (MJ/kg): 14.88 Ash content (%): 5.97	Licata et al. (2019)

(*Contd.*)

Table 7.1: (*Contd.*)

Plant species	Type of wastewater	Wastewater properties	CWs system classification	Productivity (aboveground biomass)	Energetic characteristic of biomass	References
P. australis	Municipal wastewater (Exp. 1)	Exp.1 COD: 63 mg/L TN: 27 mg/L TP: 8 mg/L	H-SSF full scale	Exp.1 56.5 ton/ha	Exp. 1 LHV (GJ/ton): 11.59 HHV (GJ/ton): 20.42 Combustion energy (GJ/ha): 1217.26	Politeo et al. (2011)
	Piggery manure wastewater (Exp. 2)	Exp. 2 COD: 309 mg/L TN: 147 mg/L TP: 21 mg/L	H-SSF full scale	Exp. 2 3.6 ton/ha	Exp. 2 LHV (GJ/ton): 10.41 HHV (GJ/ton): 16.69 Combustion energy (GJ/ha): 103.11	

Note: COD: chemical oxygen demand; TN: total nitrogen; TP: total phosphorous; H-SSF: horizontal subsurface flow; V-SSF: vertical subsurface flow; LHV: low heating value; HHV: high heating value, HCV: heating calorific value

Moreover, harvesting and regrowth of plants may also affect biomass yields (Langergraber et al. 2020), improving plant performance in the next growing season (Carballeira et al. 2016). For instance, according to Rozema et al. (2016), biomass production of *Juncus torreyi* and *T. latifolia* was enhanced by successive harvests, while a higher accumulation of Na^+ and Cl^- in plant tissues was observed. Nonetheless, Jinadasa et al. (2008) reported that successive harvests influenced differently biomass production among different plant species. Four consecutive harvests induced a decline in biomass of *T. angustifolia*, whereas *Scirpus grossus* were able to maintain the levels of aerial biomass. The time of the year for biomass harvesting is also essential to keep in consideration. Langergraber et al. (2020) reported that biomass collected in late summer has high concentrations of nutrients, which may cause corrosive problems on plants' combustion. However, if the end purpose is biogas production, a single harvest in late summer or two harvests at early growth stages would bring advantages since biomass will have lower lignin contents, which facilitates digestion and improves methane yield. For direct biomass combustion, a single annual harvest performed in late autumn is recommended, as it implies a reduction in ash and moisture contents. Indeed, the final use of the biomass and intended valorization will be drivers for the plant management options through their cycle in the CW, not compromising their primary role as clean water providers.

In order to address the suitability of the biomass as fuel, it will be adequate to perform studies on biomass composition and thermal behavior. For that purpose, it will also be important to have in consideration the wastewater composition, the plant pollutant uptake, and the intrinsic plant species characteristics (Ciria et al. 2005). Concerning biomass ash, its use may be limited to a certain extent by the presence of contaminants, as heavy metals. Bonanno et al. (2013) stated that rather than hazardous waste, this material could be considered a potential fertilizer.

7.5 Concluding Remarks

The current need for increasing biofuels production has exacerbated the competition for natural resources, such as land and water, with the agricultural sector suffering a negative impact by the land-use changes policies. This scenario is expected to be worsened by climate change; therefore, prompt sustainable alternatives for the production of biomass for energy purposes are needed.

The state of the art on CWs sets the foundation for integrated management of water, nutrient, and energy cycles. Constructed wetlands to treat wastewater can act as integrated systems for water reuse and energy production. Of particular relevance is the integration of CWs for biomass valorization in order to underpin more sustainable strategies for energy production and following a circular economy alignment.

Promoting the use of CWs as a source of materials, i.e. bioresource that can be further applied for energetic purposes is a keystone strategy. Water management strategies coupled to feedstock production for energy use directly contribute to the UN's 2030 SDGs (UN 2015), particularly Goal 6: "Ensure availability and sustainable management of water and sanitation for all" and Goal 7: "Ensure affordable, reliable, sustainable and clean modern energy for all" and should be a driver of new integrated solutions.

Acknowledgments

This research was supported by national funds through FCT - Foundation for Science and Technology within the scope of UIDB/04423/2020, UIDP/04423/2020, and UIDB/50016/2020 projects. This work was also co-funded by the Fundo Europeu Agrícola de Desenvolvimento Rural (FEADER) and the Portuguese government through the PDR2020 under the BIOCHORUME project (PDR2020-101-032094).

References

Angassa, K., S. Leta, W. Mulat, H. Kloos and E. Meers. 2019. Evaluation of pilot-scale constructed wetlands with *Phragmites karka* for phytoremediation of municipal wastewater and biomass production in Ethiopia. Environ. Proc. 6: 65-84. DOI: 10.1007/s40710-019-00358-x.

Avellán, C.T., R. Ardakanian and P. Gremillion. 2017. The role of constructed wetlands for biomass production within the water-soil-waste nexus. Water Sci. Technol. 75(10): 2237-2245. DOI: 10.2166/wst.2017.106

Avellán, C.T. and P. Gremillion. 2019. Constructed wetlands for resource recovery in developing countries. Renew. Sust. Energ. Rev. 99: 42-57.

Ballantine, D.J. and C.C. Tanner. 2010. Substrate and filter materials to enhance phosphorus removal in constructed wetlands treating diffuse farm runoff: A review. New Zeal. J. Agri. Res. 53: 71-95. DOI: 10.1080/00288231003685843.

Barbagallo, S., A.C. Barbera, G.L. Cirelli, M. Milani and A. Toscano. 2014. Reuse of constructed wetland effluents for irrigation of energy crops. Water Sci. Technol. 70(9): 1465-1472.

Barbera, A.C., G.L. Cirelli, V. Cavallaro, I. Di Silvestro, P. Pacifici, V. Castiglionea, A. Toscano and M. Milani. 2009. Growth and biomass production of different plant species in two different constructed wetland systems in Sicily. Desalination 246: 129-136.

Barnabè, D., R. Bucchi, A. Rispoli, C. Chiavetta, P.L. Porta, C.L. Bianchi, C. Pirola, D.C. Boffito and G. Carvoli. 2013. Land use change impacts of biofuels: A methodology to evaluate biofuel sustainability. pp. 1-37. *In*: Z. Fang. [ed.]. Biofuels – Economy, Environment and Sustainability. IntechOpen Limited, London, UK.

Bonanno, G., G.L. Cirelli, A. Toscano, R. Lo Giudice and P. Pavone. 2013. Heavy metal content in ash of energy crops growing in sewage-contaminated natural wetlands: Potential applications in agriculture and forestry. Sci. Total Environ. 452-453: 349-354.

Brisson, J. and F. Chazarenc. 2009. Maximizing pollutant removal in constructed wetlands: Should we pay more attention to macrophyte species selection? Sci. Total Environ. 407(13): 3923-3930. DOI: 10.1016/j.scitotenv.2008.05.047

Calheiros, C.S.C., V.S. Bessa, R.B. Mesquita, H. Brix, A.O.S.S. Rangel and P.M.L. Castro. 2015. Constructed wetland with a polyculture of ornamental plants for wastewater treatment at a rural tourism facility. Ecol. Eng. 79: 1-7. DOI: 10.1016/j.ecoleng.2015.03.001.

Calheiros, C.S.C., C.M.R. Almeida and A.M. Mucha. 2018. Multiservices and functions of constructed wetlands. pp. 269-298. *In*: W. Halicki [ed.]. Wetland Function, Services, Importance and Threats. Nova Science Publishers, Inc. New York. USA.

Calheiros, C.S.C., S.I.A. Pereira and P.M.L. Castro. 2020. Constructed wetlands as nature-based solutions. pp. 97-142. *In*: A.N. Jespersen [ed.]. An Introduction to Constructed Wetlands. Nova Science Publishers, New York, USA.

Carballeira, T., I. Ruiz and M. Soto. 2016. Effect of plants and surface loading rate on the treatment efficiency of shallow subsurface constructed wetlands. Ecol. Eng. 90: 203-214.

Chauhan, S.K. and A. Shukla. 2011. Environmental impacts of production of biodiesel and its use in transportation sector. pp. 1-18. *In*: M.A.S. Bernardes [ed.]. Environmental Impact of Biofuels. IntechOpen Limited, London, UK.

Ciria, M.P., M.L. Solano and P. Soriano. 2005. Role of macrophyte *Typha latifolia* in a constructed wetland for wastewater treatment and assessment of its potential as a biomass fuel. Biosyst. Eng. 92(4): 535-544. DOI:10.1016/j.biosystemseng.2005.08.007.

Dale, B.E. and M. Holtzapple. 2015. The need for biofuels. SBE Supplement: Lignocellulosic Biofuels. American Institute of Chemical Engineers (AIChE). www.aiche.org/cep.

Dragone G., B. Fernandes, A.A. Vicente and J.A. Teixeira. 2010. Third generation biofuels from microalgae. pp. 1355-1366. *In*: A. Méndez-Villas [ed]. Current Research, Technology and Education Topics in Applied Microbiology and Microbial Biotechnology. Formatex Research Center. Spain.

European Commission. 2019. Press release: The European Green Deal sets out how to make Europe the first climate neutral continent by 2050, boosting the economy, improving people's health and quality of life, caring for nature and leaving no one behind. Accessible at: file:///C:/Users/cristina/AppData/Local/Temp/The_European_Green_Deal_sets_out_how_to_make_Europe_the_first_climate_neutral_continent_by_2050__boosting_the_economy__improving_people_s_health_and_quality_of_life__caring_for_nature_and_leaving_no_one_behind.pdf. Accessed in: 12/07/2020.

European Commission. 2020. Communication from the commission to the European parliament, the council, the European economic and social committee and the committee of the regions. A new Circular Economy Action Plan. For a cleaner and more competitive Europe, Brussels, 11.3.2020. COM(2020) 98 final.

Evangelou, M.W.H., E.G. Papazoglou, B.H. Robinson and R. Schulin. 2014. Phytomanagement: Phytoremediation and the production of biomass for economic revenue on contaminated land. pp: 115-132. *In*: A.A. Ansari, S.S. Gill, R. Gill, G.R. Lanza and L. Newman [eds.]. Phytoremediation: Management of Environmental Contaminants. Springer International Publishing, Switzerland.

FAO. 2008. Biofuels and agriculture – A technical overview. The State of Food and Agriculture. Rome, Italy.

Fattahi, M.M., M. Darvish, H.R. Javidkia and M. Adnani. 2011. Assessment and mapping of desertification total risk using FAO-UNEP method (Case study: Qomroud watershed). Iranian Journal of Range and Desert Research, 17(4), 575-588.

Fingerman, K.R., M.S. Torn, M.H. O'Hare and D.M. Kammen. 2010. Accounting for the water impacts of ethanol production. Environ. Res. Lett. 5: 014020.

Gaurav, N., S. Sivasankari, G.S. Kiran, A. Ninawe and J. Selvin. 2016. Utilization of bioresources for sustainable biofuels: A review. Renew. Sust. Energ. Rev. 73: 205-214.

Gizińska-Górna, M., W. Czekała, K. Jóźwiakowskia, A. Lewicki, J. Dach, M. Marzec, A. Pytka, D. Janczak, A. Kowalczyk-Juśkoa and A. Listosz. 2016. The possibility of using plants from hybrid constructed wetland wastewater treatment plant for energy purposes. Ecol. Eng. 95: 534-541.

Gregersen, P. and H. Brix. 2001. Zero-discharge of nutrients and water in a willow dominated constructed wetland. Water Sci. Technol. 44(11-12): 407-412.

Giuntoli, J., 2018. Final recast renewable energy directive for 2021-2030 in the European Union, ICCT: Washington, DC. https://www.theicct.org/publications/final-recast-renewable-energy-directive-2021-2030-european-union.

Idris, S.M., P.L. Jones, S.A. Salzman, G. Croatto and G. Allinson. 2012. Evaluation of the giant reed (*Arundo donax*) in horizontal subsurface flow wetlands for the treatment of recirculating aquaculture system effluent. Environ. Sci. Pollut. Res. 19: 1159-1170. DOI: 10.1007/s11356-011-0642-x.

IEA. 2019. World Energy Outlook 2019, IEA, Paris https://www.iea.org/reports/world-energy-outlook-2019.

Jenssen, P.D., T. Mæhlum, T. Krogstad and L. Vråle. 2005. High performance constructed wetlands for cold climates. J. Environ. Sci. Health 40(6-7): 1343-1353. DOI: 10.1081/ESE-200055846.

Jinadasa, K.B.S.N., N. Tanaka, S. Sasikala, D.R.I.B. Werellagama, M.I.M. Mowjood and W.J. 2008. Impact of harvesting on constructed wetlands performance – A comparison between *Scirpus grossus* and *Typha angustifolia*. J. Environ. Sci. Health, Part A 43: 664-671, DOI: 10.1080/10934520801893808.

Kasak, K., J. Truu, I. Ostonen, J. Sarjas, K. Oopkaup, P. Paiste, M. Kõiv-Vainik, Ü. Mander and M. Truu. 2018. Biochar enhances plant growth and nutrient removal in horizontal subsurface flow constructed wetlands. Sci. Total Environ. 639: 67-74.

Kouki, S., N. Saidi, F. M'hiri, A. Hafiane and A. Hassen. 2016. Co-composting of macrophyte biomass and sludge as an alternative for sustainable management of constructed wetland by-products. Clean – Soil, Air, Water. 44(6): 694-702.

Langergraber, G., G. Dotro, J. Nivala, A. Rizzo and O.R. Stein. 2020. Wetland technology practical information on the design and application of treatment wetlands. Scientific and Technical Report Series No. 27. IWA Publishing. London. UK ISBN: 9781789060171.

Laurent, A., E. Pelzer, C. Loyce and D. Makowski. 2015. Ranking yields of energy crops: A meta-analysis using direct and indirect comparisons. Renew. Sustain. Energy Rev. 46: 41-50.

Licata, M., M.C. Gennaro, T. Tuttolomondo, C. Leto and S. La Bella. 2019. Research focusing on plant performance in constructed wetlands and agronomic application of treated wastewater – A set of experimental studies in Sicily (Italy). PLoS ONE 14(7): e0219445. DOI:10.1371/ journal.pone.0219445

Lin, Y., Y. Zhao, X. Ruan, T.J. Barzee, Z. Zhang, H. Kong and X. Zhang. 2020. The potential of constructed wetland plants for bioethanol production. BioEnergy Res. 13: 43-49.

Liu, D., X. Wu, Gu J. Chang, Y. Min, Y. Ge, Y. Shi, H. Xue, C. Peng and J. Wu. 2012. Constructed wetlands as biofuel production systems. Nature Clim. Change. 2: 190-194. DOI: 10.1038/nclimate1370.

Liu, D., C. Zou and M. Xu. 2019. Environmental, ecological, and economic benefits of biofuel production using a constructed wetland: A case study in China. Int. J. Environ. Res. Public Health. 16: 827. DOI:10.3390/ijerph16050827.

Masi, F., A. Rizzo and M. Regelsberger. 2018. The role of constructed wetlands in a new circular economy, resource oriented, and ecosystem services paradigm. J. Environ. Manage. 216: 275-284.

Milani, M., S. Consoli, A. Marzo, A. Pino, C. Randazzo, S. Barbagallo and G.L. Cirelli. 2020. Treatment of winery wastewater with a multistage constructed wetland system for irrigation reuse. Water. 12: 1260. DOI:10.3390/w12051260.

Nielsen, S. and J.D. Larsen. 2016. Operational strategy, economic and environmental performance of sludge treatment reed bed systems based on 28 years of experience. Water Sci. Technol. 74: 1793-1799.

OECD/FAO. 2019. OECD-FAO Agricultural Outlook 2019-2028, OECD Publishing, Paris/ Food and Agriculture Organization of the United Nations, Rome. DOI: 10.1787/agr_outlook-2019-en.

Pandey, V.C., O. Bajpai and N. Singh. 2016. Energy crops in sustainable phytoremediation. Renew. Sust. Energ. Rev. 54: 58-73.

Perbangkhem, T. and C. Polprasert. 2010. Biomass production of papyrus (*Cyperus papyrus*) in constructed wetland treating low-strength domestic wastewater. Biores. Technol. 101: 833-835. DOI: 10.1016/j.biortech.2009.08.062

Politeo, M., M. Borin, M. Milani, A. Toscano and G. Molari. 2011. Production and energy value of *Phragmites australis* obtained from two constructed wetlands. 19th European Biomass Conference and Exhibition, 6-10 June, Berlin, Germany.

Popp, J., Z. Lakner, M. Harangi-Rákos and M. Fári. 2014. The effect of bioenergy expansion: Food, energy, and environment. Renew. Sust. Energ. Rev. 32: 559-578. DOI: 10.1016/j.rser.2014.01.056

Rodionova, M.V., R.S. Poudyal, I. Tiwari, R.A. Voloshin, S.K. Zharmukhamedov, H.G. Nam, B.K. Zayadan, B.D. Bruce, H.J.M. Hou and S.I. Allakhverdiev. 2017. Biofuel production: Challenges and opportunities. Int. J. Hydrog. Energy. 42: 8450-8461. DOI: 10.1016/j.ijhydene.2016.11.125

Rozema, E.R., R.J. Gordon and Y. Zheng. 2016. Harvesting plants in constructed wetlands to increase biomass production and Na^+ and Cl^- removal from recycled greenhouse nutrient solution. Water Air Soil Pollut. 227: 136. DOI: 10.1007/s11270-016-2831-1.

Shao, L., Z. Wu, L. Zeng, Z.M. Chen, Y. Zhou and G.Q. Chen. 2013. Embodied energy assessment for ecological wastewater treatment by a constructed wetland. Ecol. Model. 252: 63-71. DOI: 10.1016/j.ecolmodel.2012.09.004.

Sönnichsen, N. 2020. Biofuels production in selected European countries 2019. https://www.statista.com/statistics/332510/biofuels-production-in-selected-countries-in-europe/#statisticContainer.

Srivastava, P., R. Abbassi, A.K. Yadav, V. Garaniya and M. Asadnia. 2020. A review on the contribution of electron flow in electroactive wetlands: Electricity generation and enhanced wastewater treatment. Chemosphere 254: 126926. DOI: 10.1016/j.chemosphere.2020.126926.

Stefanakis, A.I. 2019. The role of constructed wetlands as green infrastructure for sustainable urban water management. Sustainability. 11: 6981. DOI: 10.3390/su11246981.

Tripathi, V., S.A. Edrisi and P.C. Abhilash. 2016. Towards the coupling of phytoremediation with bioenergy production. Renew. Sust. Energ. Rev. 57: 1386-1389. DOI: 10.1016/j.rser.2015.12.116

United Nations. 2015. A/RES/70/1-Transforming our world: The 2030 Agenda for Sustainable Development. Resolution adopted by the General Assembly on 25 September 2015. General Assembly. Accessible at: https://sdgs.un.org/2030agenda Accessed in: 19/12/2020.

United Nations. 2016. International Decade for Action, "Water for Sustainable Development", 2018-2028: Revised draft resolution. General Assembly. A/C.2/71/L.12/Rev.1. Accessible at: https://digitallibrary.un.org/record/849767/files/A_C-2_71_L-12_Rev-1-EN.pdf. Accessed in: 12/07/2020.

United Nations. 2019. Report of the Secretary-General. Special edition: Progress towards the Sustainable Development Goals. Economic and Social Council. E/2019/68. Accessible at: https://undocs.org/E/2019/68. Accessed in: 22/06/2020.

Vymazal, J. 2013. Emergent plants used in free water surface constructed wetlands: A review. Ecol. Eng. 61P: 582-592. DOI: 10.1016/j.ecoleng.2013.06.023

WWAP (United Nations World Water Assessment Programme)/UN-Water. 2018. The United Nations World Water Development Report 2018: Nature-Based Solutions for Water. Paris, UNESCO.

Yang, Z., Q. Wang, J. Zhang, H. Xie and S. Feng. 2016. Effect of plant harvesting on the performance of constructed wetlands during summer. Water. 8: 24. DOI:10.3390/w8010024.

CHAPTER
8

Wastewater Remediation and Biomass Production: A Hybrid Technology to Remove Pollutants

Palliyath Sruthi[1], A.K. Sinisha[2] and K.V. Ajayan[3]*

[1] PG Department of Botany, CPA College of Global Studies, Puthanathani P.O. Malappuram, Kerala – 676510, India

[2] Department of Botany, PSMO College, Malappuram, Kerala – 676306, India

[3] Biomass Laboratory, Environmental Science Division, Department of Botany, University of Calicut, Calicut University P.O., Kerala – 673635, India

8.1 Introduction

Clean and fresh water is the unavoidable requirement of life equal to food and shelter, by surface and ground water sources. Due to increasing urbanization and industrialization, water sources are becoming polluted and getting destroyed. The natural water bodies get destroyed by the discharge of organic and inorganic waste, causing detrimental impacts on aquatic ecosystems (Gandhi et al. 2013, Safauldeen et al. 2019). Among the conventional methods, the bioremediation technique is an efficient and low cost method (phytoremediation and phycoremediation techniques) being used to cleanup the deadly polluted environment. Phytoremediation and phycoremediation are the branches of bioremediation that employ the application of plants and algae, respectively, for the remediation of wastewater. Aquatic plants have the capacity to absorb contaminants such as organic and inorganic substances, heavy metals, and pharmaceutical pollutants present in agricultural, domestic and industrial wastewater (Mustafa and Hayder 2021).

The world has become increasingly reliant on renewable energy rather than fossil fuels, with a goal of reducing carbon emissions in society. Among various natural renewable energy sources available, plant and microbe based biomass energy are an attractive option. Biomass resources can be categorized into first, second and third generation crops, where urban waste can be useful for energy production, but it requires a tremendous amount of heat to dry sewage sludge. First-generation feedstocks include sugarcane, barley, potato wastes, sugar beets and corn used

Corresponding author: ajukasa@gmail.com

to produce biofuels that include ethanol and biodiesel using yeast strains such as *Saccharomyces cerevisiae* (Lee and Lavoie 2013), whereas the combined wastewater treatment with these edible crops may increase the partitioning of heavy metals into the food chain. Production of ethanol from sugarcane is in close competition with the sugar market, leading to a reduction in biofuel production.

Aquatic plants like *Potamogeton, Eichhornia, Azolla, Lemna,* and *Spirodela* are good phytoremediators and universal feedstock of biofuels (Miranda et al. 2016); also, they have high efficiency in remediating aquatic contamination. Among them, water hyacinth (*Eichhornia*) is a good bioaccumulator of various metals, acidic and alkaline forms of chemicals even in their high concentrations (Ansari et al. 2020). Industrial and domestic wastewaters are the important nutrient rich water for the cultivation of energy crops, which would reduce the cost extensively and would meet all energy requirements for biofuel production (Miranda et al. 2016). In recent days, terrestrial crops reduced the usage of wastewater and all the attention shifted to aquatic plants and microalgae production for the potent use of wastewater and biomass production. Second generation biomass production for biofuel are lignocellulosic crops. The major part of biomass is the residual nonfood parts of current crops, such as stems, leaves, and husks (Bhatia and Goyal 2014).

The most accepted third-generation biofuels would be formed from the algal biomass because of the incredible growth yield as compared with other conventional crops (Brennan and Owendea 2010). Algae are the most diverse category of organisms, ranging from basic blue-green algae to sophisticated seaweeds (Shahid et al. 2017). Generally, algae are high sources of lipids (70%), carbohydrates (60%) and proteins (65%) (Afzal et al. 2017). Research efforts investigating the effects that different growth conditions have on increasing useful biochemical content have been made. For the higher production of algal biomass, various researchers in the world designed and developed different cultivation systems. Every cultivation system development focuses on at least one product from each alga; it may be in lab scale or as in large scale. From the single cell of multitask genus, *Chlorella* can produce a huge amount of hydrogen in the form of fuel energy; simultaneously, each cell can be consumed as single cell protein. Moreover, the same algae can uptake high amount of CO_2 from atmosphere and accumulate pollutants from different industrial wastewaters such as dairy, paper and pulp, pesticide, textile, fertilizer, petroleum, tannery, electroplating etc. (Shashirekha et al. 2005, Swamy 2011, Rehman 2011, Ajayan et al. 2012, 2015). Recently, algal researchers attempted to integrate the microalgal cultivation system with various wastewater treatment plants and industrial CO_2 emission point sources. These are the novel potential strategies to produce large quantities of biomass to make algal biofuels with less cost and more effective method (Wiley et al. 2011).

Algae proffer vast benefits as in terms of a feedstock for various biofuels and also they have significant rate of areal biomass productivity due to well-organized solar energy conversion and nutrient acquisition strategies. Many species can grow in various industrial wastewater treatment plants, which can help to remediate nutrients and metals. The incorporation of algal biomass production with wastewater treatment would allow to achieve better economic performance as well as improved environmental sustainability. Other than biofuel production from the residual biomass, commodities with high market value products such as nutriceuticals, therapeutics, algae-based fertilizer and animal feeds can be formulated (Mc Ginn et al. 2011).

8.2 Sources of Wastewaters

Various sources of wastewater contaminate the earth's fresh water supplies, with home wastewater, industrial wastewater, and agricultural wastewaters being the most important concerns (Table 8.1). Domestic sewage carries used water from houses and apartments; it is also called sanitary sewage. Industrial sewage is used water from manufacturing or chemical processes.

Table 8.1: Characteristics of major sources of wastewater

Sources from various industries	Characteristics
Domestic sewage	TS, BOD, total nitrogen, total phosphorus, alkalinity, oils and grease
Dairy and ice cream	TDS, TSS, hardness, COD, BOD, nitrogen, sodium, potassium, chloride
Textiles and leather	BOD, solids, sulfates and chromium
Pulp and paper	BOD, COD, solids, chlorinated organic compounds
Petrochemicals and refineries	BOD, COD, mineral oils, phenols, and chromium
Chemicals	COD, organic chemicals, heavy metals, SS, and cyanide
Non-ferrous metals	Fluorine and SS
Microelectronics	COD and organic chemicals
Mining	SS, metals, acids and salts
Fertilizer	BOD, urea, residual chlorine, PO_4, NO_3, P, K, Fe, and Zn
Iron and steel	BOD, COD, oil, metals, acids, phenols, and cyanide

Notes: TS – Total solids; TSS – Total suspended solids; TDS – Total dissolved solids; BOD – Biochemical oxygen demand; COD – Chemical oxygen demand; P – Phosphorus; K – Potassium; Zn – Zinc, Fe-Iron.

The industrial wastewater may itself differ according to the contaminants and it makes the combination of different pollutants according to the sector. The highly pollutant compounds discharged from the metal working industries are chromium, nickel, zinc, cadmium, lead, iron and titanium compounds; among them, the electroplating industry is an important pollution distributor (Basiglini et al. 2018). High lethal chlorine based substances are released from the pulp and paper industry, which contributes to the increase of high organic wastes. A high amount of phenoic compounds and mineral oils are released from the petrochemical industry and leads to the source of wastewater on Earth (Gandhi et al. 2013).

Inorganic industrial wastewater is produced largely in the coal and steel industry, nonmetallic minerals industry, commercial enterprises and iron and electroplating industries. The wastewater runoffs from these industries have a large quantity of suspended pollutant. Organic industrial wastewaters from the chemical industries mainly use organic substances for chemical reactions for the production of compounds. The majority of organic industrial wastewaters are discharged from the factories developed for the pharmaceuticals, cosmetics, soaps, synthetic detergents, glue and adhesives, organic dye-stuffs, pesticides and herbicides. Moreover, the industries such as tanneries , textile, paper industry and oil refinaries contribute enormous level of pollutants to the ecosystem (Srivastava and Majumder 2008).

8.3 Phytoremediation

8.3.1 Role of aquatic plants in wastewater treatment

Phytoremediation technique is one of the branches of bioremediation in which plants are used for remediation of wastewater. Aquatic plants can be used as a cost-effective green emerging technology with long-lasting applicability and resourceful clean up technique which can be used for phytoremediation of a large contaminated area since these plants absorb heavy metals and other contaminants. They possess extraordinary capacity to remove various pollutants, organic and inorganic. Inorganic pollutants include heavy metals, macronutrients and radio nuclides. Organic pollutants include pesticides, herbicides etc. The mechanisms of uptake of organic and inorganic contaminants are also different among plants. Through active and passive uptake, inorganic compounds (ionic or complexed form) are absorbed by the plant whereas uptake of organic compounds is by hydrophobic interactions and polarity (Dhir 2013). Heavy-metal (HM) pollution is considered as a major source of environmental contamination. Heavy metal raises several problems to both plants and animals on soil and water. Toxic heavy metal in ground water poses a serious problem to human health and also to the aquatic ecosystem (Ali et al. 2020).

There are different types of aquatic plants such as free-floating, which occupy water surface (*Lemna*, *Hydrocharis* and *Nymphaea*), submerged macrophytes (*Ceratophyllum*), which grow primarily below the water surface and may be anchored to the substrate, and emergent macrophytes (*Typha* and *Phragmites*), which can be seen on the margins of water bodies and have roots attached to substratum along with considerable shoot growth above the water level. The uptake and removal of contaminant also varies for above three categories (Nichols et al. 2016). Aquatic plants *Micranthemum umbrosumhas* can be used for the phytofiltration of arsenic and cadmium. Also, *Oenothera picensis* plant has been studied for phytoextraction of copper (Gonzalez et al. 2014, Islam et al. 2015). According to Singh and coworker, in addition to water hyacinth, *Lemna minor* (Duckweed) and *Hydrilla verticillata* (Hydrilla) can be used to remove lead in water bodies (Singh et al. 2012). Boron, chromium, and manganese can be phytoremediated by *Lemna gibba, Ipomonea aquatica* and *Azolla pinnata,* respectively (Chen et al. 2010, Jasrotia et al. 2017).

8.3.2 Aquatic plants as the source of biomass production and energy

Excess nutrients in wastewater can be filtered out by aquatic plants, resulting in significant amounts of energy biomass (Fig. 8.1). Species like *Lemna*,

Hydrocharis, *Nymphaea*, *Callitriche stagnalis* Scop., *Potamogeton natans* L. and *Potamogeton pectinatus* L. show high accumulation levels of uranium and/or high biomass production (Pratas et al. 2014). Among them, lemnaceae or duckweed took more than twenty years to pull through the nutrients and conversion of the generated biomass into bofuels from wastewaters (Mohedano et al. 2012). 39.2–44 t dw/ha-year is the average yield of duckweed and is much more elevated than many other main bioenergy grasses: switchgrass (5.2–26 t/ha-year), poplar (9–15 t/ha-year) and miscanthus (5.0–44 t/ha-year) (Miranda et al. 2016). Azolla is another fast growing aquatic plant able to double its biomass every five to six days. The biofuel composition (n C10–C21 alkanes) of Azolla comprises a uniqueness with that of lignocellulosic, starch- and oil-producing terrestrial bioenergy crops. The neutral lipid (8 t/ha-year) content of Azolla biomass is higher than soybean, sunflower, rapeseed and its oil palm (Brouwer et al. 2016). Another important perennial grass, vetiver, commonly known as Kasa, is a hydrophyte, often dominant in fresh-water swamps, flood plains and on stream banks; they also grow in wet and dry condition (Seaforth and Tikasingh 2002). Major quantity of vetiver grass leaves is burnt by rural people for cooking and oil extraction industry.

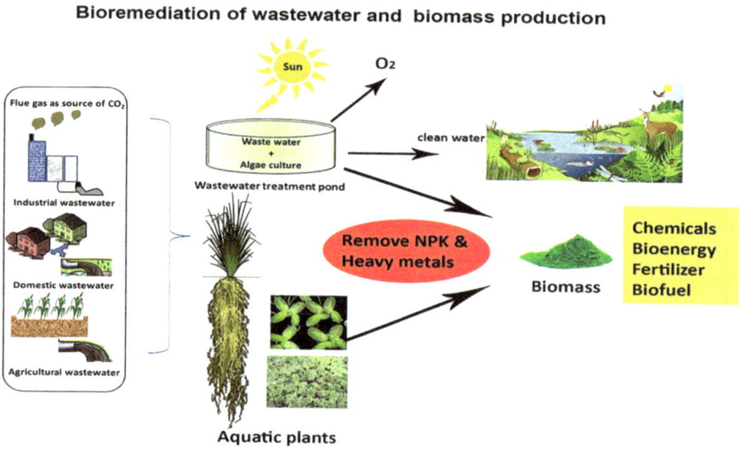

Figure 8.1: Process of phyto and phycoremediation and biomass generation.

8.4 Phycoremediation

8.4.1 Macro and micro algae – A general perspective

Algae are photosynthetic organisms that grow in salt or fresh water. Based on growth size, algae are classified into two different categories, namely multicellular macroalgae or seaweed and unicellular microalgae (Brodie and Lewis 2007). Microalgae are small aquatic photosynthetic plants that include eukaryotes such as green, red, brown and golden algae, diatoms, dinoflagellates, and also prokaryotic cyanobacteria (blue-green algae) (Möller and Clayton 2007). Due to their extensive application potential in the renewable energy, biopharmaceutical, and nutraceutical industries, microalgae have considerable interest worldwide and they are the renewable, sustainable, and economical sources of biofuels, bioactive medicinal products, and food ingredients (Khan et al. 2018). The biological CO_2 fixation through photosynthesis with microalgae is more efficient than with land plants (Williams and Laurens 2010).

Algae can be used as a raw material for the production of biofertilizers, biopesticides, feeds, and feed additives due to their high growth rate, a high ability to bind carbon dioxide and the potential to accumulate biogenic elements and light metals (Piwowar and Harasym 2020). The major algae which are cultivated in large scale for the above purposes include *Laminaria japonica* and *Undaria pinnatifida* of Phaeophyta, *Eucheuma, Gracilaria, Porphyra* and *Kappaphycus* of Rhodophyta, and *Enteromorpha* and *Monostroma* of Chlorophyta (Luning and Pang 2003).

8.4.2 Phycoremediation of wastewaters

Remediation by algae, known as phycoremediation, is considered as a viable option for nutrients and heavy metal remediation. Though many conventional physicochemical methods are currently being practiced, biotechnological methods are becoming attractive alternatives, as they are economical and eco-friendly (Shashirekha et al. 2008). Finding inexpensive or underutilized sources of nutrients will be an important factor in algal cultivation. Currently, algae are used in some wastewater treatment facilities because of their ability to provide oxygen for the bacterial breakdown of organic materials and to sequester nitrogen and phosphorous into biomass for water cleanup. Because of their growth potentiality, they are extensively used in wastewater treatment to remove heavy metals such as Fe, Mn, Zn, Cu, Mo, Ni, Co, Pb, Cd, A1, Cr, Hg, Ag, As and Sn from wastewaters generated by different sources (Hoffman 1998, Ajayan et al. 2012, 2015).

Microalgae-based phycoremediation as well as biorefinery based approaches of wastewater were studied and reported by several authors and from this, microalgae have been considered for the study of industrial and domestic wastewater treatment (Calicioglu and Demirer 2019, Ferreira et al. 2017), pharmaceutical waste streams (Xie et al. 2019), palm oil mill effluents (Hariz and Takrif 2017), textile wastewater (Wu et al. 2017a), slaughterhouse industry (Aziz et al. 2019), heavy metal-containing wastewater (Khan et al. 2017), starch-containing textile wastewater (Lin et al. 2017), agro-industrial wastewater (Jayakumar et al. 2017) and tannery wastewater (Ajayan et al. 2015, 2018). Wastewater treatment using algae has many advantages. It offers the feasibility to recycle these nutrients into algae biomass as a fertilizer and can thus offset treatment cost. Oxygen rich effluent is released into water bodies after wastewater treatment using algae (Becker 2004). The addition of carbon is not required to remove nitrogen and phosphorus from wastewater. *Chlorella* (Gonzalez et al. 1997), *Scenedesmus* (Martinez et al. 2000), and *Spirulina* (Olguin et al. 2003) are the most widely used algae for nutrient removal.

8.4.3 Phycoremediation coupled energy production

Among the researchers, there is keen interest in algae due to significant production of biomass as well as the good source of feedstock for the manufacture of renewable liquid fuels (Fig. 8.1). Also, perhaps the most comprehensive and detailed program based on microalgae biofuels research ever conducted was the Aquatic Species Program (ASP) sponsored by the U.S. Department of Energy from 1978–1996 (Sheehan et al. 1998). The importance of the research tests done under this initiative is that the United States currently imports two-thirds of its petroleum from just a few countries across the world. Energy demand is rising in several of the world's fast developing countries, such as China and India. Additionally, the unremitting burning of fossil fuels has raised serious environmental concerns, which results in the increased release of greenhouse gases and thereby leads to global warming (Shahid et al. 2019).

The algal cell can be considered an apparatus for solar energy capture, conversion, and storage of valuable molecules. The photosynthesis in algal cell performed by the solar energy in the form of packet of photons is biologically transduced to ATP energy and reductant NADPH. Both are mandatory to convert CO_2 to reduced 3-carbon sugar compounds by Calvin cycle (Reinfelder et al. 2000). Various algal cell strains have been exposed in the lab for the generation of biomass, with the lipid biomass primarily triacylglycerides (TAGs), also known as triglycerides, being the anticipated starting material for biofuels (Hu et al. 2008).

Biofuels are one of the potential options to reduce the world's dependence on fossil fuels. The availability of land is the recent concern to increase the biofuels' production rate and it is found that the greenhouse gases will be benefited from the production of biofuels if land with existing high carbon intensity is cleared for the production of biofuel feedstock (Lam et al. 2019). Biofuels that could be produced without large increases in arable land or reductions in tropical rainforests could be very attractive in the future. Starch, glucose, cellulose/hemicelluloses, and a range of polysaccharides are the main algal carbohydrates and out of these, the conventionally used carbohydrate for biofuel production, particularly for bioethanol and hydrogen, is algal starch/glucose (Chochois et al. 2009, John et al. 2011).

Algal biomass mediated biodiesel production is much significant and considered as the best substitute than any other biofuel feedstock. It is mainly because of the environmental friendly manner of the microalgae; it will never alter the food chain and leads to lesser pressure on the arable lands as well as the environment (Srivastava et al. 2020, Pavithra et al. 2020). Algae, particularly microalgae, are considered as an admirable source for biodiesel production due to the enhanced growth rates as compared to the terrestrial crops. An approximate 20000 to 80000 L/acre/year oil is consumed from the microalgae which are about 7–31 times higher than the most widely used source (Demirbas 2009, Demirbas and Demirbas 2011). Moreover, by using different chemical and biological methods, the algal biomass can be converted into bioethanol and biohydrogen (Demirbas 2010). Macroalgae are the first to be studied for the various extraction methods and biomass production with the good usability like the ease of handling and visible features. But subsequently, the research trend shifted towards the use of microalgae because of the high potential for oil production (Scragg et al. 2003, Mondal et al. 2017).

Recently, the coupling of various wastewaters (like domestic, industrial, municipal and agricultural) with microalgae cultivation provides an effective means of utilizing nitrogen and phosphorus with lipid accumulation for biodiesel production

(U.S. DOE 2010). Among macroalgae, the *Laminaria* sp and *Ulva* sp are the most important prospects from an energy perspective. The majority of Asian seaweed resources are cultivated.

8.5 Conclusion and Future Prospects

Wastewater is a major global problem, particularly in heavily populated areas, and must be recycled or treated before disposal. On the other hand, wastewater is a rich source of nutrients and phycoremediation of this wastewater will enhance the production of biomass and other value added products (lipids, pigments, proteins, carbohydrates, etc.). Remediation is a promising technique to remove or recover excess nutrients from wastewater and simultaneously utilize the biomass for various industrial uses. The application of energy plants, aquatic plants and microalgae in remediation of wastewater are beneficial because they have tremendous capacity to absorb and degrade pollutants and convert into energy. The effective cost and benefits of biomass production may be regulated with optimum growing technologies of plants with high oil content, which leads to a lot of potential in bioenergy production.

Algal biomass provides evidence to be the most capable and potential source of bioenergy to cope with high energy demands. It can create improved quantities of bioenergy products, like biogases, biodiesel, biofuels etc. apart from the conventional prospects. With the latest and more advance technologies, it can successfully replace the conventional feedstock. The biofuels from the algal biomass is more environmentally friendly and it can also contribute to reducing global warming. Aquatic plants include bioenergy compounds that are similar to those found in lignocellulosic and starch-producing energy crops. Aquatic species' ability to grow on wastewaters and their high growth and productivity rates make them ideal feedstock for low-cost, low-energy-demanding, near-zero-maintenance biofuel production systems.

Acknowledgments

KVA greatly acknowledges University Grants Commission - Dr. D.S. Kothari Postdoctoral Fellowship (DSKPDF), Grant No. F. 4-2/2006 (BSR)/BL/18-19/0559, (2019-2022).

References

Afzal, I., A. Shahid, M. Ibrahim, T. Liu, M. Nawaz and M.A. Mehmood. 2017. Microalgae: A promising feedstock for energy and high-value products. pp. 55-75. *In*: Zia, K.M., M. Zuber and M. Ali [eds.]. Algae Based Polymers, Blends, and Composites. Elsevier. Amsterdam, Netherlands.

Ajayan, K.V. and M. Selvaraju. 2012. Heavy metal induced antioxidant defense system of green microalgae and its effective role in phycoremediation of tannery effluent. Pak. J. Biol. Sci. 15(22): 1056.

Ajayan, K.V., M. Selvaraju and K. Thirugnanamoorthy. 2012. Enrichment of chlorophyll and phycobiliproteins in *Spirulina platensis* by the use of reflector light and nitrogen sources: An in-vitro study. Biomass Bioenerg. 47: 436–441.

Ajayan, K.V., M. Selvaraju, P. Unnikannan and P. Sruthi. 2015. Phycoremediation of tannery wastewater using microalgae *Scenedesmus* species. Int. J. Phytoremediation 17: 907–916.

Ajayan, K.V., C.C. Harilal and M. Selvaraju. 2018. Phycoremediation resultant lipid production and antioxidant changes in green microalgae *Chlorella* sp. Int. J. Phytoremediation 20(11): 1144-1151.

Ali, S., Z. Abbas, M. Rizwan, I.E. Zaheer, İ. Yavaş, A. Ünay, M.M. Abdel-Daim, M. Bin-Jumah, M. Hasanuzzaman and D. Kalderis. 2020. Application of floating aquatic plants in phytoremediation of heavy metals polluted water: A review. Sustainability, 12: 1927.

Ansari, A.A., M. Naeem, S.S. Gill and F.M. AlZuaibr. 2020. Phytoremediation of contaminated waters: An eco-friendly technology based on aquatic macrophytes application. Egypt. J. Aquat. Res. 46: 371-376.

Aziz, A., F. Basheer, A. Sengar, S.U. Khan and I.H. Farooqi. 2019. Biological wastewater treatment (anaerobic-aerobic) technologies for safe discharge of treated slaughterhouse and meat processing wastewater. Sci. Total Environ. 686: 681-708.

Basiglini, E., M. Pintore and C. Forni. 2018. Effects of treated industrial wastewaters and temperatures on growth and enzymatic activities of duckweed (*Lemna minor* L.). Ecotoxicol. Environ. Saf. 153: 54-59.

Becker, W. 2004. 21 Microalgae for aquaculture. Handbook of Microalgal Culture: Biotechnology and Applied Phycology, pp. 380. John Wiley & Sons.

Bhatia, M. and D. Goyal. 2014. Analyzing remediation potential of wastewater through wetland plants: A review. Environ. Prog. Sustain Energy. 33(1): 9-27.

Brennan, L. and P. Owende. 2010. Biofuels from microalgae – A review of technologies for production, processing, and extractions of biofuels and co-products. Renew. Sust. Energy. Rev. 14(2): 557-577.

Brodie, J. and J. Lewis. 2007. Introduction in unravelling the algae: The past, present, and future of algal systematics. pp. 1-6. *In*: Brodie, J. and J. Lewis [eds.]. Unravelling the Algae: The Past, Present, and Future of Algal Systematics, 1st ed. CRC Press: Boca Raton, FL, USA.

Brouwer, P., A. van der Werf, H. Schluepmann, G.J. Reichart and K.G.J. Nierop. 2016. Lipid yield and composition of *Azolla filiculoides* and the implications for biodiesel production. Bioenerg Res. 9: 369-377.

Calicioglu, O. and G.N. Demirer. 2019. Carbon-to-nitrogen and substrate-to-inoculum ratio adjustments can improve co-digestion performance of microalgal biomass obtained from domestic wastewater treatment. Environ. Technol. 40: 614-624.

Chen, J.C., K.S. Wang, H. Chen, C.Y. Lu, L.C. Huang, H.C. Li, T.H. Peng and S.H. Chang. 2010. Phytoremediation of Cr(III) by *Ipomonea aquatica* (water spinach) from water in the presence of EDTA and chloride: Effects of Cr speciation. Bioresour. Technol. 101: 3033-3039.

Chochois, V., D. Dauvillee, A. Beyly, D. Tolleter, S. Cuine, H. Timpano, S. Ball, L. Cournac and G. Peltier. 2009. Hydrogen production in *Chlamydomonas*: Photosystem II-dependent and independent pathways differ in their requirement for starch metabolism. Plant Physiol. 151: 631-640.

Demirbas, A. 2010. Use of algae as biofuel sources. Energy Convers. Manag. 51(12): 2738-2749.

Demirbas, A. and M.F. Demirbas. 2011. Importance of algae oil as a source of biodiesel. Energy Convers. Manag. 52(1): 163-170.

Demirbas, M.F. 2009. Biorefineries for biofuel upgrading: A critical review. Appl. Energy 86: 51-61.

Dhir, B. 2013. Aquatic plant species and removal of contaminants. pp. 21-50. *In*: Dhir, B. [eds.]. Phytoremediation: Role of Aquatic Plants in Environmental Clean-Up. Springer, India.

Doe, U.S. 2010. National algal biofuels technology roadmap. US Department of Energy, Office of Energy Efficiency and Renewable Energy, Biomass Program.

Ferreira, A., B. Ribeiro, P.A. Marques, A.F. Ferreira, A.P. Dias, H.M. Pinheiro, A. Reisand, and L. Gouveia. 2017. *Scenedesmus obliquus* mediated brewery wastewater remediation and CO_2 biofixation for green energy purposes. J. Clean. Prod. 165: 1316-1327.

Gandhi, N., D. Sirisha and K.C. Sekhar. 2013. Adsorption studies of chromium by using low cost adsorbents. Our Nature 11(1): 11-16.

González, I., A. Neaman, A. Cortés and P. Rubio. 2014. Effect of compost and biodegradable chelate addition on phytoextraction of copper by *Oenothera picensis* grown in Cu-contaminated acid soils. Chemosphere 95: 111-115.

Gonzalez, L.E., R.O. Cañizares and S. Baena. 1997. Efficiency of ammonia and phosphorus removal from a Colombian agroindustrial wastewater by the microalgae *Chlorella vulgaris* and *Scenedesmus dimorphus*. Bioresour. Tech. 60: 259-262.

Hariz, H.B. and M.S. Takriff. 2017. Palm oil mill effluent treatment and CO_2 sequestration by using microalgae—sustainable strategies for environmental protection. Environ. Sci. Pollut. Res. 24: 20209-20240.

Hoffman, J.P. 1998. Wastewater treatment with suspended and nonsuspended algae. J. Phycol. 34: 757-763.

Hu, C. and M.X. He. 2008. Origin and offshore extent of floating algae in Olympic sailing area. Eos Trans. A.G.U. 89(33): 302-303.

Islam, M.S., T. Saito and M. Kurasaki. 2015. Phytofiltration of arsenic and cadmium by using an aquatic plant, *Micranthemum umbrosum*: Phytotoxicity, uptake kinetics, and mechanism. Ecotoxicol. Environ. Saf. 112: 193-200.

Jasrotia, S., A. Kansal and A. Mehra. 2017. Performance of aquatic plant species for phytoremediation of arsenic-contaminated water. Appl. Water Sci. 7: 889-896.

Jayakumar, S., M.M. Yusoff, M.H.A. Rahim, G.P. Maniam and N. Govindan. 2017. The prospect of microalgal biodiesel using agro-industrial and industrial wastes in Malaysia. Renewable Sustainable Energy Rev. 72: 33-47.

John, R.P., G.S. Anisha, K.M. Nampoothiri and A. Pandey. 2011. Micro and macroalgal biomass: A renewable source for bioethanol. Bioresour. Technol. 102(1): 186-193.

Khan, M.I., J.H. Shin and J.D. Kim. 2018. The promising future of microalgae: Current status, challenges, and optimization of a sustainable and renewable industry for biofuels, feed, and other products. Microb. Cell. Fact 17: 1-21.

Khan, Z.I., S. Iqbal, F. Batool, K. Ahmad, M.S. Elshikh, A. Al Sahli, M. El-Zaidy, H. Bashir, I.R. Noorka, M. Sher and A. Muneeb. 2017. Evaluation of heavy metals uptake by wheat growing in sewage irrigated soil: Relationship with heavy metal in soil and wheat grains. Fresenius Environ. Bull. 26: 7838-7848.

Lam, M.K., C.G. Khoo and K.T. Lee. 2019. Scale-up and commercialization of algal cultivation and biofuels production. Biofuels from Algae, pp. 475-506. Elsevier.

Lee, R.A. and J.M. Lavoie. 2013. From first- to third-generation biofuels: Challenges of producing a commodity from a biomass of increasing complexity. Anim. Front. 3: 6–11. doi: 10.2527/af.2013-0010

Lin, C.Y., M.L.T. Nguyen and C.H. Lay. 2017. Starch-containing textile wastewater treatment for biogas and microalgae biomass production. J. Clean. Prod. 168: 331-337.

Lüning, K. and S. Pang. 2003. Mass cultivation of seaweeds: Current aspects and approaches. J. Appl. Phycol. 15: 115-119.

Martınez, M.E., S. Sánchez, J.M. Jimenez, F. El Yousfi and L. Munoz. 2000. Nitrogen and phosphorus removal from urban wastewater by the microalga *Scenedesmus obliquus*. Bioresour. Technol. 73: 263-272.

McGinn, P.J., K.E. Dickinson, S. Bhatti, J.C. Frigon, S.R. Guiot and S.J. O'Leary. 2011. Integration of microalgae cultivation with industrial waste remediation for biofuel and bioenergy production: Opportunities and limitations. Photosynth. Res. 109(1): 231-247.

Miranda, A.F., B. Biswas, N. Ramkumar, R. Singh, J. Kumar, A. James, F. Roddick, B. Lal, S. Subudhi, T. Bhaskar and A. Mouradov. 2016. Aquatic plant Azolla as the universal feedstock for biofuel production. Biotechnol. Biofuels. 9: 1-17.

Mohedano, R.A., R.H. Costa, F.A. Tavares and P. Belli Filho. 2012. High nutrient removal rate from swine wastes and protein biomass production by full-scale duckweed ponds. Bioresour. Technol. 112: 98-104.

Möller, R. and D. Clayton. 2007. Micro- and macro-algae: Utility for industrial applications. *In*: A.S Carlsson, J.B. van Beilen, R. Möller and D. Clayton [eds.]. Outputs from the EPOBIO Project. CPL Press. Science Publishers. UK.

Mondal, M., S. Goswami, A. Ghosh, G. Oinam, O.N. Tiwari, P. Das and G.N. Halder. 2017. Production of biodiesel from microalgae through biological carbon capture: A review. 3 Biotech. 7(2): 1-21.

Mustafa, H.M. and G. Hayder. 2021. Cultivation of *S. molesta* plants for phytoremediation of secondary treated domestic wastewater. Ain Shams Eng. J. doi.org/10.1016/j.asej.2020.11.028

Nichols, P., T. Lucke, D. Drapper and C. Walker. 2016. Performance evaluation of a floating treatment wetland in an urban catchment. Water 8(6): 244.

Olguín, E.J., S. Galicia, G. Mercado and T. Pérez. 2003. Annual productivity of Spirulina (Arthrospira) and nutrient removal in a pig wastewater recycling process under tropical conditions. J. Appl. Phycol. 15: 249-257.

Pavithra, K.G., P.S. Kumar, V. Jaikumar, K.H. Vardhan and P. Sundar Rajan. 2020. Microalgae for biofuel production and removal of heavy metals: A review. Environ. Chem. Lett. 18: 1905-1923.

Piwowar, A. and J. Harasym. 2020. The importance and prospects of the use of algae in agribusiness. Sustainability, 12: 5669.

Pratas, J., C. Paulo, P.J. Favas and P. Venkatachalam. 2014. Potential of aquatic plants for phytofiltration of uranium-contaminated waters in laboratory conditions. Ecol. Eng. 69: 170-176.

Rehman, A. 2011. Heavy metals uptake by *Euglena proxima* isolated from tannery effluents and its potential use in wastewater treatment. Russ. J. Ecol. 42(1): 44-49.

Reinfelder, J.R., A.M. Kraepiel and F.M. Morel. 2000. Unicellular C4 photosynthesis in a marine diatom. Nature 407(6807): 996-999.

Safauldeen, S.H., H. Abu Hasan and S.R.S. Abdullah. 2019. Phytoremediation efficiency of water hyacinth for batik textile effluent treatment. Ecol. Eng. 20(9).

Scragg, A.H., J. Morrison and S.W. Shales. 2003. The use of a fuel containing *Chlorella vulgaris* in a diesel engine. Enzyme Microb. Technol. 33(7): 884-889.

Seaforth, C., T. Tikasingh, M. Gupta and Gilbertha St. Rose. 2002. A study for the development of a handbook of selected Caribbean herbs for industry. CTA, Wageningen, The Netherlands.

Shahid, A., M. Ishfaq, M.S. Ahmad, S. Malik, M. Farooq, Z. Hui and M.A. Mehmood. 2019. Bioenergy potential of the residual microalgal biomass produced in city wastewater assessed through pyrolysis, kinetics and thermodynamics study to design algal biorefinery. Bioresour. Technol. 289: 121701.

Shahid, M., S. Shamshad, M. Rafiq, S. Khalid, I. Bibi, N.K. Niazi and M.I. Rashid. 2017. Chromium speciation, bioavailability, uptake, toxicity and detoxification in soil-plant system: A review. Chemosphere 178: 513-533.

Shashirekha, V., M. Pandi and S. Mahadeswara. 2005. Bioremediation of tannery effluent sand chromium containing wastes using cyanobacterial species. J. Amer. Leather Chem. Asso. 11: 419-426.

Shashirekha, V., M.R. Sridharan and M. Swamy. 2008. Biosorption of trivalent chromium by free and immobilized blue green algae: Kinetics and equilibrium studies. J. Environ. Sci. Health A 43(4): 390-401.

Sheehan, J., T. Dunahay, J. Benemann and P. Roessler. 1998. Look back at the US department of energy's aquatic species program: Biodiesel from algae; close-out report (No. NREL/TP-580-24190). National Renewable Energy Lab., Golden, CO (US).

Singh, D., A. Tiwari and R.Gupta. 2012. Phytoremediation of lead from wastewater using aquatic plants. J. Agric. Technol. 8: 1-11.

Srivastava, N.K. and C.B. Majumder. 2008. Novel biofiltration methods for the treatment of heavy metals from industrial wastewater. J. Hazard. Mater. 151(1): 1-8.

Srivastava, R.K., N.P. Shetti, K.R. Reddy and T.M. Aminabhavi. 2020. Biofuels, biodiesel and biohydrogen production using bioprocesses. A review. Environ. Chem. Lett. 18(4): 1049-1072.

Swamy, M.A. 2011. Marine algal sources for treating bacterial diseases. Adv. Food Nutr. Res. 64: 71-84.

Wiley, P.E., J.E. Campbell and B. McKuin. 2011. Production of biodiesel and biogas from algae: A review of process train options. Water Environ. Res. 83(4): 326-338.

Williams, P.J.L.B. and L.M. Laurens. 2010. Microalgae as biodiesel & biomass feedstocks: Review & analysis of the biochemistry, energetics & economics. Energy Environ. Sci. 3: 554-590.

Wu, J.Y., C.H. Lay, C.C. Chen and S.Y. Wu. 2017. Lipid accumulating microalgae cultivation in textile wastewater: Environmental parameters optimization. J. Taiwan. Inst. Chem. Eng. 79: 1-6.

Xie, J., Y. Chen, X. Duan, L. Feng, Y. Yan, F. Wang, X. Zhang, Z. Zhang and Q. Zhou. 2019. Activated carbon promotes short-chain fatty acids production from algae during anaerobic fermentation. Sci. Total Environ. 658: 1131-1138.

CHAPTER
9

Waste to Bioenergy: A Sustainable Approach

Monika Yadav[1], Gurudatta Singh[2], Jayant Karwadiya[2], Akshaya Prakash Chengatt[3], Delse Parekkattil Sebastian[3] and R.N. Jadeja[1]*

[1] Department of Environmental Studies, Faculty of Science, The Maharaja Sayajirao University of Baroda, Vadodara - 390002, India

[2] Institute of Environment and Sustainable Development, Banaras Hindu University, Varanasi - 221005, India

[3] Climate Change Division, Centre for Post Graduate Studies and Research, Department of Botany, St. Joseph's College (Autonomous), Devagiri, Calicut, Kerala - 673008, India

9.1 Introduction

Increasing population growth requires more energy that automatically results in enhanced utilization of non-renewable energy resources; thereby, search of novel sustainable energy resource is highly recommended (Yi et al. 2018, Khalil et al. 2019). High consumption of fossil related fuels enhances global environmental problems such as global warming and air pollution that adversely affect the environment (Abdeshahian et al. 2016). After energy crisis of the 1970s, many countries showed their interest in developing biomass as a fuel source. Up until recently, biomass energy interest has been lessened due to the technological breakthrough that makes fossil energy relatively inexpensive. However, enhanced greenhouse gas emissions, severe air pollution, unstable fossil-based energy prices as well as intense growth of global transportation of fuel demand have raised extensive research efforts in development of bioenergy. Bioenergy is a form of energy derived from any fuel that has been originated from biomass. Since biomass is a renewable resource, it has been considered as an alternative feedstock for production of sustainable energy in future. Historically, biomass has been traditionally used as form of firewood for providing energy to humans through direct combustion.

In industrialized and developed countries, wide and varied ranges of feedstocks are available in excess for production of biofuel, including forestry and agricultural waste, building and industrial waste and municipal solid waste (MSW). The biofuels that have been generated from these feedstocks are categorized as second generation

Corresponding author: rjadeja-chem@msubaroda.ac.in

biofuels. The first generation biofuels have been derived from edible food crops (i.e. sunflower, potato, barley, sugarcane, soybean, wheat, corn, and coconut), whereas second generation biofuels are generated from lignocellulosic materials (i.e. jatropha, cassava, switchgrass, wood, and straw) and biomass residues. The utilization of biomass residues and waste as primary resource for production of biofuels is a promising approach towards reduction of environmental issues concerning the waste disposal. This method will convert the wastes that would otherwise have been left to decompose into useful bioenergy. Another biomass such as algae has been introduced as feedstock for third generation biofuels, due to their high potential to produce large amounts of lipids that have been utilized for biodiesel production. Besides, this fast-growing biomass could be applied directly for generation of wide range of biofuels.

The present chapter provides an in-depth overview of the potential organic sources and technological details of biomass residues conversion techniques, waste to biofuels as well as bioelectricity generation. More specifically, the chapter presents a list of the waste-to-energy technological options along with sources. Conversion technologies covered in this chapter include gasification, liquefaction, pyrolysis, anaerobic digestion, alcoholic fermentation, photobiological hydrogen production, transesterification, supercritical fluid processing, combustion, and photosynthetic microbial fuel cells (MFC). The chapter serves to encompass the up-to-date information related to bioenergy production from biomass residues and waste in the rapidly expanding bioenergy field.

In several developing countries, bioenergy plays a crucial role in production of useful renewable energy. In emerging nations, bioenergy provides approximately 35% of power needs, bringing global power demand up to total of 13%. In several nations, bioenergy is considered as primary power source (e.g. Bhutan 86%, Nepal 97%, Asia 16%, East Sahelian Africa 81%, and Africa 39%). Biomass produced firewood has been utilized for cooking and heating in these nations (Balat 2006). As an evidence of the continuous degradation of the amount, the quality and cost of these resources are getting increased. The need towards awareness of environmental problems is also getting increased. The social and economic function of natural ecosystems will become clearer, and it will be important to reduce or eliminate the negative impacts on ecosystems. Thus, a country's economic strategies will work for the development that has been bounded by natural ecosystems. With the advancement of economic development observed recently, China is the world's major energy consumer, at present. China's energy security is very critical for heavy dependency on imported petroleum; this energy production causes serious economic issues (Wu et al. 2010). Biomass mainly consists of carbon dioxide and lignin that has been generated by the photosynthesis process; therefore, it helps in easy absorption of solar energy from plant crops. Biofuels define biomass that could be produced in a more comfortable way for domestic applications such as biogas. It is widely used for the domestic usage, but also included for producing crude oil and strong fuel-like pellets of timber (Rajmohan and Varjani 2019). Although various kinds of biomass could be transformed directly into heat or other forms of energy, certain types of biomass are only transformed into an advanced organic fuel with more effective manner. These materials have several useful characteristics such as improved retention, ease of operation, higher functionality and increased energy density.

Chemical or thermal conversion process has been used for conversion of organic waste to its bioenergy form (Cucchiella et al. 2017, Ramos et al. 2018, Sun et al. 2018, Wang et al. 2020). The organic waste has been burned to produce bioenergy in thermal conversion process. Thermal conversion includes conventional incineration and advanced thermal method (Dong et al. 2018, Makarichi et al. 2018). The bioenergy production varies on the basis of provided feedstock. Advanced thermal method includes pyrolysis and gasification (Wang et al. 2012). In order to increase energy recovery, a new technique called plasma gasification has been applied (Mazzoni et al. 2017, Mazzoni and Janajreh 2017, Sanlisoy and Carpinlioglu 2017, Perna et al. 2018). Methanation is proved to be one of the chemical process for generation of bioenergy. The development of biofuel generation with highlights on the second generation biofuels produced by biomass residues is presented in Fig. 9.1.

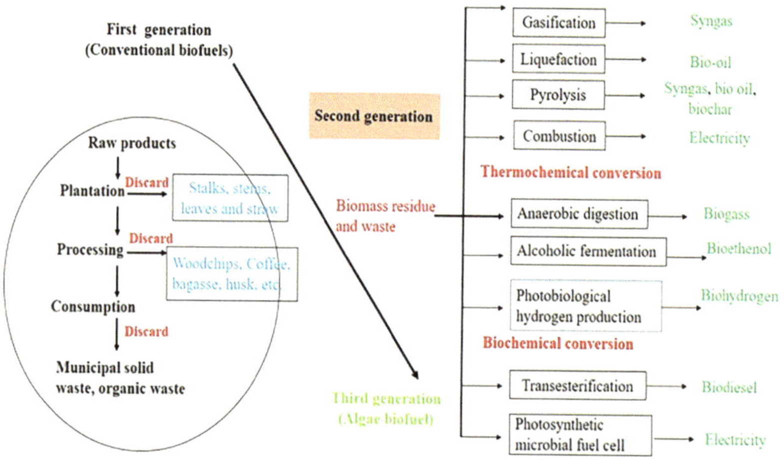

Figure 9.1: Diagram of the development of biofuel generation with highlights on the second generation biofuels produced by biomass residues and waste and their conversion pathways to produce a wide variety of bioenergy (Adopted from Lee et al. 2019).

9.2 Potential or Significant Organic Waste Sources

Bioenergy can be defined as the energy generated from the fuel that is produced from biomass. That means, biomass is a source of bioenergy. Biomass can be an important raw material for the production of energy in the future. Production of biofuels or bioenergy from biomass residues and wastes has been found to be a wonderful measure for reducing environmental problems caused due to the disposal of such wastes. Moreover, the conversion of a useless product into a useful product is an added advantage of this process (Lee et al. 2019). Any wastes containing biomass can be used to generate bioenergy. The constituents of biomass may include carbohydrates, lipids, proteins, lignin, water, ash and a variety of other compounds. Since these constituents are present in palm oil mill effluent, paper mill effluent, agricultural wastes, microalgal biomass, animal waste and textile waste in good amounts, these wastes can be considered as promising sources of bioenergy.

9.2.1 Palm oil mill effluents (POME)

Palm oil is an important product obtained from oil palms for which there is an increasing demand due to its various health benefits. The use of palm oil has been found to reduce the cholesterol level and oxidative stress, improve skin and hair health etc. (May and Nesaretnam 2014). This increased demand has resulted in an increased production of biomass waste. The various waste substances produced during the processing of palm oil include palm kernel cake, kernel shells, fibres, empty bunches of fruits, fronds and trunks of oil palms and the liquid waste palm oil mill effluent (Singh et al. 2010). Of these, the most important waste produced is palm oil mill effluent (POME). POME is produced as a result of sterilization, clarification and extraction procedures palm oil mills (Onyia et al. 2001). POME is a colloidal mixture of oil, water and suspended solids, which is thick and brownish in colour (Wu et al. 2009). The POME has an acidic pH due to the presence of free fatty acids and organic acids (Din et al. 2006). It has high chemical oxygen demand and biochemical oxygen demand and is several folds more polluting in comparison with municipal sewage. The phosphorus and organic nitrogen concentration are also higher in POME (Hadiyanto and Soetrisnanto 2013). Other constituents of POME include oil, grease, suspended solids, carbohydrates, lipids, proteins and nutrients like potassium, calcium and magnesium (Habib et al. 1997, Iwuagwu and Ugwuanyi 2014, Kamyab et al. 2016). POME itself is not a hazardous waste but it can cause adverse effects in aquatic environments because of its oxygen depleting capacity due to the presence of organic or natural substances and nutrients in it (Kamyab et al. 2014). So, it should be either treated to make it less polluting or it should be used as a raw material for other purposes like bioenergy production.

Tan et al. (2018) has reported that palm oil mill effluent is a good material for the generation of bioenergy. Anaerobic decomposition of POME can result in the production of methane, water and carbon dioxide through a number of reactions like hydrolysis, acidogenesis and methanogenesis (Bitton 2005, Nwuche and Ugochi 2010). The complex molecules like lipids, carbohydrates and proteins are broken down into fatty acids, sugars and amino acids, respectively, by the enzymes of anaerobic bacteria in the hydrolysis process. Acidogenic bacteria work upon these molecules and convert these fatty acids, sugars and amino acids to organic acids in the process of acidogenesis. Acetogens convert organic acids to acetate, carbon dioxide and hydrogen. Acetoclastic methanogens use this carbon dioxide and convert it into methane (Demirel and Scherer 2008, Weiland 2010). Methane can also be produced from lipids present in POME by the action of methanogenic archaea and acetogenic bacteria (Ahmad et al. 2011). In these ways, biogas, which is a biofuel, can be produced from POME.

9.2.2 Paper mill effluent

There are about 5000 paper and pulp mills all over the world, which produce about 400 million tonnes of paper annually (Skogsindustrierna 2010). Thus, paper production has become a fast-growing business in the recent years (Mensink 2007). A huge quantity of water is required by the paper industry for breaking down the raw materials, transporting fibres and for the formation of paper. Therefore, large quantity of wastewater is produced after the process (Büyükkamaci and Koken 2010). This discharge from pulp and paper mills, which contains a large quantity of suspended solids, phenols, aldehydes, ketones, cyanide, corrosive alkalis, oils, greases, proteins, carbohydrates etc., is called paper mill effluent (PME) (Sharma and Kaur 2000). Paper mill effluent is characterized by high chemical oxygen demand and biological oxygen demand, dark colour, high organic content, foul odour and also high pH (Sharma and Kaur 2000, Pokhrel 2004). These wastes or paper mill effluents are often discharged into water bodies, thereby contaminating them (Afzal et al. 2008). The use of such effluents for bioenergy production can reduce the problem of environmental pollution.

Bioenergy can be produced through biological processes from the huge quantity of paper mill effluents having high chemical oxygen demand (values between 1 and 80 g/L) formed from the paper and pulp industry (Lin et al. 2013, Kamali and Khodaparast 2015). Anaerobic digestion of paper mill effluents can reduce the chemical oxygen demand as well as produce renewable energy like biogas (Habets and Driessen 2007). A series of dark fermentation followed by anaerobic digestion of paper mill effluent was found to produce biohydrogen and biomethane at the expense of COD reduction (Vaez and Zilouei 2020). Zwain et al. (2013) reported that the use of modified anaerobic baffled reactor could produce methane efficiently along with COD removal of paper mill effluent. According to Pontual et al. (2015), anaerobic digestion is the better method of pre-treatment of paper mill effluents, as this process produces less amount of sludge, is less space consuming, utilizes less energy for aeration and easier running of the process without the bulking of sludge.

9.2.3 Agricultural waste

The harvesting and processing of various crops result in the production of carbon containing matter as by-products. These are often called agricultural residues. Agricultural residues can be classified as primary and secondary residues. Primary residues are the wastes generated during the process of harvesting while the wastes generated during the processing of products are called secondary residues. These wastes are heterogeneous with different moisture content, bulk density, particle size and are fibrous with low nitrogen content. But the characteristics may vary with different geographical location (Smith 1987). The species of crop, period of harvest, duration of storage of the crop residues etc. influence the chemical composition of the agricultural residues (Cooper and Laing 2007, OECD/IEA 2010). Nowadays, improper agricultural practices and high usage of chemical fertilizers and pesticides affect the agricultural waste quality. Sometimes these wastes contain toxic contaminants including heavy metals. The deposition on soil or burning of these crop residues by the farmers results in extreme land and water pollution locally and regionally (Kumar and Joshi 2013, Sarath and Puthur 2020).

The conversion of agricultural residues to fertilizers, bioenergy etc. can help in the growth of economy, prevents exerting pressure on land and improve human well-being (UNEP 2011). For the conversion of agricultural wastes into bioenergy and other bio-based substances, anaerobic digestion has been found to be one of the most effective and grown-up technology (Merlin and Boileau 2013). Anaerobic digestion of agricultural wastes can yield biogas as well as biofertilizer. Biochar, syngas and bio-oil can also be obtained after the pyrolysis of the solid anaerobic digestate and these can be utilized for the generation of electricity (Gontard et al. 2018). But the anaerobic digestion of lignocellulosic rich residues results in low yield of bioenergy (Bolzonella et al. 2017). This limitation can be overcome by the integration of pyrolysis with anaerobic digestion (Fabbri and Torri 2016).

9.2.4 Food waste

Food waste is defined as the pre- or post-cooked biodegradable waste that is discarded from different sources like households, hotels and restaurants, food processing industries etc. (FAO 2012). The main components of food waste include carbohydrates, lipids and proteins but the composition may vary according to the type of food waste (Parithosh et al. 2017). Approximately 1.3 billion tonnes of food is wasted along the food supply chain according to the reports of Food and Agricultural Organization. This includes fruits, vegetables, bakery, meat, dairy products etc. (FAO 2012). Due to the increasing economic growth and population growth, the amount of food waste would increase in the coming 25 years with Asian countries as the main contributors. An increment from 278 million tonnes to 416 million tonnes of food waste is expected by the year 2025 (Melikoglu et al. 2013). Nearly 28% of world's farm land is required for the cultivation of wasted food, which means this land is also wasted. Food wastes also contribute to the emission of greenhouse gases along with wastage of land resources. Dumping of food waste in open area or incineration of food waste containing moisture can cause several environmental issues (Agarwal et al. 2005, Talyan et al. 2008, Kumar and Goel 2009, Kumar et al. 2009, Pattnaik and Reddy 2010). Therefore, food waste management has become necessary.

Anaerobic digestion of food waste is an eco-friendly method of food waste management and is a good choice since biogas can be generated in addition to nutrient cycling and waste management. The process of conversion of food waste to biogas occurs through three phases which are enzymatic hydrolysis, acid formation (acidogenesis and acetogenesis) and methanogenesis. Enzymatic hydrolysis breaks down polymers to oligomers or monomers (Parithosh et al. 2017). These are fermented to volatile fatty acids during acidogenesis, of which acetate, carbon dioxide and hydrogen can be used for methane generation directly (Bryant 1979, Mittal 1997, Schink 1997). During acetogenesis, acetogenic bacteria convert the products of acidogenesis into hydrogen and acetates (Schink 1997). During methanogenesis, methanogens convert acetic acid to methane or reduce carbon dioxide to methane (Griffin et al. 1998, Karakashev et al. 2005). Using food waste as a feedstock for the production of biogas can produce higher yields as compared to animal waste (Curry and Pillay 2012).

9.2.5 Microbial biomass

Microalgae can be defined as oxygenic photosynthetic microorganisms that are either prokaryotic or eukaryotic. Photosynthesis in microalgae results in the production of proteins, carbohydrates and lipids (Richmond 2004). It is quite easy to cultivate microalgae since it does not require arable lands like other food crops, can grow in wastewaters, has high growth rates etc. Its ability to grow in wastewaters makes it an efficient tool to remediate industrial wastewaters (Montingelli et al. 2015, Peng and Colosi 2016). The most widely used microalgae for industrial purposes include *Haematococcus*, *Dunaliella* and *Chlorella* (Benedetti et al. 2018). Due to the high accumulation of lipids in microalgae, it is a promising tool for the production of bio-oil (Lee et al. 2019). Besides, the spent microalgal biomass is an alternative source for the production of bioenergy. Biohydrogen, bioethanol, biodiesel and biomethane can be produced using microalgal biomass (Nguyen and Hoang 2016).

High quantities of polysaccharides are accumulated by green algae in their cell wall as well as in the form of storage molecules. The fermentation of this can generate bioethanol (Shokrkar et al. 2017). Photofermentation of cyanobacteria and microalgae, which is an anaerobic process, can result in the production of biohydrogen. This occurs by the activity of hydrogenase enzyme that oxidises ferredoxin (Sharma and Arya 2017, Khetkorn et al. 2017). Microalgal biomass consists of 20-60% of oil fraction. The transesterification of this oil fraction can generate biodiesel (Rodolfi et al. 2009, Ghasemi Naghdi et al. 2016). Anaerobic digestion of microalgal biomass can produce biogas. Energetically, anaerobic digestion has been found to be the more favourable process for biogas production (Gonzalez-Fernandez et al. 2015). The effectiveness of production of biogas is affected by the degradability of the cell walls (Santos-Ballardo 2016). Chemical methods, physical methods or enzymatic hydrolysis pre-treatment methods can help in the disruption of cell walls and enhance the yield of biomethane (Mahdy et al. 2014, Passos et al. 2014, Mahdy et al. 2016).

9.2.6 Animal waste

Animal wastes are commonly the excreta of livestock animals (Duku et al. 2011). Animals produce large amount of waste daily (Simonyan and Fasina 2013). The amount of waste produced by the animal depends on the quality and quantity of the feed consumed and the weight of the animal (Duku et al. 2011). Organic material, ash and moisture are the main components of livestock waste. Being an important source of pollution, greenhouse gases, pathogens and some livestock wastes (EPA 1998) require to be converted to other useful and less polluting products. Since it has been reported to be a good source of energy and fertilizers, it can be used for their production. The conversion of animal waste into bioenergy helps farmers to have the additional benefits of generating revenues annually, expands income from farms and reduces the effect of commodity costs (Cantrell et al. 2008). In developing countries, people use dung cakes to cook food in rural areas (Burton and Turner 2003).

Livestock waste can be used for the production of biogas. Nearly 20-25 m^3 of biogas can be generated from 1 tonne of manure, which in turn can be used to produce 35-40 kWh electricity (Burton and Turner 2003). Aerobic decomposition of livestock wastes results in the production of carbon dioxide and stable organic materials, while anaerobic decomposition results in the production of methane, carbon dioxide and stable organic materials (Duku et al. 2011). The biogas thus produced can be filled in cylinders and used for cooking as well as vehicular fuel, provided it contains

98% methane (Sorathiya et al. 2014). Thermochemical conversion methods like liquefaction, pyrolysis and gasification are also capable of transforming animal waste into combustible oils and biofuels (Cantrell et al. 2008).

9.2.7 Textile waste

Textile wastes are those substances that become useless during or after the manufacturing or use of textile products. Textile wastes are capable of polluting land, water as well as air. Most of the textile wastes are landfilled (Rago et al. 2018). Greenhouse gases, which cause air pollution, will be formed during the decomposition of textile wastes. The large amount of chemicals that are used during textile manufacturing can pollute the water bodies once these wastes reach the environment (Woolard 2009). This necessitates the proper management of textile wastes. Since these wastes are produced in large amounts and consist of huge amounts of cellulose having high calorific value, renewable energy production is possible (Bansal et al. 2016, Leonel et al. 2018).

Yousef et al. (2019) developed a technology involving pyrolysis of textile wastes along with several other steps for the production of bio-oil, bio-gases and char. They could produce 37.5% high quality bio-oil, 44.7% biogas and 17.8% char from the textile waste. The heavy metals present in the textile dyes played the role of catalysts and accelerated the process. The sludge from textile industries could also be used for the production of biogas efficiently by anaerobic digestion with the addition of cow dung to the sludge (Kumar et al. 2020).

9.3 Bioenergy Generation

Currently, in most of the countries, the primary and major source of energy is fossil fuels (Caetano et al. 2017, Sugiawan and Managi 2019). This is because they are relatively cheap and useful by-products can be produced. But the emission of large quantities of greenhouse gases, contribution of fossil fuels to air pollution and the fast depletion of resources, persuades us for a shift from the use of non-renewable fossil fuels to the use of bioenergy. Bioenergy, being a renewable resource, can be an alternative source of energy in the future (Lee et al. 2019). The various sources of biomass that can be used to produce bioenergy include palm oil mill effluents, animal waste, agricultural waste, paper mill effluents, microalgal biomass etc. Different technologies are adopted for the conversion of biomass to bioenergy. The choice of adopting a specific technology for this conversion is based on several factors, which are the quality of the raw material or feedstock, its quantity, end products required and environmental problems caused (Matsumura 2015).

The technologies used for biomass conversion can be classified into biochemical methods and thermochemical methods. Biochemical methods make use of microorganisms or enzymes for the conversion of biomass into bioenergy. This conversion can occur via anaerobic digestion, photobiological reaction or alcoholic fermentation. Thermochemical conversion utilizes high temperature for breaking bonds and converting the organic matter into bioenergy. It includes the processes of pyrolysis, gasification, liquefaction and combustion (Goyal et al. 2008). Use of thermochemical methods is advantageous over the biochemical methods for bioenergy generation. This is because of the highly improved industrial infrastructure that is available, less water and time utility, production of energy from even plastic wastes, which is not possible by biological methods, and no dependency on the environmental conditions for bioenergy production (Uzoejinwa et al. 2018). The various biomass to bioenergy conversion techniques are explained below.

9.3.1 Gasification

Gasification is a process of incomplete combustion which involves conversion of carbonaceous material (biomass) into syngas (CO_2, CO, and H_2) by heating it to a very high temperature (> 700 °C). A controlled amount of oxygen (oxygen-deficient atmosphere) is supplied to carry out the reaction. The quality of the produce (syngas) depends on the composition of the biomass and the mechanism used for gasification. The end result of gasification contains energy content up to about 70–80% of what was in the raw material (Devi and Kamaraj 2017).

There are various types of gasification techniques. Common gasification is carried out in oxygen-deficient environment. The temperature required is around 700-900°C. Hydrothermal gasification (HTG) is carried out in supercritical aqueous environment. In this process, the feedstock is gasified to syngas and methane under supercritical and non-oxidative conditions (Toor et al. 2014). The temperature requirement for HTG is comparatively lower than the non-aqueous one but a very high pressure is required to build a supercritical atmosphere. When gasification is carried out in air, the syngas produced has a lower heating value. But, when the same process occurs in presence of pure oxygen, products with medium heating value are obtained. When a catalyst is used, the resultant syngas has a high heating value.

Gasification of waste material has certain benefits:

1. The products of gasification are CO_2, CO, and H_2, together known as syngas. It is effectively utilized for generation of heat and power.
2. The generated syngas is a clean fuel, which can be easily transferred than the waste itself.
3. High energy recovery and heating capacity.
4. It is more advantageous when compared with pyrolysis and liquefaction due to their complex technique and operating conditions (Sansaniwal et al. 2017).

Along with the benefit one reaps, there is one limitation to the process. Together with the syngas, tar is also produced, thereby limiting its direct usage in internal combustion engines. To make it fit for usage, biomass is first pyrolyzed and then gasification is carried out (Haydary 2017).

9.3.2 Pyrolysis

Pyrolysis is a process of thermal decomposition (typically between ~300 and 800 °C) of organic matter which occurs under anoxygenic conditions (Czajczyńska et al. 2017). In this process, high density biomass is converted into energy rich fuel. This high-density biomass is nothing but lignocellulosic biomass, which is mainly composed of three components: cellulose, hemicellulose and lignin (Roy and Dias

2017). The process of pyrolysis typically depends on factors like temperature, rate of heating and residence time.

Pyrolysis is broadly classified into three types, i.e. slow, intermediate, fast and flash pyrolysis. Slow pyrolysis, also known as conventional pyrolysis, requires lower temperature (300–650 °C) at a low heating rate of 0.1–10 °C/s and a longer residence time (it takes several hours to complete). For many centuries, humans have been using this technique for the production of charcoal from wood (Williams and Nugranad 2000). Intermediate pyrolysis requires temperature in the range of 400 to 500 °C at a heating rate of 1 and 1000 °C/s with a residence time of 5-10 minutes. Such processes are mostly suitable for pyrolysis of food, wood and sewage sludge (Yang et al. 2013). Pyrolysis process is presented in Fig. 9.2.

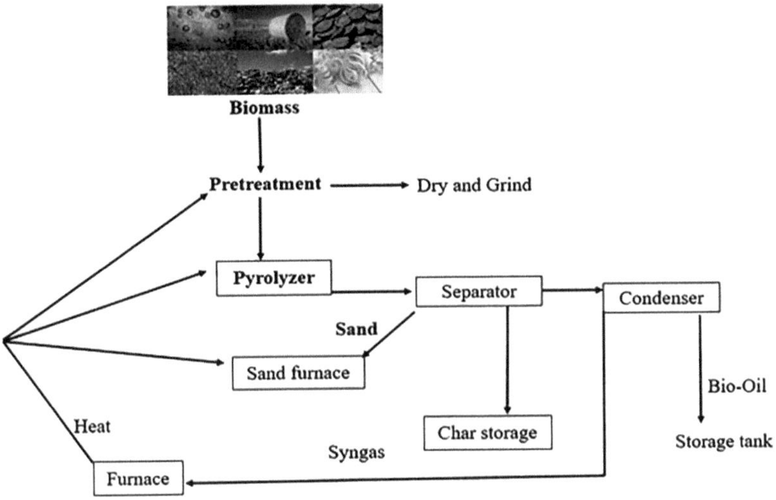

Figure 9.2: Schematic diagram representing the pyrolysis process. (Source: U.S. Department of Agriculture)

Fast pyrolysis is generally employed for the production of bio-oil where the feedstock is directly converted to vapours and later condensed to produce bio-oil. It takes seconds to complete this process. It requires high temperature ranging from 450–650°C at a rate of 100–1000°C/s and for a very short residence time (0.5–5 s). Bio-oil is a dark coloured viscous fluid composed mainly of organic compounds, tars such as benzene and toluene and a quarter amount of water released during fast pyrolysis of feedstock (Li et al. 2013). It results in almost 60-65% of bio-oil. In addition, it gives 20% biochar and 20% syngas.

Flash pyrolysis, also known as very fast pyrolysis, requires high temperatures range of 900-1200°C at a high heating rate of over 1000°C with a very short residence time (~ 0.5 s.). Bio-oil produced using flash pyrolysis has very low water content with efficiency more than 70 percent (Li et al. 2013). There are numerous potential uses of pyrolysis product. For example, biochar has multiple applications such as in purification of water, used as fertilizer to increase the productivity and also used for energy production (Roy and Dias 2017).

9.3.3 Transesterification

Transesterification, also known as alcoholysis, is the process in which an ester reacts with alcohol and the displacement of alcohol in an ester occurs by another alcohol (Srivastava and Prasad 2000). The reaction occurs in the presence of a catalyst, which may be an acid or a base or a biocatalyst (enzymes) and leads to the formation of fatty acid alkyl esters and alcohol.

$$\underset{\text{Ester}}{RCOOR^1} + \underset{\text{Alcohol}}{R^2OH} \Leftrightarrow \underset{\text{Ester}}{RCOOR^2} + \underset{\text{Alcohol}}{R^1OH}$$

There lie numerous challenges in using the desired biomass for producing biofuel. This comes from the fact that the properties and performance of the extracted oil should be modulated according to the properties of hydrocarbon-based fuels. The oils or fats thus obtained when converted to biofuels should be able to adequately replace the conventional fuels. When lignocellulosic materials are converted to biofuels, the major concern arises from high viscosity, low vitality and polyunsaturated characteristics. Different pretreatment techniques, most important being transesterification, are required to resolve the above issue (Clark et al. 1984). Thus, fatty acid methyl esters (known as FAME or biodiesel fuel) obtained by transesterification can be used as an alternative fuel for diesel engines. Transesterification process for production of bioenergy is discussed in Fig. 9.3.

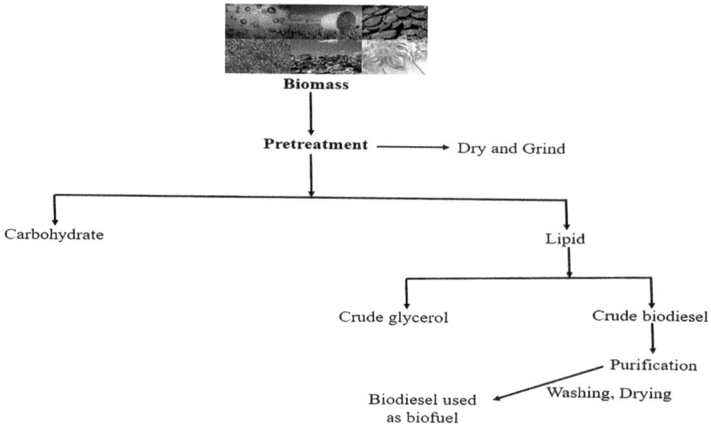

Figure 9.3: Representation of transesterification process for production of bioenergy.

9.3.3.1 Transesterification mechanism and kinetics

When triglycerides (ester) undergo the process of transesterification, they produce fatty acid alkyl esters and glycerol, and diglycerides and monoglycerides are produced as the intermediate's products. The transesterification reaction is reversible in nature.

The first step is the conversion of triglycerides to diglycerides, which is followed by the conversion of diglycerides to monoglycerides and of monoglycerides to glycerol, yielding one methyl ester molecule from each glyceride at each step (Freedman et al. 1984, Kudsiana and Saka 2001).

General equations for transesterification of glycerides are shown below:

$$\text{Triglycerides} + R^1OH \Leftrightarrow \text{Diglycerides} + RCOOR^1$$
$$\text{Diglycerides} + R^1OH \Leftrightarrow \text{Monoglycerides} + RCOOR^1$$
$$\text{Monoglycerides} + R^1OH \Leftrightarrow \text{Glycerol} + RCOOR^1$$

Several researchers have reported the kinetics for both acid- (Eckey 1956, Dufek et al. 1972, Freedman et al. 1986, Noureddini and Zhu 1997) and alkali-catalyzed (Noureddini and Zhu 1997, Freedman et al. 1986) transesterification reactions. In transesterification, the use of the alkali as a catalyst is preferred over acid since it yields faster results and higher fatty acid methyl ester (FAME). Contrarily, enzyme catalysts, though slow in the reaction, are more environmentally friendly and can produce high-quality products (Freedman et al. 1986).

In alkali catalysed transesterification, the alkoxide ion attacks the carbonyl carbon of the triglyceride, thereby forming a tetrahedral intermediate. The intermediate reacts with alcohol to produce the alkoxide ion again. The tetrahedral intermediate then rearranges to give rise to an ester and a diglyceride (Ma and Hanna 1999).

Acid-catalysed transesterification is carried out preferably by sulfonic and sulfuric acids. The reaction, though slow, produces very high yields in presence of alkyl esters (Schuchardt et al. 1998).

Triglyceride transesterification in presence of a biocatalyst occurs conforming with the principle of successive reaction mechanism (Kaieda et al. 1999). In this method, triglycerides and partial glycerides are first hydrolyzed to partial glycerides and free fatty acids in presence of lipase, respectively. Methyl esters are then synthesized from free fatty acids and methanol.

9.3.3.2 Supercritical fluid method

Supercritical transesterification is the process of biofuel production that utilizes the supercritical operating conditions, and works in the absence of a catalyst (Deshpande et al. 2017). This technique was utilised to overcome limitations of catalysis like long reaction time, catalyst poisoning, catalyst regeneration, high operation cost, saponification and biodiesel washing. But, having said that, high temperature and pressure are compulsory to build a supercritical condition. So, this requires a heavy infrastructure set up to sustain high temperature and pressure. But still non-catalytic transesterification is advantageous in commercial biodiesel production.

9.3.3.3 Factors affecting transesterification reaction

Various factors affecting the process of transesterification are described below:

1. Effect of free fatty acid and moisture
2. Catalyst type and concentration
3. Molar ratio of alcohol to oil and type of alcohol
4. Effect of reaction time and temperature

5. Mixing intensity
6. Effect of using organic cosolvents

9.3.4 Anaerobic digestion: Biogas and methane production

Due to growing demands of increasing population and depletion of fossil fuels for energy, it is crucial to shift to our energy choices based on renewable sources (Laperriere et al. 2017), such as biogas production in the near future. In biogas production, conversion of complex organic compounds into CH_4 and inorganic compounds such as CO_2, N_2, NH_3, and H_2S through anaerobic digestion is essential (Pavlostathis and Giraldo-Gomez 1991). Over the past decade, biogas has been effectively used as a renewable energy source, with major raw material as agriculture produce (Sturmer 2017). Flow chart of anaerobic digestion process is presented in Fig. 9.4.

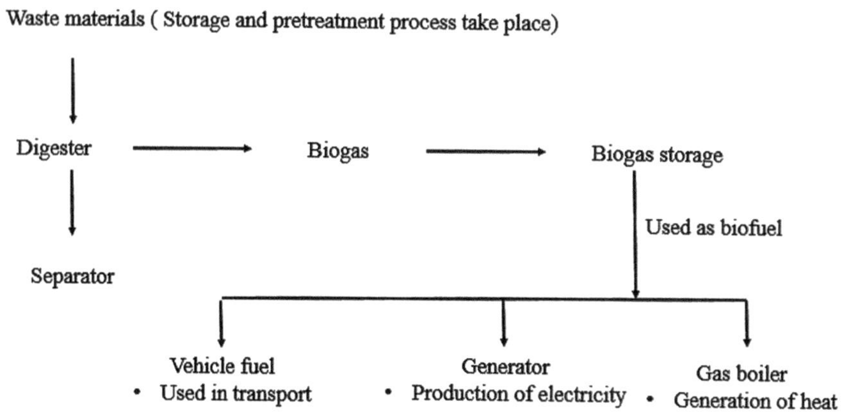

Figure 9.4: Flow chart of anaerobic digestion for production of bioenergy. (Adopted from e-inst.com)

Anaerobic digestion refers to microbial degradation of complex organic matter that occurs in the absence of oxygen. It mainly involves four main steps: hydrolysis, acidogenesis, acetogenesis, and methanogenesis (Bremond et al. 2018). But, biogas generation is a time-consuming process because the bacterial consortia, before being able to degrade, require time to adapt to the new environment (Poh and Chong 2009).

For anaerobic digestion to occur, microbes play an important role in biochemical pathways (Fitamo et al. 2017). Bacteria can be categorised into three different metabolic groups, which are fermentative, acidogenic and methanogenic (Bryant 1979). Fermentative bacteria hydrolyze lipids, protein, and polysaccharides to simpler materials. Acidogenic bacteria produce acetate and H_2 from the end-products of the fermentative bacteria. The methanogenic bacteria break down the end products produced jointly by the fermentative and acidogenic bacteria, to the final products.

Some fundamental steps involved in the anaerobic digestion process, i.e. hydrolysis, acidogenesis, acetogenesis and methanogenesis, are shown in Fig. 9.5.

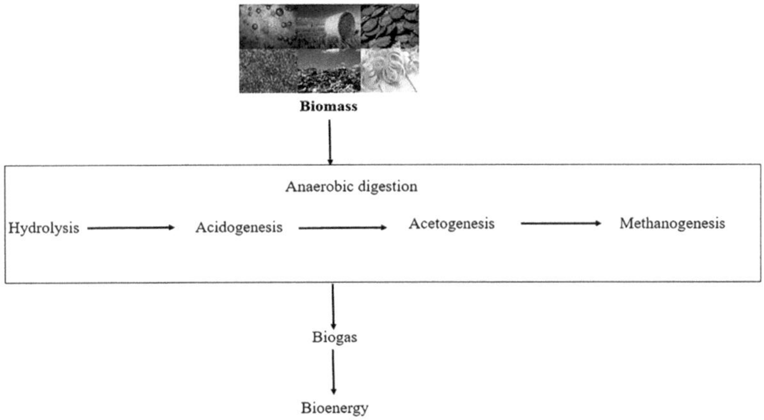

Figure 9.5: Fundamental steps involved in anaerobic digestion process.

Firstly, the hydrolytic or the fermentative bacteria, through the process of hydrolysis or liquefaction, breaks down complex organic compounds, such as lipids, proteins, and polysaccharides, into soluble monomers or oligomers like amino acids, long-chain fatty acids, sugars, and glycerol. Secondly, through acidogenesis, the oligomers and monomers produced in the first step are then fermented by acidogenic bacteria to release a mixture of carbon dioxide (CO_2), hydrogen (H_2), alcohol, and low molecular weight volatile fatty acids like propionic and butyric acids (Surendra et al. 2014). Thirdly, during the process of acetogenesis, acetogenic bacteria convert fatty acids produced by acidogenesis into hydrogen and acetic acid, which can be further used by methanogenic bacteria (Schink 1997). During the process of methanogenesis, methanogenic bacteria act on acetate and carbon dioxide, transforming them into methane (CH_4). Hydrolysis or methanogenesis are often considered as a limiting step in the case of complex organic substrates because the production rate, quantity, and variety of hydrolytic enzymes released by hydrolytic microorganisms are often not sufficient to properly degrade a given substrate (Bremond et al. 2018). The composition of biogas varies with the type of raw material and the operational condition of the digester. In general, biogas consists of 50-75% CH_4 and 25-50% CO_2 together with other traces of components, such as water vapour (H_2O), hydrogen sulfide (H_2S), and ammonia (NH_3) (Surendra et al. 2014).

9.3.4.1 Use of anaerobic bio-digested substrate

The solid waste generated from different sources such as municipal, industrial, and agricultural wastes varies widely and thereby affects anaerobic digestion (Li et al. 2011). Sugarcane bagasse and Agave tequilana bagasse have been employed effectively for bioenergy generation (Silva et al. 2014, Siddique and Wahid 2018). Agricultural waste can successfully be employed in bioenergy generation due to its plentiful supply and low cost (Kapoor et al. 2020).

According to various studies, it was concluded that the residue left after the oil extraction of Jatropha species (*J. curcas* L.) acts as an excellent substrate for the production of biogas, yielding up to 70% methane. Pig manure was used as an inoculum in a concentration of 20% (vol./vol.) (Staubmann et al. 1997). Jatropha (esp. *J. curcas*), due to its fast growth, high biomass production, abundant root system, hardy nature, and inedibility, is economically viable and the most suitable raw material (Khalil et al. 2019). But the main limitation in bioenergy production from biomass is the economic viability of the raw material production and conversion process (Chen et al. 2007, Carlson et al. 2008). The waste cooking oils can be used effectively for biodiesel production due to the low cost of raw material, thereby reducing the overall manufacturing cost and low environmental pollution (Balat 2011). Alternatively, microalgae can be preferred as raw material for bioenergy production for their high productivity per unit area, and reduced land use (Li et al. 2008). But microalgae offer a wide range of usage for other commercial purposes such as a protein source for human and animal, as corant astaxanthin (Zhao et al. 2019), for use as single-cell protein, as dietary supplements (Jacob-Lopes et al. 2019, Yamaguchi et al. 2019) and animal food (Wild et al. 2019).

Multipurpose modeling methods were very viable use to produced biogas. It involves using multiple technologies to produce a range of products such as biofuels, biochemicals, and bioenergy using biomass feedstocks. But to enable the sustainability of bioenergy production, economic viability alone is not enough. The expectation of the development of bioenergy technologies is based on the conversion of fast-growing plant materials. These energy crops such as *Miscanthus, Panicum* (switchgrass), poplar, willow microalgae, and *Jatropha* are less dependent on favourable soil and weather conditions and require fewer agrochemical inputs, thus reducing their direct competition with food production, ensuring safety to feed.

9.3.5 Alcoholic fermentation

Alcoholic fermentation is a biotechnological process accomplished by yeast, some kinds of bacteria, or a few other microorganisms to convert sugars into ethyl alcohol and carbon dioxide. In this fermentation process, yeast is mostly used as a bio-culture and aqueous solution of monosaccharide (raw materials) as the culture media for the production of beverages. In the alcoholic fermentation process, yeast generally carries out the aerobic fermentation process, but it may also ferment the raw materials under anaerobic conditions. In the absence of oxygen, alcoholic fermentation occurs in the cytosol of yeast (Sablayrolles 2009, Stanbury et al. 2013). Alcoholic fermentation begins with the breakdown of sugars by yeasts to form pyruvate molecules, which is also known as glycolysis. Glycolysis of a glucose molecule produces two molecules of pyruvic acid. The two molecules of pyruvic acid are then reduced to two molecules of ethanol and $2CO_2$ (Huang et al. 2015). Bioethanol is used as a bioenergy sources for transportation and other uses.

9.4 Conclusion

In this chapter, various organic sources and methods of bioenergy generation have been discussed. It can be concluded that bioenergy could be generated from organic waste, which no longer remains as waste but the rich source of substrate and nutrients for energy production. The most preferable, economic and environmental friendly approach to get valuable energy products is anaerobic digestion. Due to increasing population growth, the world has to enhance the efforts towards future development and advancement of valuable renewable energy sources. Biofuels can significantly substitute the fuels for future on the basis of versatile technology, convenience of transportation, and economic advantages. The cost of bioenergy production has to be improved extensively in manufacturing as well as transformation techniques. The waste generated by food industry, microalgal waste, agricultural waste, palm oil mill effluents etc. could be effectively used for generation of bioenergy in a sustainable way. Overcoming several challenges by enforcing strict regulations and implementation of government policies, the active bioenergy production from waste organic sources via several methods will definitely provide a renewable and techno-economical solution for recycling waste biomass and establishing sustainable and continuous source of clean and green energy.

References

Abdeshahian, P., J.S. Lim, W.S. Ho, H. Hashim and C.T. Lee. 2016. Potential of biogas production from farm animal waste in Malaysia. Renew. Sustain. Energy Rev. 60: 714-723.

Afzal, M., G. Shabir, I. Hussain and Z.M. Khalid. 2008. Paper and board mill effluent treatment with the combined biological–coagulation–filtration pilot scale reactor. Bioresour. Technol. 99(15): 7383-7387.

Agarwal, A., A. Singhmar, M. Kulshrestha and A.K. Mittal. 2005. Municipal solid waste recycling and associated markets in Delhi, India. Resour. Conserv. Recycl. 44(1): 73-90.

Ahmad, A., R. Ghufran and Z.A. Wahid. 2011. Bioenergy from anaerobic degradation of lipids in palm oil mill effluent. Rev. Environ. Sci. Biotechnol. 10(4): 353-376.

Balat, M. 2006. Biomass energy and biochemical conversion processing for fuels and chemicals. Energ. Source. Part A 28(6): 517-525.

Balat, M. 2011. Potential alternatives to edible oils for biodiesel production – A review of current work. Energy Convers. Manag. 52(2): 1479-1492.

Bansal, A., P. Illukpitiya, F. Tegegne and S.P. Singh. 2016. Energy efficiency of ethanol production from cellulosic feedstock. Renew. Sust. Energ. Rev. 58: 141-146.

Benedetti, M., V. Vecchi, S. Barera and L. Dall'Osto. 2018. Biomass from microalgae: The potential of domestication towards sustainable biofactories. Microb. Cell Factories 17(1): 1-18.

Bitton, G. 2005. Anaerobic digestion of wastewater and biosolids. Wastewater Microbiology. 3: 1-729. Wiley, Hoboken, New Jersey.

Bolzonella, D., M. Gottardo, F. Fatone and N. Frison. 2017. Nutrients recovery from anaerobic digestate of agro-waste: Techno-economic assessment of full-scale applications. J. Environ. Manage. 216: 111-119.

Brémond, U., R. de Buyer, J.P. Steyer, N. Bernet and H. Carrere. 2018. Biological pretreatments of biomass for improving biogas production: An overview from lab scale to full-scale. Renew. Sustain. Energy Rev. 90: 583-604.

Bryant, M.P. 1979. Microbial methane production-theoretical aspects. J. Anim. Sci. 48(1): 193-201.

Burton, C.H. and C. Turner. 2003. Manure management. Treatment strategies for sustainable agriculture (2nd edn). *In*: Proceedings of the MATRESA, EU Accompanying Measure Project Silsoe Research Institute, Wrest Park, Silsoe, Bedford, UK. (2), 1-449.

Buyukkamaci, N. and E. Koken. 2010. Economic evaluation of alternative wastewater treatment plant options for pulp and paper industry. Sci. Total Environ. 408(24): 6070-6078.

Caetano, N.S., T.M. Mata, A.A. Martins and M.C. Felgueiras. 2017. New trends in energy production and utilization. Energ. Proced. 107: 7-14.

Cantrell, K.B., T. Ducey, K.S. Ro and P.G. Hunt. 2008. Livestock waste-to-bioenergy generation opportunities. Bioresour. Technol. 99: 7941-7953.

Carlson, T.R., T.P. Vispute and G.W. Huber. 2008. Green gasoline by catalytic fast pyrolysis of solid biomass derived compounds. ChemSusChem: Chem. Sustaina. Energ. Materia. 1(5): 397-400.

Chen, M., L. Xia and P. Xue. 2007. Enzymatic hydrolysis of corncob and ethanol production from cellulosic hydrolysate. Int. Biodeterior. Biodegrad. 59(2): 85-89.

Clark, S.J., L. Wagner, M.D. Schrock and P.G. Piennaar. 1984. Methyl and ethyl soybean esters as renewable fuels for diesel engines. J. Am. Oil Chem. Soc. 61(10): 1632-1638.

Cooper, C.J. and C.A. Laing. 2007. A macro analysis of crop residue and animal wastes as a potential energy source in Africa. J. Energy South. Afr. 18: 10-19.

Cucchiella, F., I. D'Adamo and M. Gastaldi. 2017. Sustainable waste management: Waste to energy plant as an alternative to landfill. Energy Convers. Manag. 131: 18-31.

Curry, N. and P. Pillay. 2012. Biogas prediction and design of a food waste to energy system for the urban environment. Renew. Energy 41: 200-209.

Czajczyńska, D., T. Nannou, L. Anguilano, R. Krzyżyńska, H. Ghazal, N. Spencer and H. Jouhara. 2017. Potentials of pyrolysis processes in the waste management sector. Energ. Proced. 123: 387-394.

Demirel, B. and P. Scherer. 2008. The roles of acetotrophic and hydrogenotrophic methanogens during anaerobic conversion of biomass to methane: A review. Rev. Environ. Sci. Biotechnol. 7(2): 173-190.

Deshpande, S.R., A.K. Sunol and G. Philippidis. 2017. Status and prospects of supercritical alcohol transesterification for biodiesel production. Wiley Interdiscip. Rev. Energy Environ. 6(5): 252.

Devi, R.P. and S. Kamaraj. 2017. Design and development of updraft gasifier using solid biomass. Int. J. Curr. Microbiol. Appl. Sci. 6: 182-189.

Din, M.M., Z. Ujang, M.C.M. Van Loosdrecht, A. Ahmad and M.F. Sairan. 2006. Optimization of nitrogen and phosphorus limitation for better biodegradable plastic production and organic removal using single fed-batch mixed cultures and renewable resources. Water Sci. Technol. 53: 15-20.

Dong, J., Y. Tang, A. Nzihou, Y. Chi, E. Weiss-Hortala and M. Ni. 2018. Life cycle assessment of pyrolysis, gasification and incineration waste-to-energy technologies: Theoretical analysis and case study of commercial plants. Sci. Total Environ. 626: 744-753.

Dufek, E.J., R.O. Butterfield and E.N. Frankel. 1972. Esterification and transesterification of 9(10)-carboxystearic acid and its methyl esters. Kinetic studies. J. Am. Oil Chem. Soc. 49(5): 302-306.

Duku, M.H., S. Gu and E.S. Hagan. 2011. A comprehensive review of biomass resources and biofuels potential in Ghana. Renew. Sustain. Energy Rev. 15: 404-415.

Eckey, E.W. 1956. Esterification and interesterification. J. Am. Oil Chem. Soc. 33(11): 575-579.

EPA. 1998. National Air Pollutant Emission Trends 1990-1998.

Fabbri, D. and C. Torri. 2016. Linking pyrolysis and anaerobic digestion (Py-AD) for the conversion of lignocellulosic biomass. Curr. Opin. Biotechnol. 38: 167-173.

FAO. 2012. Towards the Future We Want: End Hunger and Make the Transition to Sustainable Agricultural and Food Systems, Food and Agriculture Organization of the United Nations Rome.

Fitamo, T., L. Treu, A. Boldrin, C. Sartori, I. Angelidaki and C. Scheutz. 2017. Microbial population dynamics in urban organic waste anaerobic co-digestion with mixed sludge during a change in feedstock composition and different hydraulic retention times. Water Res. 1(118): 261-271.

Freedman, B., E.H. Pryde and W.F. Kwolek. 1984. Thin layer chromatography/flame ionization analysis of transesterified vegetable oils. J. Am. Oil Chem. Soc. 61(7): 1215-1220.

Freedman, B., R.O. Butterfield and E.H. Pryde. 1986. Transesterification kinetics of soybean oil 1. J. Am. Oil Chem. Soc. 63(10): 1375-1380.

Ghasemi Naghdi, F., L.M. González González, W. Chan and P.M. Schenk. 2016. Progress on lipid extraction from wet algal biomass for biodiesel production. Microb. Biotechnol. 9(6): 718-726.

Gontard, N., U. Sonesson, M. Birkved, M. Majone, D. Bolzonella, A. Celli, H. Angellier-Coussy, G.W. Jang, A. Verniquet, J. Broeze and B. Schaer. 2018. A research challenge vision regarding management of agricultural waste in a circular bio-based economy. Crit. Rev. Environ. Sci. Technol. 48(6): 614-654.

Gonzalez-Fernandez, C., B. Sialve and B. Molinuevo-Salces. 2015. Anaerobic digestion of microalgal biomass: Challenges, opportunities and research needs. Bioresour. Technol. 198: 896-906.

Goyal, H.B., D. Seal and R.C. Saxena. 2008. Bio-fuels from thermochemical conversion of renewable resources: A review. Renew. Sustain. Energy Rev. 12: 504-517.

Griffin, M.E., K.D. McMahon, R.I. Mackie and L. Raskin. 1998. Methanogenic population dynamics during start-up of anaerobic digesters treating municipal solid waste and biosolids. Biotechnol. Bioeng. 57: 342-355.

Habets, L. and W. Driessen. 2007. Anaerobic treatment of pulp and paper mill effluents – Status quo and new developments. pp. 223-230. *In*: [Mark Hammond] Forest Industry Wastewaters VIII. IWA Publishing, London, United Kingdom.

Habib, M.A.B., F.M. Yusoff, S.M. Phang, K.J. Ang and S. Mohamed. 1997. Nutritional values of chironomid larvae grown in palm oil mill effluent and algal culture. Aquac. 158: 95-105.

Hadiyanto, M.C. and D. Soetrisnanto. 2013. Phytoremediation of palm oil mill effluent (POME) by using aquatic plants and microalgae for biomass production. Environ. Sci. Technol. 6: 79-90.

Haydary, J. 2017. Modelling of two stage gasification of waste biomass. Chem. Eng. Trans. 61: 1465-1470.

https://www.ars.usda.gov/northeast-area/wyndmoor-pa/eastern-regional-researchcenter/docs/biomass-pyrolysis-research-1/what-is-pyrolysis/

https://www.e-inst.com/training/biomass-to-biogas/

Huang, H., N. Qureshi, M.H. Chen, W. Liu and V. Singh. 2015. Ethanol production from food waste at high solids content with vacuum recovery technology. J. Agric. Food Chem. 63(10): 2760-2766.

Iwuagwu, J.O. and J.O. Ugwuanyi. 2014. Treatment and valorization of palm oil mill effluent through production of food grade yeast biomass. Journal of Waste Management 1-10. Hindawi Publishing Corporation.

Jacob-Lopes, E., M.M. Maroneze, M.C. Deprá, R.B. Sartori, R.R. Dias and L.Q. Zepka. 2019. Bioactive food compounds from microalgae: An innovative framework on industrial biorefineries. Curr. Opin. Food Sci. 25: 1-7.

Kaieda, M., T. Samukawa, T. Matsumoto, K. Ban, A. Kondo, Y. Shimada, H. Noda, F. Nomoto, K. Ohtsuka, E. Izumoto and H. Fukuda. 1999. Biodiesel fuel production from plant oil

catalyzed by *Rhizopus oryzae* lipase in a water-containing system without an organic solvent. J. Biosci. Bioeng. 88(6): 627-631.
Kamali, M. and Z. Khodaparast. 2015. Review on recent developments on pulp and paper mill wastewater treatment. Ecotoxicol. Environ. Saf. 114: 326-342.
Kamyab, H., C. Tin Lee, M.F. Md Din, M. Ponraj, S.E. Mohamad and Sohrabi, M. 2014. Effects of nitrogen source on enhancing growth conditions of green algae to produce higher lipid. Desalination Water Treat. 52: 3579-3584.
Kamyab, H., M.F.M. Din, S.E. Hosseini, S.K. Ghoshal, V. Ashokkumar, A. Keyvanfar, A. Shafaghat, C.T. Lee, A. Bavafa and M.Z.A. Majid. 2016. Optimum lipid production using agro-industrial wastewater treated microalgae as biofuel substrate. Clean Technol. Environ. Policy 18: 2513-2523.
Kapoor, R., P. Ghosh, M. Kumar, S. Sengupta, A. Gupta, S.S. Kumar, V. Vijay, V. Kumar, V.K. Vijay and D. Pant. 2020. Valorization of agricultural waste for biogas based circular economy in India: A research outlook. Bioresour. Technol. 123036.
Karakashev, D., D.J. Batstone and I. Angelidaki. 2005. Influence of environmental conditions on methanogenic compositions in anaerobic biogas reactors. Appl. Environ. Microbiol. 71: 331-338.
Khalil, M., M.A. Berawi, R. Heryanto and A. Rizalie. 2019. Waste to energy technology: The potential of sustainable biogas production from animal waste in Indonesia. Renew. Sustain. Energy Rev. 105: 323-331.
Khetkorn, W., R.P. Rastogi, A. Incharoensakdi, P. Lindblad, D. Madamwar, A. Pandey and C. Larroche. 2017. Microalgal hydrogen production – A review. Bioresour. Technol. 243: 1194-1206.
Kumar, K.N. and S. Goel. 2009. Characterization of Municipal Solid Waste (MSW) and a proposed management plan for Kharagpur, West Bengal, India. Resour. Conserv. Recycl. 53: 166-174.
Kumar, P. and L. Joshi. 2013. Pollution caused by agricultural waste burning and possible alternate uses of crop stubble: A case study of Punjab. pp. 367-385. *In*: Sunil Nautiyal, K.S. Rao, Harald Kaechele, K.V. Raju, Ruediger Schaldach [eds.]. Knowledge Systems of Societies for Adaptation and Mitigation of Impacts of Climate Change. Springer, Berlin, Heidelberg.
Kumar, P., S. Samuchiwal and A. Malik. 2020. Anaerobic digestion of textile industries wastes for biogas production. Biomass Conversion and Biorefinery, 1-10.
Kumar, S., J.K. Bhattacharyya, A.N. Vaidya, T. Chakrabarti, S. Devotta and A.B. Akolkar. 2009. Assessment of the status of municipal solid waste management in metro cities, state capitals, class I cities, and class II towns in India: An insight. J. Waste Manag. 29: 883-895.
Kusdiana, D. and S. Saka. 2001. Methyl esterification of free fatty acids of rapeseed oil as treated in supercritical methanol. J. Chem. Eng. Japan 34(3): 383-387.
Laperrière, W., B. Barry, M. Torrijos, B. Pechiné, N. Bernet and J.P. Steyer. 2017. Optimal conditions for flexible methane production in a demand-based operation of biogas plants. Bioresour. Technol. 245: 698-705.
Lee, S.Y., R. Sankaran, K.W. Chew, C.H, Tan, R. Krishnamoorthy, D.T. Chu and P.L. Show. 2019. Waste to bioenergy: A review on the recent conversion technologies. BMC Energy, 1(1): 4.
Leonel, N.J.R., G. Radu, M.C.O. Joao and C.P.S. Joao. 2018. Economic and environmental benefits of using textile waste for the production of thermal energy. J. Clean. Prod. 171: 1353-1360.
Li, L., J.S. Rowbotham, C.H. Greenwell and P.W. Dyer. 2013. An introduction to pyrolysis and catalytic pyrolysis: Versatile techniques for biomass conversion. 173-208. *In*: New and Future Developments in Catalysis: Catalytic Biomass Conversion. Amsterdam: Elsevier.

Li, Li and Rowbotham, Jack S. and Greenwell, Christopher H. and Dyer, Philip W. (2013) 'An introduction to pyrolysis and catalytic pyrolysis : versatile techniques for biomass conversion.', *In*: New and future developments in catalysis : catalytic biomass conversion. Amsterdam: Elsevier, pp. 173-208.

Li, Q., W. Du and D. Liu. 2008. Perspectives of microbial oils for biodiesel production. Appl. Microbiol. Biotechnol. 80(5): 749-756.

Li, Y., S.Y. Park and J. Zhu. 2011. Solid-state anaerobic digestion for methane production from organic waste. Renew. Sustain. Energy Rev. 15(1): 821-826.

Lin, Y., S. Wu and D. Wang. 2013. Hydrogen-methane production from pulp & paper sludge and food waste by mesophilic–thermophilic anaerobic co-digestion. Int. J. Hydrog. Energy 38(35): 15055-15062.

Ma, F. and M.A. Hanna. 1999. Biodiesel production: A review. Bioresour. Technol. 70(1): 1-15.

Mahdy, A., L. Mendez, S. Blanco, M. Ballesteros and C. González-Fernández. 2014. Protease cell wall degradation of *Chlorella vulgaris*: Effect on methane production. Bioresour. Technol. 171: 421-427.

Mahdy, A., L. Mendez, E. Tomás-Pejó, M. del Mar Morales, M. Ballesteros and C. González-Fernández. 2016. Influence of enzymatic hydrolysis on the biochemical methane potential of *Chlorella vulgaris* and *Scenedesmus* sp. J. Chem. Technol. Biotechnol. 91(5): 1299-1305.

Makarichi, L., W. Jutidamrongphan and K. Techato. 2018. The evolution of waste-to-energy incineration: A review. Renew. Sustain. Energy Rev. 91: 812-821.

Matsumura, Y. 2015. Hydrothermal gasification of biomass. pp. 251-267. *In*: Ashok Pandey, Michael Stöcker Thallada Bhaskar and Rajeev K. Sukumaran [eds.]. Recent Advances in Thermo-Chemical Conversion of Biomass. Elsevier.

May, C.Y. and K. Nesaretnam. 2014. Research advancements in palm oil nutrition. Eur. J. Lipid Sci. Technol. 116(10): 1301-1315.

Mazzoni, L. and I. Janajreh. 2017. Plasma gasification of municipal solid waste with variable content of plastic solid waste for enhanced energy recovery. Int. J. Hydrog. Energy. 42(30): 19446-19457.

Mazzoni, L., R. Ahmed and I. Janajreh. 2017. Plasma gasification of two waste streams: Municipal solid waste and hazardous waste from the oil and gas industry. Energ. Proced. 105: 4159-4166.

Melikoglu, M., C.S.K. Lin and C. Webb. 2013. Analysing global food waste problem: Pinpointing the facts and estimating the energy content. Cent. Eur. J. Eng. 3: 157-164.

Mensink, M. 2007. Speaking the same language, the way forward in tracking industrial energy efficiency and CO_2 emissions. *In*: International Council of Forest & Paper Associations, Presentation at Expert Review Workshop, International Energy Agency, Oct 1-2, 2007, Paris.

Merlin, G. and H. Boileau. 2013. Anaerobic digestion of agricultural waste: State of the art and future trends. *In*: A. Torres [ed.]. Anaerobic Digestion: Types, Processes and Environmental Impact. Nova Science Publishers, Inc., New York.

Mital, K.M. 1997. Biogas Systems: Policies, Progress and Prospects. Taylor & Francis. pp. 278.

Montingelli, M.E., S. Tedesco and A.G. Olabi. 2015. Biogas production from algal biomass: A review. Renew. Sustain. Energy Rev. 43: 961-972.

Nguyen, M.A. and A.L. Hoang. 2016. A Review on Microalgae and Cyanobacteria in Biofuel Production. USTH: Hanoi, Vietnam. 1-37.

Noureddini, H. and D. Zhu. 1997. Kinetics of transesterification of soybean oil. J. Am. Oil Chem. Soc. 74(11): 1457-1463.

Nwuche, C.O. and E.O. Ugoji. 2010. Effect of co-existing plant species on soil microbial activity under heavy metal stress. Int. J. Environ. Sci. Technol. 7(4): 697-704.

OECD/IEA. 2010. Sustainable Production of Second-Generation Biofuels, Potential and Perspectives in Major Economies and Developing Countries, Information Paper. IEA. Paris, France.

Onyia, C.O., A.M. Uyu, J.C. Akunna, N.A. Norulaini and A.K. Omar. 2001. Increasing the fertilizer value of palm oil mill sludge: Bioaugmentation in nitrification. Water Sci. Technol. 44: 157-162.

Paritosh, K., S. Kushwaha, M. Yadav, N. Pareek, A. Chawade and V. Vivekanand 2017. Food waste to energy: an overview of sustainable approaches for food waste management and nutrient recycling. Biomed. Res. Int. 1-20.

Passos, F., E. Uggetti, H. Carrère and I. Ferrer. 2014. Pretreatment of microalgae to improve biogas production: A review. Bioresour. Technol. 172: 403-412.

Pattnaik, S. and M.V. Reddy. 2010. Assessment of municipal solid waste management in Puducherry (Pondicherry), India. Resour. Conser. Recycl. 54: 512-520.

Pavlostathis, S.G. and E. Giraldo-Gomez. 1991. Kinetics of anaerobic treatment: A critical review. Crit. Rev. Environ. Sci. Technol. 21(5-6): 411-490.

Peng, S. and L.M. Colosi. 2016. Anaerobic digestion of algae biomass to produce energy during wastewater treatment. Water Environ. Res. 88(1): 29-39.

Perna, A., M. Minutillo, A.L. Lavadera and E. Jannelli. 2018. Combining plasma gasification and solid oxide cell technologies in advanced power plants for waste to energy and electric energy storage applications. J. Waste Manag. 73: 424-438.

Poh, P.E. and M.F. Chong. 2009. Development of anaerobic digestion methods for palm oil mill effluent (POME) treatment. Bioresour. Technol. 100(1): 1-9.

Pokhrel, D. and T. Viraraghavan. 2004. Treatment of pulp and paper mill wastewater – A review. Sci. Total Environ. 333: 37-58.

Pontual, L., F.B. Mainier and G.B.A. Lima. 2015. The biogas potential of pulp and paper mill wastewater: An Essay. Am. J. Environ. Engineer. 5: 53-57.

Rago, Y.P., D. Surroop and R. Mohee. 2018. Torrefaction of textile waste for production of energy-dense biochar using mass loss as a synthetic indicator. J. Environ. Chem. Eng. 6(1): 811-822.

Rajmohan, K.S. and S. Varjani. 2019. Trends and advances in bioenergy production and sustainable solid waste management. Energy & Environment, p.0958305X19882415.

Ramos, A., E. Monteiro, V. Silva and A. Rouboa. 2018. Co-gasification and recent developments on waste-to-energy conversion: A review. Renew. Sustain. Energy Rev. 81: 380-398.

Richmond, A. 2004. Handbook of microalgal culture: Biotechnology and applied phycology (Vol. 577). Oxford: Blackwell Science. pp. 556.

Rodolfi, L., G. Chini Zittelli, N. Bassi, G. Padovani, N. Biondi, G. Bonini and M.R. Tredici. 2009. Microalgae for oil: Strain selection, induction of lipid synthesis and outdoor mass cultivation in a low-cost photobioreactor. Biotechnol. Bioeng. 102(1): 100-112.

Roy, P. and G. Dias. 2017. Prospects for pyrolysis technologies in the bioenergy sector: A review. Renew. Sustain. Energy Rev. 77: 59-69.

Sablayrolles, J.M. 2009. Control of alcoholic fermentation in winemaking: Current situation and prospect. Int. Food Res. 42(4): 418-424.

Sanlisoy, A. and M. Carpinlioglu. 2017. A review on plasma gasification for solid waste disposal. Int. J. Hydrog. Energy. 42: 1361-1365.

Sansaniwal, S.K., K. Pal, M.A. Rosen and S.K. Tyagi. 2017. Recent advances in the development of biomass gasification technology: A comprehensive review. Renew. Sustain. Energy Rev. 72: 363-384.

Santos-Ballardo, D.U., S. Rossi, C. Reyes-Moreno and A. Valdez-Ortiz. 2016. Microalgae potential as a biogas source: Current status, restraints and future trends. Rev. Environ. Sci. Biotechnol. 15(2): 243-264.

Sarath, N.G. and J.T. Puthur. 2020. Heavy metal pollution assessment in a mangrove ecosystem scheduled as a community reserve. Wetl. Ecol. Manag. 1-12.

Schink, B. 1997. Energetics of syntrophic cooperation in methanogenic degradation. Microbiol. Mol. Biol. Rev. 61(2): 262-280.

Schuchardt, U., R. Sercheli and R.M. Vargas. 1998. Transesterification of vegetable oils: A review. J. Braz. Chem. Soc. 9(3): 199-210.

Sharma, A. and S.K. Arya. 2017. Hydrogen from algal biomass: A review of production process. Biotechnol. Rep. 15: 63-69.

Sharma, B.K. and H. Kaur. 2000. Environmental Chemistry. Krishna Prakashan Media (P) Ltd. Meerut, UP.

Shokrkar, H., S. Ebrahimi and M. Zamani. 2017. Bioethanol production from acidic and enzymatic hydrolysates of mixed microalgae culture. Fuel 200: 380-386.

Siddique, M.N.I. and Z.A. Wahid. 2018. Achievements and perspectives of anaerobic co-digestion: A review. J. Clean. Prod. 194: 359-371.

Silva, D.A.L., I. Delai, M.L.D. Montes and A.R. Ometto. 2014. Life cycle assessment of the sugarcane bagasse electricity generation in Brazil. Renew. Sustain. Energy Rev. 32: 532-547.

Simonyan, K.J. and O. Fasina. 2013. Biomass resources and bioenergy potentials in Nigeria. Afr. J. Agric. Res. 8: 4975-4989.

Singh, R.P., M.H. Ibrahim, N. Esa and M.S. Iliyana. 2010. Composting of waste from palm oil mill: A sustainable waste management practice. Rev. Environ. Sci. Biotechnol. 9: 331-344.

Skogsindustrierna 2010. The Swedish Forest Industries, Facts and Figures e Report. Swedish Forest Industries Federation.

Smith, O.B. 1987. Utilization of crop residues in the nutrition of sheep and goats in the humid tropics of West Africa. pp. 21-25. *In*: Proceedings of a Seminar held in Yamoussoukro, Côte d'lvoire.

Sorathiya, L.M., A.B. Fulsoundar, K.K. Tyagi, M.D. Patel and R.R. Singh. 2014. Eco-friendly and modern methods of livestock waste recycling for enhancing farm profitability. Int. J. Recycl. Org. Waste Agric. 3(1): 50.

Srivastava, A. and R. Prasad. 2000. Triglycerides-based diesel fuels. Renew. Sustain. Energy Rev. 4(2): 111-133.

Stanbury, P.F., A. Whitaker and S.J. Hall. 2013. Principles of fermentation technology. Jour 2: 1-351. Elsevier.

Staubmann, R., G. Foidl, N. Foidl, G.M. Gübitz, R.M. Lafferty, V.M.V. Arbizu and W. Steiner. 1997. Biogas production from Jatropha curcas press-cake. Biotechnol. Appl. Biochem. 63(1): 457.

Stürmer, B. 2017. Feedstock change at biogas plants – Impact on production costs. Biomass Bioenerg. 98: 228-235.

Sugiawan, Y. and S. Managi. 2019. New evidence of energy-growth nexus from inclusive wealth. Renew. Sustain. Energy Rev. 103: 40-48.

Sun, L., M. Fujii, T. Tasaki, H. Dong and S. Ohnishi. 2018. Improving waste to energy rate by promoting an integrated municipal solid-waste management system. Resour. Conserv. Recycl. 136: 289-296.

Surendra, K.C., D. Takara, A.G. Hashimoto and S.K. Khanal. 2014. Biogas as a sustainable energy source for developing countries: Opportunities and challenges. Renew. Sustain. Energy Rev. 31: 846-859.

Talyan, V., R.P. Dahiya and T.R. Sreekrishnan. 2008. State of municipal solid waste management in Delhi, the capital of India. J. Waste Manag. 28: 1276-1287.

Tan, H.M., D. Gouwanda and P.E. Poh. 2018. Adaptive neural-fuzzy inference system vs. anaerobic digestion model No. 1 for performance prediction of thermophilic anaerobic digestion of palm oil mill effluent. Process Saf. Environ. Prot. 117: 92-99.

Toor, S.S., L.A. Rosendahl, J. Hoffmann, T.H. Pedersen, R.P. Nielsen and E.G. Søgaard. 2014. Hydrothermal liquefaction of biomass. pp. 189-217. *In*: Fangming [ed.]. Application of Hydrothermal Reactions to Biomass Conversion. Springer, Berlin, Heidelberg.

UNEP. 2011. Towards a Green Economy: Pathways to Sustainable Development and Poverty Eradication.

Uzoejinwa, B.B., X. He, S. Wang, A. El-Fatah Abomohra, Y. Hu and Q. Wang. 2018. Co-pyrolysis of biomass and waste plastics as a thermochemical conversion technology for high-grade biofuel production: Recent progress and future directions elsewhere worldwide. Energ. Convers. Manag. 163: 468-492.

Vaez, E. and H. Zilouei. 2020. Towards the development of biofuel production from paper mill effluent. Renew. Energy. 146: 1408-1415.

Wang, J., G. Cheng, Y. You, B. Xiao, S. Liu, P. He, D. Guo, X. Guo and G. Zhang. 2012. Hydrogen-rich gas production by steam gasification of municipal solid waste (MSW) using NiO supported on modified dolomite. Int. J. Hydrog. Energy. 37(8): 6503-6510.

Wang, S., S. Yu, Q. Lu, Y. Liao, H. Li, L. Sun, H. Wang and Y. Zhang. 2020. Development of an alkaline/acid pre-treatment and anaerobic digestion (APAD) process for methane generation from waste activated sludge. Sci. Total Environ. 708: 34564.

Weiland, P. 2010. Biogas production: Current state and perspectives. Appl. Microbiol. Biotechnol. 85(4): 849-860.

Wild, K.J., H. Steinga and M. Rodehutscord. 2019. Variability of in vitro ruminal fermentation and nutritional value of cell disrupted and nondisrupted microalgae for ruminants. Gcb Bioenerg. 11(1): 345-359.

Williams, P.T. and N. Nugranad. 2000. Comparison of products from the pyrolysis and catalytic pyrolysis of rice husks. Energy, 25(6): 493-513.

Woolard, R. 2009. Logistical model for closed loop recycling of textile materials [thesis]. North Carolina: Textile Engineering.

Wu, C.Z., X.L. Yin, Z.H. Yuan, Z.Q. Zhou and X.S. Zhuang. 2010. The development of bioenergy technology in China. Energy, 35(11): 4445-4450.

Wu, T.Y., A.W. Mohammad, J.M. Jahim and N. Anuar. 2009. A holistic approach to managing palm oil mill effluent (POME): Biotechnological advances in the sustainable reuse of POME. Biotechnol. Adv. 27: 40-45.

Yamaguchi, S.K.F., J.B. Moreira, J.A.V. Costa, C.K. de Souza, S.L. Bertoli and L.F.D. Carvalho. 2019. Evaluation of adding spirulina to freeze-dried yogurts before fermentation and after freeze-drying. Ind. Biotechnol. 15(2): 89-94.

Yang, Y., J.G. Brammer, M. Ouadi, J. Samanya, A. Hornung, H.M. Xu and Y. Li. 2013. Characterisation of waste derived intermediate pyrolysis oils for use as diesel engine fuels. Fuel, 103: 247-257.

Yi, H., M. Jiang, D. Huang, G. Zeng, C. Lai, L. Qin, C. Zhou, B. Li, X. Liu, M. Cheng and W. Xue. 2018. Advanced photocatalytic Fenton-like process over biomimetic hemin-Bi2WO6 with enhanced pH. J. Taiwan Inst. Chem. Eng. 93: 184-192.

Yousef, S., J. Eimontas, N. Striūgas, M. Tatariants, M.A. Abdelnaby, S. Tuckute and L. Kliucininkas. 2019. A sustainable bioenergy conversion strategy for textile waste with self-catalysts using mini-pyrolysis plant. Energy Convers. Manag. 196: 688-704.

Zhao, Y., C. Yue, S. Geng, D. Ning, T. Ma and X. Yu. 2019. Role of media composition in biomass and astaxanthin production of *Haematococcus pluvialis* under two-stage cultivation. Biopro. Biosyst. Eng. 42(4): 593-602.

Zwain, H.M., S.R. Hassan, N.Q. Zaman, H.A. Aziz and I. Dahlan. 2013. The start-up performance of modified anaerobic baffled reactor (MABR) for the treatment of recycled paper mill wastewater. J. Environ. Chem. Eng. 1(1-2): 61-64.

CHAPTER

10

Reclamation and Phytoremediation of Heavy Metal Contaminated Land

E. Janeeshma[1], Akhila Sen[1], K.P. Raj Aswathi[1], Riya Johnson[1], Om P. Dhankher[2] and Jos T. Puthur[1]*

[1] Plant Physiology and Biochemistry Division, Department of Botany, University of Calicut, C.U. Campus P.O., Kerala – 673635, India

[2] Agriculture Biotechnology, Stockbridge School of Agriculture, University of Massachusetts, Amherst (U.S.A.)

10.1 Introduction

Heavy metals are elements having high density, atomic weight and number. Mercury (Hg), copper (Cu), cadmium (Cd), zinc (Zn), chromium (Cr), nickel (Ni), arsenic (As) and lead (Pb) are the common heavy metals originating from both natural or anthropogenic sources and causes pollution at a greater extent (Singh et al. 2011, Sarath and Pothur 2020). Heavy metals present in the soil for a long time are a major hazard to the environment (Suman et al. 2018). Land and water pollution due to heavy metal contamination is a significant risk at the global level. All the countries face this situation, but the intensity may vary depending on the area and severity of affecting pollution.

Major disasters like the Exxon Valdez oil spill, Minimata disease and radioactive release due to the Chernobyl accident are pointing out the necessity of preventing the contamination of the environment due to effluents from such sources (Kumar et al. 2011). To overcome heavy metal contamination, techniques that can entirely eliminate the pollutants, or convert them into biodegradable substances have a great importance in maintaining a sustainable environment. This method can be achieved through phytoremediation (Saxena et al. 2019). Phytoremediation includes eradication of elemental pollutants or reducing their bioavailability in the soil using plants. Phytoremediation is an economically feasible and eco-friendly method that can enhance soil fertility through increased release of various organic substances (Ali et al. 2013).

Phytoremediation is also known as the "Green Revolution" in the area of innovative clear-out technologies. According to EPA's Comprehensive Environmental

Corresponding author: jtputhur@yahoo.com

Response Compensation Liability Information System (CERCLIS), using phytoremediation technique, almost 30,000 waste sites contaminated due to battery manufacturers, electroplating, metal finishing, and mining companies were cleaned in US. As compared with USA and Canada, the application of phytotechnology is limited in other countries. Research on phytoremediation technology is flourishing in Europe even if there is a shortage of enough funding (Sharma and Pandey 2014). In such countries, focusing on pollution control through plant resources by forming private companies has become a trend (Lelie et al. 2001). In South Africa, "Ecological Engineering and Phytoremediation Research Programme" was introduced in 1995 by AngloGold Ashanti (then Anglo American Gold Division) and the School of Animal, Plant and Environmental Sciences (APES) of the University of the Witwatersrand, Johannesburg (Wits University). AngloGold Ashanti reduced the negative impact of company's tailings storage facilities (TSFs) by planting half a million number of trees over the last decade using phyto (plant) technology. There are several attempts put forward to prevent/reduce heavy metal contamination. Initiatives like Global Innovation Solution SRL (Romania), BIOVALA (Lithuania), MLM group, Biofuel NET, SUEZ group, Pond tech, Biofuel NET, BioRemed AB (Sweden), BioPlanta (Germany), Phytorem (France), Clean Biotec S.L.L (Spain), and Waterloo Environmental Biotechnology Inc (Canada) are focusing on delivering cost effective solutions for soil and water pollutions and thus intends to ensure environmental sustainability. These industrial sectors play a major role in the contaminated sites through phyto/bio remediation techniques.

For effective phytoremediation, active involvement and collaboration of native people, farmers, technology suppliers and advisers, remediation experts, regulatory agencies, financial sponsors, NGOs and other voluntary organizations plays a critical role (Hauptvogl et al. 2019). However, due to the lack of practical knowledge, still there are uncertainties in the public view about phytoremediation and, hence, there is limited acceptance among people. To overcome this, the government has to make strong awareness and enough investments to encourage the clean-up initiative among the public and private sectors. For this purpose, a combination of diverse methods like genetic engineering, microbe-assisted and chelate-assisted methods could be more effective in the future.

10.2 Current Scenario

In Western Europe, it was reported that 1,400,000 sites were polluted with heavy metals (McGrath et al. 2001). As per EPA report up to May 2004, the United States possesses almost greater than 40,000 contaminated areas. Furthermore, 600,000 brownfields in the USA are contaminated by heavy metals and require reclamation, due to which 1,00,000 ha of cropland, 55,000 ha of pasture and 50,000 ha of forest have been lost (Sharma and Pandey 2014).

Developing countries like India, China, Pakistan and Bangladesh are also facing soil and water pollution due to the effluent discharge from industrial units. In China, one-sixth of total arable lands are exposed to heavy metal contamination, as well as more than 40% were destroyed because of erosion and desertification (Tang et al. 2013). Industrialization results in the increased content of Ni, Zn, Fe, Cd, Cr, Cu and

Pb in surface sediments of Xiamen, China (Zhang et al. 2007). Recent reports prove that in China, industrial activities resulted in the increased heavy metal pollution of soil. As per the data of Bulletin on National Survey of Soil Contamination, almost 4 million hectares of arable lands were polluted due to Cu, Pb, Ni, As, Cd, DDT (dichloro-diphenyl-trichloroethane), Hg, and PAHs (polycyclic aromatic hydrocarbons) contamination (Su 2014).

A study conducted in Pakistan proved that the coastal sediment is severely contaminated due to Cd, Fe, Pb, Cu, Cr, Co, Zn and Ni (Siddiqui and Saher 2021). In another study conducted, it was proved that increasing industrial activities result in severe ecological concern in Pakistan due to soil and water pollution. Heavy metals such as Cr, Co, Mn, Ni, Cd, Fe, Pb and Zn have resulted in soil and groundwater contamination (Afzal et al. 2014). Similarly, Khan et al. (2010) studied the heavy metals contamination from rock sites in vegetables and soil in Gilgit, Northern Pakistan. Their study detected higher content of heavy metals such as Pb, Zn, Cu, and Ni in vegetables like *Brassica oleracea*, *Mintha sylvestris*, *B. campestris* and *Malva neglecta*. In India, irrigation using wastewater results in the increased heavy metal contamination of soil resulting in the poor quality of crop production. A study by Sharma et al. (2009) proved that wastewater irrigation resulted in heavy metal content such as Ni, Cd, and Pb in the cultivated vegetables collected from production and market sites of a tropical urban area of India. In a study conducted by Patel et al. (2018), majority part of the river Swarnamukhi, India is contaminated with Pb, Cr, Zn, and Cu, which are from various anthropogenic sources.

10.3 Importance of Industrial Developments in Metal Contamination

Rapid industrialization and concomitant increase in the number of toxic metals and metalloids has become a matter of immense concern. Sequestration of the hazardous heavy metals through the phytoremediation potential of plants reduces the risk of exposure of soil flora and fauna to a vast range of toxic materials (Ali et al. 2013, Janeeshma and Puthur 2020). A thorough theoretical and practical knowledge regarding the use of plants for mitigating the hazardous contaminants from the polluted sites will provide futuristic promise to resolve the problem of contamination (Arthur et al. 2005)

According to the national general survey of soil contamination published by the Ministry of Ecology and Environment and the Ministry of Natural Resources in 2014, 19.4 % of the agricultural land in China exceeded Level II requirements of the soil environmental quality standard. It was reported that there are about 1.7 million heavy metal contaminated sites in central and eastern European countries (Li et al. 2019). In developing countries like India, China and Pakistan, industrial effluents cause serious threats to the natural ecosystem (Sharma and Pandey 2014). Due to the enormous ecological and health risk exerted by these hazardous materials, we need to urgently plan for a green and sustainable solution for the reclamation of contaminated lands. This can be attained through various phytoremediation methods. Phytoremediation has more acceptance due to its eco-friendly and cost-effective nature (Pandey et al. 2016). Numerous plants having this remediation capability are used for the restoration of the polluted land. The accelerated rate of increase in contamination sites demands the need for large-scale phytoremediation industries.

Exceeded level in the expulsion of industrial effluents and the subsequent increase in the level of toxic materials becomes a severe problem of the era. Phytoremediation techniques have their role in such polluted sites (Luo et al. 2017, Sameena and Puthur 2021). Even though the government is spending a huge part of its economy for the reclamation of contaminated sites, the resulting outcome is not satisfactory. The world requires a cost-effective and eco-sustainable method for the renovation of the polluted areas.

Different countries around the world are taking initiatives to develop a worthy technology for the decontamination of land. Government, academicians and non-profit research groups are hardly utilizing the full potential of phytoremediation (Hauptvogl et al. 2019). To explore phytoremediation on a larger scale, multi-technique phytoremediation is being experimented (Luo et al. 2017). Modern scale phytoremediation utilizes the advantage of genetically modified organisms (GMO). Genetic engineering provides enhanced phytoremediation potential to plants through the augmented expression level of metal-chelating genes (Rai et al. 2020). Transgenic *Beta vulgaris* overexpressing StGCS-GS showed enhanced tolerance to Zn, Cd and Cu (Liu et al. 2015). Similarly, overexpression of *Arabidopsis thaliana* ACR2 gene (AtACR2) in *Nicotiana tabacum* increased the phytoremediation of As (Nahar et al. 2017).

Several plants are used as potential candidates for phytoremediation. *Helianthus annuus* proved to be an effective strategy to reduce the level of uranium (U) in contaminated soil in Jordan (Alsabbagh and Abuqudaira 2017). Likewise, *Brassica juncea* was used as a phytoremediation agent in Pb contaminated site in Trenton, New Jersey (Blaylock et al. 1999). The current trends in phytoremediation exploit the combined application of knowledge from the field of genetics, molecular biology, omics, metabolic engineering, and nanotechnology (Rai et al. 2020).

Governmental and non-governmental organizations are more focusing on the reclamation of contaminated lands. They offers job opportunities in various fields and helps in remediating the polluted land in an eco-sustainable manner. Even though the governments have invested a significant amount in conventional phytoremediation technologies, it is inadequate to clean up the contaminated sites. Furthermore, the maintenance of the decontaminated sites by preventing further exposure requires continuous input of energy and money. Leaving the contaminated space as such leads to the spreading of contaminants, and also prevents the utilization of the land for any other purposes. Creating awareness among the native people living in the contaminated sites is another difficult task faced by experts. Hence, collaborative efforts jointly implemented by government and private sectors can provide a solution for this problem.

Large-scale phytoremediation markets get immense importance in the current scenario of global pollution. At present, the two largest markets for phytoremediation are in the USA and Europe. In addition to the clearing up of the polluted lands, phytoremediation offers opportunities in the field of entrepreneurship and employment; also, it gives opportunities for applied research. Opportunities offered by phytoremediation industries are represented in Fig. 10.1.

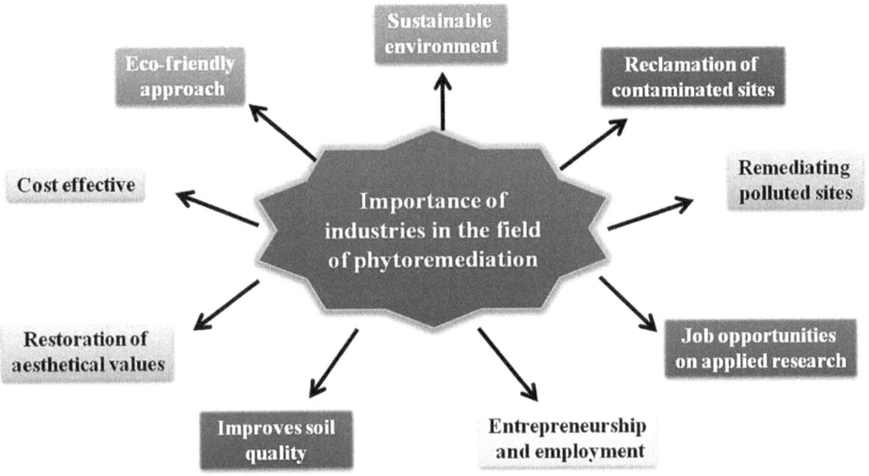

Figure 10.1: Importance and possibilities of industries in phytoremediation sector.

Field projects were successfully conducted in various contaminated sites in Europe. Lead (Pb) and Cd contaminated sites in Czechowice oil refinery in Poland were reclaimed using *B. juncea* by Phytotech, Florida State University, Institute for Ecology of Industrial Areas. Similarly, *Salix viminalis* was used by the Swiss Federal Institute of Technology for the remediation of the Zn and Cd contaminated sites in Switzerland. Implementation of innovative ideas in phytoremediation market sector provides various advantages over conventional techniques. Phytoremediation conducted by various industries for the reclamation of contaminated sites are listed in Table 10.1.

10.4 Modeling Phytoremediating Plant: Role of Public and Private Sectors

Contaminants are a global menace faced by common people of the world. Among them, heavy metal contamination is the major one and it is very important to find a solution to the current problem on an urgent basis. Phytoremediation is holistic, sustainable and cost-effective approach that involves a series of events to decontaminate heavy metals from both land as well as aquatic fields. Various researchers and companies have introduced a large number of plant candidates with phytoremediation potential. Bioenergy crops can be excellent candidates for decontamination of heavy metal contaminated lands as they possess additional advantage of producing renewable energy and hence bioenergy crops is promising to the increased demand of the non-renewable fossil fuels. Various crops such as poplars, willow, *Ricinus communis*, *Jatropha curcas*, *Acacia nilotica*, *Hibiscus cannabinus*, *Cannabis sativa*, *Azadirachta indica*, *Leucaena leucocephala*, *Millettia pinnata*, *Brassica juncea*, *Helianthus annuus* and *panicum vigratum* are successfully introduced into the heavy metal contaminated sites (EPA 1998). Similarly in water bodies, severe heavy metals pollution can be remediated and cleared using different

Table 10.1: Phytoremediation conducted by different organizations

Area contaminated	Xenobiotics	Clearing organization	Plants used	Family	References
US	Uranium	Edenspace System Corporation	*Helianthus annuus* and *Brassica juncea*	Asteraceae Brassicaceae	Singh et al. (2006)
Fort Greely, USA	Strontium	Edenspace System Corporation	*Brassica juncea*	Brassicaceae	Singh et al. (2006)
Switzerland	Zinc and cadmium	Swiss Federal Institute of Technology	*Salix viminalis*	Salicaceae	Singh et al. (2006)
University of Glasgow	Nickel, copper, zinc, cadmium	Sewage disposal site (United Kingdom)	*Salix viminalis*	Salicaceae	Singh et al. (2006)
Bayonne, New Jersey	Lead	Phytotech, Inc.	*Brassica juncea*	Brassicaceae	Watanabe (1997)
Dorchester, Maine	Lead	Edenspace System Corporation	*Brassica juncea*	Brassicaceae	Henry (2000)
U.S.A.	Arsenate	U.S. Army Corp of Engineers Edenspace Systems, Inc.	*Pteris cretica* and *Pteris vittata*	Pteridaceae	Ebbs et al. (2009)
Bayonne, New Jersey	Lead	Phytotech Inc.	*Brassica juncea*	Brassicaceae	Saxena et al. (1999)
Chernobyl	137Cs and 9OSr,	Phytotech Inc.	Helianthus annuus	Asteraceae	Saxena et al. (1999)
Holte, Denmark	Cyanide, BTEX, PAHs, oil	Technical University of Denmark	Salix viminalis and *Populus tremula*	Salicaceae	Schwitzguébe et al. (2002)

plant species such as water hyacinth, azolla, poplar, and duck weed (Liphadzi et al. 2003). In contrast to higher plant species, bioremediation via algae is a novel approach in the field of water management. As green algae are having photosynthetic machinery, they are able to produce higher biomass even under polluted environments. Moreover, being enriched with higher lipid content it can very well act as a source of biofuel production. The selection of plant and algal species having phytoremediation potential and the ability to produce higher biomass is an essential feature to be considered while preparing a water based treatment method. Different industries involved in developing phytoremediating plant candidates are:

10.4.1 Exxon Mobil Corporation

Exxon is a research based American company headquartered in Irving, Texas that works on bioremediation of contaminated hydrocarbons in soil surface. The company has found that treating the soil with microorganisms is a better technique of phytoremediation. Several laboratory studies have been conducted by Exxon for the polycyclic aromatic hydrocarbons' (PAH) biodegradability in contaminated refinery soil via phytoremediation. The incorporation of plants in the contaminated site appears to increase the rate of heavy metal degradation. In addition, when the treatment of soil with different methods is compared, the economic cost of using phytoremediation is less than half of that needed for microbial bioremediation.

10.4.2 Chevron

Chevron is a company indulged in clearing lands using phytoremediation. The company carried out a field research project in Ogden, Utah, to investigate the phytoremediation potential of poplars, to restrict the transport of solute to ground water. Generally, the ground waters are contaminated by diesel and gasoline components and further reaches deep down to eight feet below the surface of soil. Here the roots of plants absorb and forcefully uptake the contaminant by acting as a barrier. Poplars were planted in three rows with a distance of six feet apart and its roots reached the ground water. Thus, poplar was known as a plant for phytoremediation (Liphadgi et al. 2003).

10.4.3 Gas Research Institute (GRI)

Gas Research Institute, situated in Chicago, is involved in treating natural gas. According to statistical data, in United States there are around 260,000 numbers of gas wells and 700 gas processing plants. GRI is an international company concerned with identifying techniques such as phytoremediation, bioventing and land farming for clearing. Some of the well-known projects conducted by GRI include Environmentally Acceptable Project to restrict hydrocarbon mobility movements in soils.

10.4.4 Sierra Environmental Services, Inc.

Sierra Environmental Services is a company which uses plants to remediate the contaminated site in Florida. The project used non-riparian phreatophytes to clear out the groundwater contaminants of a disposal lagoon area. These non-riparian phreatophytes are commonly called as water loving plants, which develop deep roots to absorb water from the deep down surface. For the study, a number of phreatophytes were evaluated such as alders, ash, aspen, river birch, cottonwoods, mesquite, bald cypress, eucalyptus, greasewood, salt cedar, willows and poplar. These plants restrict the movement of heavy metals into water table and improve the intrinsic bioremediation. This method of bioremediation was considered as an efficient remedy for the contaminated lagoon.

10.4.5 Monsanto

Monsanto is a company that has keen interest in issues of phytoremediation. Monsanto identified various contaminated field sites and investigated the possibility of phytoremediation and its potential effects in contaminated sites. A novel technique developed and introduced by Monsanto was Lasagna TM technology. The new Lasagna TM technology includes a series of events, mainly electroosmosis, which are applied directly in contaminated soils with a remedial approach. This process can also hold an advantage of moving water from ground to plant root zones (Lambert et al. 2000).

10.4.6 Occidental Chemical Corporation

Occidental Chemical Corporation is an agency where Poplar trees are used to decontaminate trichloroethylene (TCE) contaminated groundwater. Several experiments were performed, which indicates that poplars absorb and take up the contaminants from the decontaminated soil. State of Washington provided the permission to Occidental to conduct field trials using applied TCE. Field experiment of two years was conducted to assess the rate of decontamination by hybrid poplars from groundwater. It was also noted that poplars possess the potential of high transpiration through which TCE can be removed by poplar trees (Wood 1997).

10.4.7 Phytotech

Phytotech is commonly involved in phytoremediation technologies by engineering plants to treat water and soil contaminated with heavy metals. The main treatment method used in their technology was phytoextraction. It is a simple, cost effective, sustainable and friendly approach towards removal of heavy metals. Here, plants are grown in contaminated soil and harvested after accumulation of toxic metals in tissue of the plant. The rate of accumulation of heavy metal varies for different plant species and thus cleans up the whole land. Statistical data suggests that from 10-acre site phytoextraction can remove 400 ppm of Pb from the top one foot, which requires 500 tons of biomass and $1/4^{th}$ of the soil was cleaned. Phytotech has undertaken several programs and among them, SITE and rhizofiltration technology are widely known. Under the SITE program, phytoremediation technology was done in the contaminated soil at metal-plating facility in Findlay, Ohio. Several other successful field remediations were carried out in U.S and abroad by this company. Rhizofiltration was done to remove cesium/stromium (Cs/Sr) at Chernobyl and uranium (U) from polluted groundwater in Ashtabula, Ohio. At Chernobyl, sunflowers filtered and absorbed the radionuclides from small ponds within 10 days. Similarly in Ashtabula site, a nine-month project by Phytotech cleaned up the U containing groundwater.

10.4.8 National Risk Management Research Laboratory (NRMRL)

NRMRL is an environmental protection agency situated in America, with headquarters in Cincinnati, Ohio and divisions in North Carolina, Oklahoma, and New Jersey. The agency is involved in risk management research and specifically solves environmental issues (NRMRL, 2000). They are recognized as leader of scientific expertise, which directly supports the action to reduce contaminations of air, water and land, cleanup of hazardous heavy metals and improve water quality. Their main aim is to prevent and reduce air, water and land pollution and thus to restore ecosystem (EPA 2000).

10.4.9 Remediation Technology Development Forum (RTDF)

The RTDF is a development forum which discusses the collaborative projects and works to be conducted and performed to decontaminate, purify and protect the environment from hazardous waste and heavy metal accumulation. It was started in 1992 with collaboration between two parties, i.e. Monsanto and EPA with an aim to hold hands between government and industry together with developing innovative ideas to overcome contamination problems. The RTDF was introduced to progress the development of more cost-effective, economically feasible approach for the remediation of contaminated sites.

10.4.10 National Exposure Research Laboratory in Athens, Georgia

NERL is a protection agency that looks for benefits and limitations of phytoremediation and phytodegradation. The agency described and investigated different processes involved in phytoremediation such as phytoextraction, phytoaccumulation, rhizofiltration, phytodegradation, and phytovolatilization. The advantage of using plants are that they have aesthetic values, balanced water system, advanced enzymatic machinery, are nutrients rich, and have potential of completely breaking down hazardous contaminants, and are economically feasible. Now research focuses on identifying the pathways and signalling mechanisms of hazardous material degradation by vascular plants and microorganisms. A collaborative work of Athens laboratory and the Army at the Iowa Army Ammunition Plant to phytoremediate heavy metals and radionuclides is under progress.

10.5 Clearing of Contaminated Land

Restoring the contaminated land is very essential for a nation to fulfill the basic needs of the increasing population. The task of clearing contaminated land is being achieved with the help of different firms and government agencies. Governments of different countries always made efforts for the remediation of contaminated land.

But due to different socio-economic crises, the participation of private industries in the field of phytoremediation is the need of the hour.

The generation of industrial waste is the major contribution of mankind to increase the contamination of soil and groundwater. So it is essential to arrange a waste management plan for each such industry. But these industries are also facing difficulties to obtain proper support and infrastructure for cleanup of the contaminated area. Moreover, groundwater contamination is also a growing concern for the human population. Unfortunately, conventional pump-and-treat method is being used for the cleanup of groundwater at 93 percent of contaminated sites (EPA 1996). This is a high cost and low performance method as compared to the other technologies. Thus, it is essential to develop new industries in the field of hazardous material management, phytoremediation, and land rehabilitation.

10.5.1 Government agencies in contaminated land cleanup programs

10.5.1.1 National Research Council (NRC) Canada

The scientific research of the country is chiefly dependent on this institute, started by President John McDougall in 2011. Algal carbon conversion flagship program is one of the best strategies of waste management and bioenergy production introduced by NRC with the involvement of Canadian Natural Resources Limited (Canadian Natural) and Pond Technologies within a budget of \$19 million. In this project, 25,000 L photobioreactor was planned to be constructed for the transformation of CO_2 in the environment to the algal biomass. The infrastructure and the photobioreacter is by the support of the Pond Technologies and the algal strains were supplied by the Government.

The Government of Canada introduced a 15-year program, the Federal Contaminated Sites Action Plan (FCSAP), which was executed in 2005 with a budget of \$4.54 billion. Due to the demand for environmental sustainability, FCSAP was renewed for another 15 years. The major aim of this program is to analyze the highest priority sites, federal contaminated sites and associated federal financial liabilities, and the formulation of appropriate strategies to avoid environmental and human health risks due to the toxic xenobiotics contaminated sites (FCSAP 2012).

The National Research Council (NRC), associated with Department of Fisheries and Oceans (DFO), worked on other important project, and proved that the microbial assisted biodegradation of oil at 1°C in the high Arctic marine region is possible. Program of Energy Research and Development (PERD) of NRC significantly contributed to the funding of this project. This work mainly focused on the kinetics of microbial actions.

10.5.1.2 EPA (Environmental Protection Agency)

The Environmental Protection Agency of US introduced "Initiatives to Promote Innovative Technology in Waste Management Programs" (EPA 1996) for the establishment of partnerships between government and different industries. The Technology Innovation Office of EPA mainly focuses on the innovation of novel remediation technologies that increase research on remediation technologies. Different methods to analyze sediment cleanup programs, on-site and off-site methods to assess the intensity of metal transfer to humans, improved models for predicting the biomagnification of xenobiotics to the fishes, and faster and cheaper tools that detect changes in sediment toxicity are the major outcomes of EP soil clearing program. EPA Brownfield programs and Land Revitalization Programs were the milestones in the land clearing programs of the nation. Evaluation of site condition and the hazardous material presented in the soil give primary information regarding the impact of this toxicity on the human population. The evaluation was based on the Hazard Ranking System (HRS) and sites with HRS scores of 28.50 or greater were included in the National Priorities List (NPL). Further remedial investigation and actions will be carried out at these sites.

10.5.1.3 European Commission

The European Commission (EC) was established on 16 January 1958 and this institute implemented laws to protect the environment and proposed strategies to clear contaminated sites (EC 2018). The contaminated land clearing was a major agenda of EC and with the help of European Soil Data Centre (ESDAC) by JRC, the collection of soil data and monitoring of the heavy metal pollution was possible (Tóth et al. 2016).

The research conducted in 2016 on the topic "Progress in the management of contaminated sites in Europe" with the help of National Reference Centers (NRCs) aided in providing extensive knowledge on the rate of clearing contaminated land. According to the report, 650,000 sites were remediated from the 2.5 million contaminated sites and it is a great improvement in the land clearing process of the country. According to the data obtained by EC, the mineral oil and heavy metal toxicity are the crucial issues in the contamination reported sites. Based on the information from 39 countries, the major remediation technique implemented in these land clearing is the "dig-and-dump", an *ex-situ* technique, which implicates the digging and off-site disposal of polluted sediments. But the real concern over this clearing purpose is the budget. Approximately €4.3 billion was utilized for the same and a major portion of the expenditure came from the public budget (EC 2018). It was finally decided that the responsible industries or firms have to clear the land, soil, and water, which had been contaminated by them and for this purpose extensive investments in the field of land clearing is essential.

10.5.1.4 The Ministry of Environment, Forest and Climate Change (MoFE)

The Ministry of Environment, Forest and Climate Change was established in 1985, helping with the execution of India's environmental and forestry policies and programs. The annual budget of MoFE was US$430 million for the year 2020-21. According to the remediation strategies of MoFE, site specific remediation target levels (SSTLs) were calculated for each contaminated site by analyzing the impact of contamination on humans. "Creation of Management Structure for Hazardous Substances" is another program established by the Indian government. The sub-schemes of this program are:

- SAMPATTI - 'Sustainable Management of Pre-owned Asset through Trade Initiatives':
- Capacity building of government agencies/organizations/department/civil society/institute with respect to environmentally sound management of chemicals and wastes.
- Organizing awareness program with various stakeholders for implementation of various wastes and chemicals management rules.
- Innovative technologies for environmentally sound management of chemicals and wastes
- Setting up facilities for the management of biomedical waste (CBMWTF) and treatment, storage and disposal of hazardous waste (TSDF).

"Capacity Building for Industrial Pollution Management (CBIPM) Project" from MoFE supported by World Bank from October 2010 to March 2018 is a historical move in the land clearing programs of India. The contaminated sites of Andhra Pradesh (Kadapa) and West Bengal (Dhapa) were selected for this work. West Bengal State Pollution Control Board is implementing clearing of dumpsite (12.14 hectare) at Dhapa with a budget of Rs. 57.44 Crore. Similarly, the Kadapa (10.38 acre) of Andhra is being cleared by Andhra Pollution Control Board with a project cost of Rs. 30.17 Crore.

10.5.2 Private organizations

The introduction of different private industries and stand-up programs in the field of land cleanup is really essential to overcome the shortage of experts, equipment, money and other recourses. Figure 10.2 represents different phytoremediation agencies.

Figure 10.2: Various agencies in the field of phytoremediation.

10.5.2.1 Pond tech

It is a Canadian company investing in algal bioprocess with a vision to transform algae as the basis of a low carbon economy. Algae-based superfoods like *Spirulina* and *Chlorella*, protein enriched animal feeds, biofuels and cosmetics products are the major pond tech products. This industry will help to reduce the extensive carbon emission from other firms.

10.5.2.2 Biofuel NET

BioFuelNet is another Canadian company established for the mobilization of biofuel research aiding in the utilization of non-food biomass as biofuel feedstock.

10.5.2.3 SUEZ group

It is an American company implemented to remediate polluted soil and water tables and they exploit different phytotechniques in the field. They follow standard LNE NF X 31-620-1-2-3-4 for clearing a contaminated land. They showed excellence in providing human and technical resources, technical and financial feasibility, and conducting laboratory tests and field tests for a better resolution. According to the characteristics of the contaminated sites, SUEZ is proposing different methods of treatments:

In situ remediation including venting, bioventing, sparging, simultaneous treatment of the soil, groundwater and floating materials and the treatment of chlorinated pollution.

Off-site remediation (soil excavation) was done by the transferring of contaminated soil to different certified biocenters for treatment, usage of waste tracking documents, and the optimization and recovery of tonnages on the Neoter® platform (this platform is utilized for various treatments and pretreatments).

10.5.3 MLM group

An American company that aids to remediate brownfield sites with a detailed investigation of the regulatory framework and has the familiarity of working with developers, principal contractors, engineering consultants, regulators, and other stakeholders to achieve appropriate solutions in a short duration. More than the removal of contaminants, this firm provides legal support.

MLM undertakes appropriate investigation and assessment of brownfield sites such as former gasworks, chemical works and other industrial sites which often have complex soil, gas and groundwater contamination conditions. The progressive work involves inception, providing case studies, designing appropriate site investigations and monitoring programs, undertaking risk assessments and developing a robust conceptual site model for remediation. Work may be required to clear planning conditions and enable future development, understand risks and liability for site divestment of the Environmental Protection Act. Whether it involves long-term monitoring or active remediation, the clean-up and management of contaminated sites can be complex and costly. MLM has a reputation for deliverable solutions focussed on end use. One of the projects is:

Haulbowline Island East Tip Phytoremediation Project

Haulbowline Island east tip is a project that involved phytoremediation of the East Tip located in Cork Harbour, a location in Ireland. The contaminants released by the steel works processing industry was deposited on an area of reclaimed foreshore, which became known as the 'East Tip'. This project cleared the area of the East Tip and transformed it into a public recreational amenity.

10.5.3.1 Argyll environmental

This company was founded in UK (2002) to find strategies to solve different environmental crisis and contamination issues. The collaboration with UK's top law firms makes this company a more appropriate candidate for clearing contaminated land.

10.5.3.2 Veolia's

New investments are being implemented in India to clear contaminated lands and Veolia's waste management technology is a pioneer in this field. Veolia offers remediation within an industry, large-scale restoration project for municipal governments by different techniques (physical, chemical, biological methods) and also provides the clearing equipment. This industry also implemented different techniques to treat methane, leachate and industrial wastewater.

10.6 Conclusion

Heavy metal contamination is a major threat to human health. Restoration of contaminated regions has a crucial role in attaining the needs of the exponentially growing population. Phytoremediation has great importance in the ecological aspect. It completely removes contaminants in a safe manner. There are several efforts to accomplish this task by governments of different countries. At the same time, industries that focus on phytoremediation are emerging in different countries. However, due to the lack of enough knowledge, still there are uncertainties in the public view about phytoremediation. The introduction of different industries in land reclamation is a good strategy of the nation and it reduces the efforts of the government sector in land restoration.

Acknowledgments

EJ, AS, ARKP and RJ gratefully acknowledge the financial assistance from University Grants Commission (UGC) and Council of Scientific and Industrial Research (CSIR), India through JRF fellowships. JTP acknowledges the financial assistance provided by Kerala State Council for Science, Technology and Environment in the form of KSCSTE Research Grant (KSCSTE/5179/2017-SRSLS). The authors extend their sincere thanks to the Department of Science & Technology (DST), Government of India for granting fund under Fund for Improvement of S&T Infrastructure (FIST) programme (DST-FIST/15-16/28.05.2015).

References

Afzal, M., G. Shabir, S. Iqbal, T. Mustafa, Q.M. Khan and Z.M. Khalid. 2014. Assessment of heavy metal contamination in soil and groundwater at leather industrial area of Kasur, Pakistan. CLEAN – Soil, Air, Water 42(8): 1133-1139.

Ali, H., E. Khan and M.A. Sajad. 2013. Phytoremediation of heavy metals – Concepts and applications. Chemosphere 91(7): 869-881.

Alsabbagh, A.H. and T.M. Abuqudaira. 2017. Phytoremediation of Jordanian uranium-rich soil using sunflower. Water Air Soil Pollut. 228(6): 219.

Arthur, E.L., P.J. Rice, P.J. Rice, T.A. Anderson, S.M. Baladi, K.L. Henderson and J.R. Coats. 2005. Phytoremediation—An overview. Crit. Rev. Plant Sci. 24(2): 109-122.

Blaylock, M.J., M.P. Elless, J.W. Huang and S.M. Dushenkov. 1999. Phytoremediation of lead-contaminated soil at a New Jersey brownfield site. Remed. Jour. 9(3): 93-101.

Ebbs, S., S. Hatfield, V. Nagarajan and M. Blaylock. 2009. A comparison of the dietary arsenic exposures from ingestion of contaminated soil and hyperaccumulating Pteris ferns used in a residential phytoremediation project. Int. J. Phytoremediation 12(1): 121-132.

Environmental Protection Agency of United States (EPA). 1996. National Water Quality Inventory Report to Congress.

Environmental Protection Agency of United States (US EPA). 2001. Citizen's Guide to Phytoremediation.

European Commission (EC). 2018. Annual Report of European Commission.

Hauptvogl, M., M. Kotrla, M. Prčík, Ž. Pauková, M. Kováčik and T. Lošák. 2019. Phytoremediation potential of fast-growing energy plants: Challenges and perspectives – A review. Pol. J. Environ. Stud. 29(1): 505-516.

Henry, J.R. 2000. An overview of the phytoremediation of lead and mercury. NNEMS Report 2000: 3-9.

Janeeshma, E. and J.T. Puthur. 2020. Direct and indirect influence of arbuscular mycorrhizae on enhancing metal tolerance of plants. Arch. Microbiol. 202(1): 1-16.

Khan, S., S. Rehman, A.Z. Khan, M.A. Khan and M.T. Shah. 2010. Soil and vegetables enrichment with heavy metals from geological sources in Gilgit, northern Pakistan. Ecotoxicol. Environ. Saf. 73(7): 1820-1827.

Kumar, A., B.S. Bisht, V.D. Joshi and T. Dhewa. 2011. Review on bioremediation of polluted environment: A management tool. Int. J. Environ. Sci. 1: 1079-1093.

Lambert, M., Leven, B.A. and R.M. Green. 2000. New methods of cleaning up heavy metal in soils and water. Environmental Science and Technology Briefs for Citizens 7(4): 133-163.

Lelie, D.V.D., J.P Schwitzguébel, D.J. Glass, J. Vangronsveld and A. Baker. 2001. Peer reviewed: Assessing phytoremediation's progress in the United States and Europe. Environ. Sci. Technol. 35: 446-452.

Li, T., Y. Liu, S. Lin, Y. Liu and Y. Xie. 2019. Soil pollution management in China: A brief introduction. Sustainability, 11(3): 556.

Liphadzi, M.S., M.B. Kirkham, K.R. Mankin and G.M. Paulsen. 2003. EDTA-assisted heavy-metal uptake by poplar and sunflower grown at a long-term sewage-sludge farm. Plant and Soil, 257(1), 171-182.

Liu, D., Z. An, Z. Mao, L. Ma and Z. Lu. 2015. Enhanced heavy metal tolerance and accumulation by transgenic sugar beets expressing *Streptococcus thermophilus* StGCS-GS in the presence of Cd, Zn and Cu alone or in combination. PLoS One, 10(6): e0128824.

Luo, J., L. Cai, S. Qi, J. Wu and X.S. Gu. 2017. A multi-technique phytoremediation approach to purify metals contaminated soil from e-waste recycling site. J. Environ. Manage. 204: 17-22.

Nahar, N., A. Rahman, N.N. Nawani, S. Ghosh and A. Mandal. 2017. Phytoremediation of arsenic from the contaminated soil using transgenic tobacco plants expressing ACR2 gene of *Arabidopsis thaliana*. J. Plant Physiol. 218: 121-126.

National Risk Management Research Laboratory (US). 2000. Introduction to Phytoremediation.

Pandey, V. C., O. Bajpai, and N. Singh. 2016. Energy crops in sustainable phytoremediation. Renewable Sustainable Energy Rev. 54: 58-73.

Patel, P., N.J. Raju, B.S.R. Reddy, U. Suresh, D.B Sankar and T.V.K. Reddy. 2018. Heavy metal contamination in river water and sediments of the Swarnamukhi River Basin, India: Risk assessment and environmental implications. Environ. Geochem. Health 40(2): 609-623.

Rai, P.K., K.H. Kim, S.S. Lee and J.H. Lee. 2020. Molecular mechanisms in phytoremediation of environmental contaminants and prospects of engineered transgenic plants/ microbes. Sci. Total Environ. 705: 135858.

Sameena, P.P. and J.T. Puthur. 2021. Cotyledonary leaves effectively shield the true leaves in *Ricinus communis* L. from copper toxicity. Int. J. Phytoremediation 23: 492-504.

Sarath, N.G. and J.T. Puthur. 2020. Heavy metal pollution assessment in a mangrove ecosystem scheduled as a community reserve. Wet. Ecol. and Manag. 1-12.

Saxena, G., D. Purchase, S.I. Mulla, G.D. Saratale and R.N. Bhargava. 2019. Phytoremediation of heavy metal-contaminated sites: Eco-environmental concerns, field studies, sustainability issues, and future prospects. Rev. Environ. Contam. T. 249: 71-131.

Saxena, P.K., R.S. Krishna, T. Dan, M.R. Perras and N.N. Vettakkorumakankav. 1999. Phytoremediation of heavy metal contaminated and polluted soils. pp. 305-329. *In*: Prasad, M.N.V. and J. Hagemeyer [eds.]. Heavy Metal Stress in Plants – From Molecules to Ecosystems. Springer, Berlin, Heidelberg.

Schwitzguébel, J.P., D. van der Lelie, A. Baker, D.J. Glass and J. Vangronsveld. 2002. Phytoremediation: European and American trends, successes, obstacles and needs. J. Soils and Sediments 2(2): 91-99.

Sharma, P. and S. Pandey. 2014. Status of phytoremediation in world scenario. Int. J. Environ. Bioremediat. Biodegrad. 2(4): 178-191.

Sharma, R.K., M. Agrawal and F.M. Marshall. 2009. Heavy metals in vegetables collected from production and market sites of a tropical urban area of India. Food Chem. Toxicol. 47(3): 583-591.

Siddiqui, A.S. and N.U. Saher. 2021. Distribution profile of heavy metals and associated contamination trend with the sedimentary environment of Pakistan coast bordering the Northern Arabian Sea. Environ. Sci. Pollut. Res. 1-18.

Singh, P., Rana, D. Geeta, S. Asha and K.J Pawan. 2006. Biotechnological approaches to improve phytoremediation efficiency for environment contaminants. pp. 223-258. *In*: Singh, S.N. and R.D. Tripathi [eds.]. Environmental Bioremediation Technologies. Springer Publication, NY.

Singh, R., N. Gautam, A. Mishra and R. Gupta. 2011. Heavy metals and living systems: An overview. Indian J. Pharmacol. 43(3): 246.

Su, C. 2014. A review on heavy metal contamination in the soil worldwide: Situation, impact and remediation techniques. Environ. Skeptics Critics 3(2): 24.

Suman, J., O. Uhlik, J. Viktorova and T. Macek. 2018. Phytoextraction of heavymetals: A promising tool for clean-up of polluted environment. Front. Plant Sci. 9: 1476.

Tang, W., Y. Zhao, C. Wang, B. Shan and J. Cui. 2013. Heavy metal contamination of overlying waters and bed sediments of Haihe Basin in China. Ecotoxicol. Environ. Saf. 98: 317-323.

The Federal Contaminated Sites Action Plan (FCSAP). 2012. Canada, environment and climate change: "action plan for contaminated sites." Program results.

Tóth, G., T. Hermann, G. Szatmári and L. Pásztor. 2016. Maps of heavy metals in the soils of the European Union and proposed priority areas for detailed assessment. Sci. Total Environ. 565: 1054-1062.

Van der Lelie, D., J.P. Schwitzguebel, D.J. Glass, J. Vangronsveld and A.J.M Baker. 2001. Assessing phytoremediation's progress in the United States and Europe. Environ. Sci. Technol. 35: 446A-452A.

Watanabe, M.E. 1997. Phytoremediation on the brink of commercialization. Environ. Sci. Technol. 31(4): 182A-186A.

Wood, P. 1997. Remediation methods for contaminated sites. pp. 47-71. *In*: R. Hester and R. Harrison [eds.]. Contaminated Land and Its Reclamation. The Royal Society of Chemistry, Cambridge.

Zhang, L., X. Ye, H. Feng, Y. Jing, T. Ouyang, X. Yu and W. Chen. 2007. Heavy metal contamination in western Xiamen Bay sediments and its vicinity, China. Mar. Pollu. Bull. 54(7): 974-982.

CHAPTER

11

Global Perspective and Challenges in Utilization of Bioenergy Crops for Phytoremediation

A.M. Shackira[1]*, P.P. Mirshad[2] and Misbah Naz[3]

[1] Department of Botany, Sir Syed College, Taliparamba, Kannur, Kerala – 670142, India.
[2] Department of Botany, St. Josephs College, Devagiri, Calicut, Kerala – 673008, India.
[3] State Key Laboratory of Crop Genetics and Germplasm Enhancement, Nanjing Agricultural University, Nanjing 210095, China

11.1 Introduction

Environmental remediation of toxic compounds adversely affects the ecosystem structure and function. Removal of contaminants from soil, air and water in a sustainable manner is crucial for the existence of biodiversity. Polluted soils, air, water, coal mine spoil, and other dumpsites are a worldwide problem, contaminating nearby communities and these are the major source of environmental pollution too. Conventional methods of soil decontamination have several disadvantages both in terms of environmental and financial stability. This leads to the search for an alternative approach of remediation which are eco-friendly and cost effective. Phytoremediation is one of the promising strategies as it has many advantages as compared to conventional methods and has gained increased attention for removing pollutants due to its cost-effectiveness and environmental sustainability. In addition, compared to other remediation strategies, phytoremediation technique provides aesthetic beauty and there are no air emissions or generation of harmful waste (Gomes 2012).

The biomass produced after successful phytoremediation can be reused in the form of bioenergy if energy crops are selected for the process, which also symbolizes the eco-friendly nature of this technique. Thus, employment of energy crops for remediation of contaminated soil is now getting more attention due to the multiuse of biomass after decontamination process in addition to the renewable energy production. A desirable bioenergy crop should have the ability to capture and convert the available solar energy into harvestable biomass with maximal efficiency.

Corresponding author: shackimajeed@gmail.com

Biomass appears to be an attractive feedstock for three main reasons. First, it is a renewable resource that could be sustainably developed in the future. Second, it appears to have positive environmental properties resulting in no net releases of CO_2. Third, it appears to have significant economic potential provided that fossil fuel prices increase in the future (Cadenas and Cabezudo 1998). Biomass energy potential is addressed to be the most promising among various renewable energy sources, due to its spread and its availability worldwide. Biomass also have a unique advantage among the renewable energy sources because it can be stored, transported and utilized, far away from the point of origin (Demirbas 2008).

In general, energy crops can be broadly classified into two categories which are annuals (e.g. sorghum, rapeseed, kenaf, etc.) and perennials (e.g. wheat, sugar beet, willows, poplars, etc.) (Zabaniotou et al. 2008, Simpson et al. 2009). There is growing interest in perennial bioenergy crops because of their potential for reducing the CO_2 levels either by high level carbon sequestration and/or by reduced CO_2 emission on combustion of biofuels generated from it (Rowe et al. 2009, Mleczek et al. 2010, Gomes 2012). In this scenario, the use of perennial bioenergy crops has gained significant attention because of its potential for bioethanol production along with phytoremediation of contaminated soil. Now, the task is to find out a potential perennial bioenergy crop that can accumulate high biomass and at the same time can be grown in various types of contaminated soils. According to Pandey et al. (2012), there are four main energy crops in the world which can be exploited as ideal candidate of phytoremediation and are also serving as carbon sinks as well as energy source, i.e. *Populus, Miscanthus*, Castor and *Jatropha*. In addition, growing energy crops on contaminated soil is also a means of sustainable utilization of these types of soils. Nowadays, the use of bioenergy crops for the phytoremediation has gained more attention because of its abundancy, availability and regeneration capacity. Therefore, bioenergy crops have a significant role in ecosystem functioning and balancing the carbon economy, they are considered as a prime candidate for sustainable management practices and ecosystem services.

11.2 Bioenergy Crops as Tool for Phytoremediation

The high biomass yield and non-food utilization of bioenergy crops make it to be the best candidate for phytoremediation of contaminated soil. The basic mechanisms behind the detoxification of various contaminants operating in the bioenergy crops are the following:

(i) Chemical transformation of the pollutant through the interaction with root exudates and microbes
(ii) Mineralisation of pollutants into non-toxic forms like nitrate, CO_2, chlorine, etc. along with the association of mycorrhizal fungi
(iii) Bioaccumulation of uptaken pollutants within the plant tissue
(iv) Volatilisation of pollutants through the aerial parts of the plant
(v) Detoxification of pollutant through complexation/sequestration in the vacuole

Hence, energy crops are widely employed for removing pollutants from soil and are considered as a good source of non-renewable energy. So, a plant that removes

Table 11.1: Bioenergy crops widely exploited for phytoremediation of various pollutants

Sl. No.	Plant species	Type of pollutant	References
1	*Salix viminalis* L.	Cd	Greger and Landberg (1999)
2	*Cannabis sativa* L.	Ni, Pb, Cr, Cu and Cd	Linger et al. (2002), Citterio et al. (2003), Arru et al. (2004)
3	*Arundo donax* L.	Cd, As and Ni	Mirza et al. (2010)
4	*Populus* spp.	Pollutants including fertilizers, inorganic metals and metalloids	Licht and Isebrands (2005)
5	*Cynodon dactylon* L.	Cr	Sampanpanish et al. (2010)
6	*Ricinus communis* L.	DDT and Cd	Huang et al. (2011)
7	*Poa annua* L	Zn, Pb and Cd	Barbafieri et al. (2011)
8	*Salix* sp.	Cd, Cr, Cu, Ni, Pb and Zn	Miller et al. (2011)
9	*Cannabis sativa* L.	Cd, Zn and Fe	Mihoc et al. (2012)
10	*Helianthus annuus* L.	Pb, Zn and Cd	Angelova et al. (2005)
11	*Jatropha curcas* L.	Al, Fe, Cd, Cr, Mn, Ar, Zn, Cd and Pb	Pandey et al. (2012), Marques and Nascimento (2013)
12	*Arundo donax* and *Miscanthus sacchariflorus*	Zn and Cr	Li et al. (2014)
13	*Miscanthus* sp.	As, Sn, Cd, Cr, Cu, Ni, Pb, Zn and Al	Pidlisnyuk et al. (2014)
14	*Phalaris arundinacea* L	Cd and Hg	Polechonska and Klink (2014)
15	*Arundo donax* L.	Zn and Cr	Xiao et al. (2010)
16	*Miscanthus* spp. and *Arundo donax*L.	Zn, Cr and PB	Barbosa et al. (2015)
17	*Populus* sp.	Cd, Cr, Cu, Se, Pb and Zn	Chen et al. (2015)
18	*Ricinus communis* L.	Cu, Zn, Mn, Pb and Cd	Bauddh et al. (2016)
19	*Salix klara* and *Salix inger*	Cr, Cu, Ni, Pb and Zn	Enell et al. (2016)
20	*Ricinus communis* L.	Cd, Pb, Zn and As	Kiran and Prasad (2017)
21	*Arundo donax* L.	Ni, Pb and Hg	Cristaldi et al. (2020)
22	*Pennisetum purpureum*	Cd and Zn	Hou et al. (2020)
23	*Festuca rubra* L.	As, B, Cu, Zn, Mn, Mo, and Se	Gajić et al. (2020)

contaminants from soil and are also excellent in energy production would be a good choice for the ever-increasing demand of energy and environmental remediation (Table 11.1). Bioenergy crops such as poplars (*Populus* spp.), willows (*Salix* spp.), elephant grass (*Miscanthus giganteus*), castor bean (*Ricinus communis*), and switchgrass (*Panicum virgatum*) can tolerate high concentrations of heavy metal and they can grow well on contaminated soils. Likewise, maize plants were successfully employed by Gomes (2012) to decontaminate the Campine region in Belgium, which were polluted heavily with toxic metal ions like cadmium (Cd), zinc (Zn) and lead (Pb) deposited through smelting activities. Phytoremediation properties as well as the low metal concentrations in consumable parts make it an ideal candidate as energy crop (Gomes 2012). Similarly, bana grass (*Pennisetum americanum* × *Pennisetum purpureum*) and vetiver grass (*Vetiveria zizanioides*) can be used as successful candidates of phytoremediation as it has showed significant accumulation of Cd in their biomass (Zhang et al. 2014). Two promising bioenergy crops such as *Arundo donax* and *Miscanthus sacchariflorus* exhibited enhanced accumulation of Zn/Cr, and therefore, they were regarded as capable candidates for the phytoremediation of soil contaminated with Zn/Cr (Xiao et al. 2010).

In addition to the above, *A. donax* showed good bioaccumulation capability of heavy metals such as Cd, nickel (Ni), Pb and mercury (Hg) (Cristaldi et al. 2020). *A. donax* var. *versicolor* is considered as a prime candidate for the phytoextraction of Zn, whereas *P. purpureum* is utilised for phytoextraction and phytostabilization of Cd and phytostabilization of Zn (Hou et al. 2020). Moreover, *A. donax* withstands high levels of soil salinity (Nackley and Kim 2015) and drought (Sanchez et al. 2015) so that it can be cultivated in marginal soils experiencing heavy metal toxicity, salinity and drought. Likewise, *Festuca rubra* L. is considered as a perennial, bioenergy crop which tolerates different ecological conditions and grows in various habitats. This plant also possesses vigorous growth, large biomass, and high tolerance to unfavourable environmental conditions such as salinity, drought and heavy metal contaminated soils. *F. rubra* has exhibited higher phytoremediation potential for the toxic metal ions like arsenic (As), boron (B), copper (Cu), Zn, manganese (Mn), molybdenum (Mo) and selenium (Se) (Gajić et al. 2020).

11.3 Sustainable Applications of Bioenergy Crops: Global Perspective

As far as energy production from biomass is considered, the major positive feature of perennial crops for biomass energy is the potential for high yields on marginal lands (Mehmood et al. 2016). Grasses and short rotation crops are chief source of lignocellulose, especially grasses like *M. giganteus* and *P. virgatum* (Tan et al. 2008). Per year, *Miscanthus* can produce up to 40 tonnes of biomass from a single hectare (Harvey 2007). Perennial crops have many ecological advantages such as requirement of low soil management, reducing soil erosion and need of low nutrient inputs due to the recycling of nutrients by their rhizome system (Lewandowski et al. 2003, Heaton et al. 2004). In the following sections, discussions are made on the role of energy crops which are phytoremediation agent and their employment in other attributes such as source of biofuels, agents of CO_2 sequestration, etc.

11.3.1 Bioenergy crops as source of bioethanol

Depending on the origin and production technology of biofuels, they are generally categorized as the first, second and third generation biofuels. Bioethanol produced from lignocellulosic materials of plant biomass is referred to as second-generation bioethanol and is considered as a renewable alternative to petroleum based liquid fuels (Shields and Boopathy 2011). Lignocellulosic biomass including wood, grass, weeds, forestry waste, and agricultural residues are the largest source of hexose and pentose sugars which can be converted into ethanol by fermentation (Chandel et al. 2010). Bioethanol derived from lignocellulosic biomass is regarded as an eco-friendly biofuel. The excreted CO_2 is carried back during the respiratory process of the growing organism, so it does not contribute to the total CO_2 in the atmosphere (Wyman 2007).

Energy crops utilized for the production of second-generation bioethanol comprise primarily perennials (Oliver et al. 2009). Second-generation bioethanol crops, also known as cellulosic bioethanol, include dedicated energy crops (e.g. switchgrass and miscanthus) that are grown exclusively for fuel production (Singh and Saraswat 2016). The feedstock of second-generation bioethanol should be inexpensive, cellulosic in nature, much abundant and easily available throughout the world. The food-fuel dispute can be rectified by the second-generation bioethanol production. Extensive use of first-generation bioethanol from agricultural crops such as sugarcane and corn can be overcome by second-generation bioethanol and the plant biomass can be used as feedstock (Tan et al. 2008).

P. virgatum and *Miscanthus* are used as sustainable energy crops for the production of bioethanol. *Miscanthus,* which is a C4 grass cultivated in East Asia, has high carbon sequestration capacity (Sanderson et al. 1996, Chung and Kim 2012). Besides these, bioethanol from energy crops such as *Cannabis sativa, Populus grandidentata* etc. was found to reduce the CO_2 emission and also increase energy efficiency (Zalensy et al. 2009, Prade et al. 2012). Till date, a number of energy crop plants have been characterised for the increased biomass yield and promising source of renewable energy, which includes Giant Napier grass (Ma et al. 2012), *Saccharum arundinaceum* (Feng et al. 2015), *Arundo donax* (Cristaldi et al. 2020), *Festuca rubra* (Gajić et al. 2020), etc.

11.3.2 Bioenergy crops as efficient carbon sequesters

Energy production from renewable energy sources such as biomass will play a major role in mitigation of climate change effects by reducing the release of CO_2 (Swinton et al. 2011). Moreover, energy crops are excellent in sequestering soil organic carbon, especially woody energy crops. This is because woody species of energy crops has high input of carbon from root as well as leaf litter as compared to that of herbaceous energy crops. On the other hand, in the case of herbaceous crops when the aboveground part is removed, the carbon input source will be only the root system (Tolber et al. 2002, Hangs et al. 2014, Chimento et al. 2016). The only exception to this is *Miscanthus*, as being a herbaceous crop it sequesters almost equal CO_2 as that of woody crops because of the partial defoliation of leaves during the winter period (Clifton-Brown et al. 2007, Meehan et al. 2013, Chimento et al. 2016).

Besides this attribute, it is also accountable that the litter decomposition enriches the soil fauna, which influences the decaying of fresh detritus, thereby enhancing the rate of carbon sequestration in soil (Conant et al. 2001, Fonte et al. 2012, Chimento et al. 2016). In addition to the above features, the soil organic carbon stock is significantly influenced by the harvesting time of energy crop as well as residue removal from the surface of soil (Jones and Donnelly 2004, Stewart et al. 2014, Chimento et al. 2016).

On an average, bioenergy crops have the potential to sequester about 0.6 to 3.0 Mg C ha^{-1} yr^{-1} (Lemus and Lal 2005). For instance, recent quantification of carbon sequestration has been done in some selected energy crops like poplar, willow, miscanthus and switchgrass species by Bazrgar et al. (2020). They reported that among the above energy crops, willow plants had the highest carbon sequestration rate over the other crops studied, i.e. about 1.66 Mg C $ha^{-1}yr^{-1}$. Moreover, these perennial crops are highly successful in carbon sequestration due to the long time survival and constant biomass production, thereby enhancing the residual matter. In addition to this, the well developed root system of these plants also enhances the rate of carbon sequestration to a great extent (Clifton-Brown et al. 2007, Larsen et al. 2014, Agostini et al. 2015, Tiwari et al. 2017). Hence, in addition to the decontamination of pollutants, the ability to sequester carbon greatly facilitates the energy crops more suitable and promising members for phytoremediation.

11.4 Bioenergy Crops as Agents of Phytoremediation: Challenges

Even though energy crops pose several rewarding attributes for being the agents of phytoremediation, which include the potential carbon sequestering ability and a source of renewable energy, they also have certain unconstructive characteristics. The most striking challenge is the presence of pollutant in the products like biodiesel, oils, etc., which adversely affects the quality and efficiency of goods. For instance, toxic metal ion content (Cd, Cu and Pb) were detected in the oil extracted from seeds of rape plant (*Brassica napus*), which were cultivated in polluted soil (Angelova et al. 2005). Moreover, the thermal decomposition behaviour and the property of energy is influenced by the pollutants as well as the soil pH (Ginneken et al. 2007). One more challenge is the assurance of biomass production, i.e. if the energy crop fails to produce enough biomass after exposing to the environmental pollutants, the success level as well as the economic returns will be a dilemma.

It is evident that perennial energy crops are gifted members of phytoremediation as they are able to reduce the emission of greenhouse gases and thereby regulate climate change. However, to achieve this goal to a significant level, implementation of proper land use practices is crucial (Pidlisnyuk et al. 2014). Hence, in order to evaluate the cost-effectiveness and sustainable nature of using energy crops for phytoremediation, policy makers should properly address the issue of land use (Whitaker et al. 2018). Another striking delimitation is the present legislation and decontamination practices employed in the field of soil decontamination. It is purely based on the overall concentration of pollutants trapped in the soil. It must be retreated by considering soil functionality and risk assessment criteria as the process of phytoremediation is bit slower (Andersson-Sköld et al. 2009). Moreover, phytoremediation research is interdisciplinary in nature. In order to successfully implement phytoremediation measures with bioenergy crops, site owners, locals, farmers, technology providers and consultants, restoration experts, sustainability assessors, regulatory and certification agencies, bio-refineries, financial sponsors, NGOs and other volunteers must work in tandem.

11.5 Conclusion

The ever-increasing population has raised several issues like energy crisis as well as pollution of the environment due to uncontrolled industrialization and urbanization. Hence, researchers all over the world are looking for a feasible technique that is cost-effective and eco-friendly. Bioenergy plants offer supreme model to solve both the issues as they are the source of renewable energy as well as can decontaminate the polluted soil in a sustainable manner (Fig. 11.1).

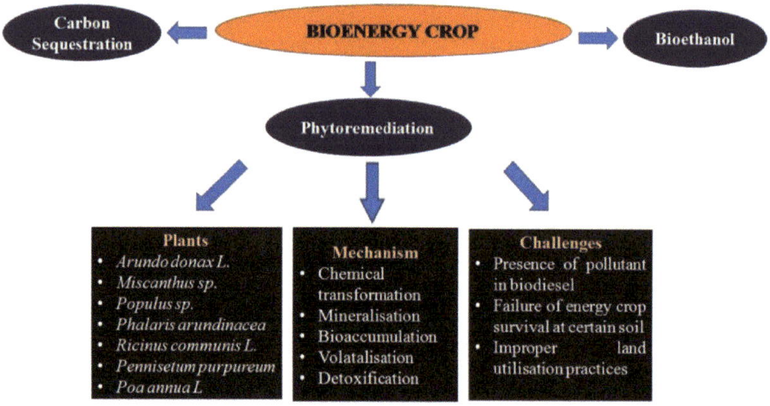

Figure 11.1: Bioenergy crops as agents of phytoremediation - promises and challenges.

In addition to this, energy crops can also mitigate climate change issue by acting as a sink for CO_2. However, attention is needed to overcome the limitations of using energy crops in the field of phytoremediation. The most important challenges are the quality of biofuels as well as the efficiency of energy. Proper assessment and implementation of the technique is essential so as to ensure the long-term success of the technique. In future, energy crisis as well as the climate change will be solved to a great extent if ideal energy crops are properly employed with proper restoration strategies.

References

Agostini, F., A.S. Gregory and G.M. Richter. 2015. Carbon sequestration by perennial energy crops: Is the jury still out? Bioenergy. Res. 8: 1057-1080.

Andersson-Sköld, Y., A. Enell, S. Blom, T. Rihm, A. Angelbratt, K. Haglund, O. Wik, P. Bardos, T. Track and S. Keuning. 2009. Biofuel and other biomass based products from contaminated sites – Potentials and barriers from Swedish perspectives. Swedish Geotechnical Institute, Linköping.

Angelova, V., R. Ivanov and K. Ivanov. 2005. Heavy metal accumulation and distribution in oil crops. Commun. Soil Sci. Plant Anal. 35: 2551-2566.

Arru, L., S. Rognoni, M. Baroncini, P.M. Bonatti and P. Perata. 2004. Copper localization in *Cannabis sativa* L. grown in a copper-rich solution. Euphytica 140: 33-38.

Barbafieri, M., C. Dadea, E. Tassi, F. Bretzel and I. Fanfani. 2011. Uptake of heavy metals by native species growing in a mining area in Sardinia, Italy: Discovering native flora for phytoremediation. Int. J. Phytoremediation. 13: 985-999.

Barbosa, B., S. Boleo, S. Sidella, J. Costa, M.P. Duarte, B. Mendes, S.L. Cosentino and A.L. Fernando. 2015. Phytoremediation of heavy metal-contaminated soils using the perennial energy crops *Miscanthus* spp. and *Arundo donax* L. Bioenerg. Res. 8: 1500-1511.

Bauddh, K., A. Kumar, S. Srivastava, R.P. Singh and R.D. Tripathi. 2016. A study on the effect of cadmium on the antioxidative defense system and alteration in different functional groups in castor bean and Indian mustard. Arch. Agron. Soil Sci. 62: 877-891.

Bazrgar, A.B., N. Aeryn, B. Coleman, M.W. Ashiq, A. Gordon and N. Thevathasan. 2020. Long-term monitoring of soil carbon sequestration in woody and herbaceous bioenergy crop production systems on marginal lands in Southern Ontario, Canada. Sustainability. 12: 3901.

Cadenas, A. and S. Cabezudo. 1998. Biofuels as sustainable technologies: Perspectives for less developed countries. Technol. Forecasting Soc. Change 58: 83-103.

Chandel, A.K., O.V. Singh, G. Chandrasekhar, L.V. Rao and M.L. Narasu. 2010. Key drivers influencing the commercialization of ethanol-based bio refineries. J. Commerc. Biotechnol. 16: 239-257.

Chen, L., X. Hu, W. Yang, Z. Xu, D. Zhang and S. Gao. 2015. The effects of arbuscular mycorrhizal fungi on sex specific responses to Pb pollution in *Populus cathayana*. Ecotoxicol. Environ. Saf. 113: 460-468.

Chimento, C., M. Almagro and A. Stefano. 2016. Carbon sequestration potential in perennial bioenergy crops: The importance of organic matter inputs and its physical protection. GCB Bioenergy 8: 111-121.

Chung, J.H. and D.S. Kim. 2012. Miscanthus as a potential bioenergy crop in East Asia. J. Crop Sci. Biotechnol. 15: 65-77.

Citterio, S., A. Santagostino, P. Fumagalli, N. Prato, P. Ranalli and S. Sgorbati. 2003. Heavy metal tolerance and accumulation of Cd, Cr and Ni by *Cannabis sativa* L. Plant Soil. 256: 243-252.

Clifton-Brown, J.C., J. Breuer and M.B. Jones. 2007. Carbon mitigation by the energy crop, Miscanthus. Global Change Biology 13: 2296-2307.

Conant, R.T., K. Paustian and E.T. Elliott. 2001. Grassland management and conversion into grassland: Effect on soil carbon. Ecol. Soc. Amer. 11: 343-355.

Cristaldi, A., G.O. Conti, S.L. Cosentino, G. Mauromicale, C. Copat, A. Grasso, P. Zuccarello, M. Fiore, C. Restuccia and M. Ferrante. 2020. Phytoremediation potential of *Arundo donax* (Giant Reed) in contaminated soil by heavy metals. Environ. Res. 185: 109427.

Demirbas, A. 2008. Biofuels sources, biofuel policy, biofuel economy and global biofuel projections. Energ. Convers. Manage. 49: 2106-2116.

Enell, A., Y. Andersson, J. Vestin and M. Wagelmans. 2016. Risk management and regeneration of brownfields using bioenergy crops. J. Soils Sediments. 16: 987-1000.

Feng, X., Y. He, J. Fang, Z. Fang, B. Jiang, M. Brancourt-Hulmel and D. Jiang. 2015. Comparison of the growth and biomass production of *Miscanthus sinensis*, *Miscanthus floridulus* and *Saccharum arundinaceum*. Span. J. Agric. Res. 13: e0703.

Fonte, S.J., D.C. Quintero, E. Velasquez and P. Lavelle. 2012. Interactive effects of plants and earthworms on the physical stabilization of soil organic matter in aggregates. Plant Soil 359: 205-214.

Gajić, G., M. Mitrović and P. Pavlović. 2020. Feasibility of *Festuca rubra* L. native grass in phytoremediation. *In*: Phytoremediation Potential of Perennial Grasses, pp. 115-164.

Ginneken, V.L., E. Meers, R. Guisson, A. Ruttens, K. Elst, F.M. Tack, J. Vangronsveld, L. Diels and W. Dejonghe. 2007. Phytoremediation for heavy metal-contaminated soils combined with bioenergy production. J. Env. Eng. Land. Manag. 15: 227-236.

Gomes, H.I. 2012. Phytoremediation for bioenergy: Challenges and opportunities. Environ. Tech. Rev. 1(1): 59-66.

Greger, M. and T. Landberg. 1999. Use of willow in phytoremediation. Int. J. Phytoremediation 1: 115-123.

Hangs, R.D., J.J. Schoenau, K.C.J. Rees, N. Belanger and T. Volk. 2014. Leaf litter decomposition and nutrient-release characteristics of several willow varieties within short rotation coppice plantations in Saskatchewan Canada. BioEnergy Research. doi 10.1007/s12155-014-9431-y.

Harvey, J. 2007. A versatile solution-growing Miscanthus for bioenergy. Renew. Energy World 10: 86.

Heaton, E.A., S.P. Long, T.B. Voigt, M.B. Jones and J. Clifton-Brown. 2004. Miscanthus for renewable energy generation: European Union experience and projections for Illinois. Mitig. Adapt. Strategies Glob. Chang. 9: 433-451.

Hou, X., W. Teng, Y. Hu, Z. Yang, C. Li, J. Scullion, Q. Guo and R. Zheng. 2020. Potential phytoremediation of soil cadmium and zinc by diverse ornamental and energy grasses. BioResources 15: 616-640.

Huang, H., N. Yu, L. Wang, D.K. Gupta, Z. He, K. Wang, Z. Zhu, X. Yan, T. Li and X. Yang. 2011. The phytoremediation potential of bioenergy crop *Ricinus communis* for DDTs and cadmium contaminated soil. Bioresour. Technol. 102: 11034-11038.

Jones, M.B. and A. Donnelly. 2004. Carbon sequestration in temperate grassland ecosystems and the influence of management, climate and elevated CO_2. New Phytol. 164: 423-439.

Kiran, B.R. and M.N.V. Prasad. 2017. *Ricinus communis* L. (Castor bean), a potential multi-purpose environmental crop for improved and integrated phytoremediation. Euro. Biotech. J. 1: 101-116.

Larsen, S.U., U. Jorgensen, J.B. Kjeldsen and P.E. Laerke. 2014. Long-term Miscanthus yields influenced by location, genotype, row distance, fertilization and harvest season. Bioenergy Res. 7: 620-635.

Lemus, R. and R. Lal. 2005. Bioenergy crops and carbon sequestration. Crit. Rev. Plant Sci. 24(1): 1-21.

Lewandowski, J.M.O., E. Scurlock, M. Lindvall and Christou. 2003. The development and current status of perennial rhizomatous grasses as energy crops in the US and Europe. Biomass Bioenerg. 25: 335-361.

Li, C., B. Xiao, H. Wang, S.H. Yao and J.Y. Wu. 2014. Phytoremediation of Zn- and Cr-contaminated soil using two promising energy grasses. Water Air Soil Pollut. 225: 20-27.

Licht, L. and J.G. Isebrands. 2005. Linking phytoremediated pollutant removal to biomass economic opportunities. Biotechnol. Bioeng. 28: 203-218.

Linger, P., J. Mussig, H. Fischer and J. Kobert. 2002. Industrial hemp (*Cannabis sativa* L.) growing on heavy metal contaminated soil: Fibre quality and phytoremediation potential. Ind. Crop. Prod. 16: 33-42.

Ma, C., R. Naidu, F. Liu, C. Lin and H. Ming. 2012. Influence of hybrid giant Napier grass on salt and nutrient distributions with depth in a saline soil. Biodegradation 23: 907-916.

Marques, M.C. and A. Nascimento. 2013. Analysis of chlorophyll fluorescence spectra for the monitoring of Cd toxicity in a bio-energy crop (*Jatropha curcas*). J. Photochem. Photobiol. Biol. 127: 88-93.

Meehan, P.G., K.P. McDonnell and J.M. Finnan. 2013. An assessment of the effect of harvest time and harvest method on biomass loss for *Miscanthus* 9 *giganteus*. Global Change Biology Bioenergy 5: 400-407.

Mehmood, M.A., M. Ibrahim, U. Rashid, M. Nawaz, S. Ali, A. Hussain and M. Gull. 2016. Biomass production for bioenergy using marginal lands. Sustainable Production and Consumption 9: 3-21.

Mihoc, M., G. Pop, E. Alexa and I. Radulov. 2012. Nutritive quality of Romanian hemp varieties (*Cannabis sativa* L.) with special focus on oil and metal contents of seeds. Chemistry Central Journal, 6(1): 1-12.

Miller, R.S., Z. Khan and S.L. Doty. 2011. Comparison of trichloroethylene toxicity, removal, and degradation by varieties of Populus and Salix for improved phytoremediation applications. J. Bioremediat. Biodegrad. 27: 231-239.

Mirza, N., Q. Mahmood, A. Pervez, R. Ahmad, R. Farooq, M.M. Shah and M.R. Azim. 2010. Phytoremediation potential of *Arundo donax* in arsenic contaminated synthetic wastewater. Bioresour. Technol. 101: 5815-5819.

Mleczek, M., P. Rutkowski, I. Rissmann, Z. Kaczmarek, P. Golinski, K. Szentner, K. Strazynska and A. Stachowiak. 2010. Biomass productivity and phytoremediation potential of *Salix alba* and *Salix viminalis*. Biomass Bioenerg. 34: 1410-1418.

Nackley, L.L. and S.H. Kim. 2015. A salt on the bioenergy and biological invasions debate: Salinity tolerance of the invasive biomass feedstock *Arundo donax* L. GCB Bioenergy 7: 752-762.

Oliver, R.J., J.W. Finch and G. Taylor. 2009. Second generation bioenergy crops and climate change: A review of the effects of elevated atmospheric CO_2 and drought on water use and the implications for yield. GCB Bioenegy 1: 97-114.

Pandey, V.C., K. Singh, J.S. Singh, A. Kumar, B. Singh and P.P. Singh. 2012. *Jatropha curcas*: A potential biofuel plant for sustainable environmental development. Ren. Sust. Energy Rev. 16: 2870-2883.

Pidlisnyuk, V., T. Stefanovska, E.E. Lewis, L.E. Erickson and L.C. Davis. 2014. Miscanthus as a productive biofuel crop for phytoremediation. Crit. Rev. Plant Sci. 33: 1-19.

Polechonska, L. and A. Klink. 2014. Trace metal bioindication and phytoremediation potentialities of *Phalaris arundinacea* L. (reed canary grass). J. Geochem. Explor. 146: 27-33.

Prade, T., M. Finell, S.E. Svensson and J.E. Mattsson. 2012. Effect of harvest date on combustion related fuel properties of industrial hemp (*Cannabis sativa* L.). Fuel 102: 592-604.

Rowe, R.L., N.R. Street and G. Taylor. 2009. Identifying potential environmental impacts of large-scale deployment of dedicated bioenergy crops in the UK. Renew. Sustain. Energy Rev. 13: 271-290.

Sampanpanish, P., K. Tippayasak and P. Chairat-Utai. 2010. Chromium accumulation by phytoremediation with monocot weed plant species and a hydroponic sand culture system. J. Environ. Res. Development 4: 654-666.

Sanchez, E., D. Scordia, G. Lino, C. Arias, S.L. Cosentino and S. Nogues. 2015. Salinity and water stress effects on biomass production in different *Arundo donax* L. clones. BioEnerg Res. 8: 1461-1479.

Sanderson, M.A., R.L. Reed, S.B. McLaughlin, S.D. Wullschleger, B.V. Conger, D.J. Parrish and D.D. Wolf. 1996. Switchgrass as a sustainable bioenergy crop. Bioresour. Technol. 56: 83-93.

Shields, S. and R. Boopathy. 2011. Ethanol production from ligno cellulosic biomass of energy cane. Int. Biodet. Biodegrad. 65: 142-146.

Simpson, J.A., G. Picchi, A.M. Gordon, N.V. Thevathasan, J. Stanturf and I. Nicholas. 2009. Task 30 Short rotation crops for bioenergy systems. Environmental benefits associated with short-rotation woody crops. Technical Review No. 3, IEA Bioenergy.

Singh, G. and D. Saraswat. 2016. Development and evaluation of targeted marginal land mapping approach in SWAT model for simulating water quality impacts of selected second-generation bio feedstock. Environ. Modell. Softw. 81: 26-39.

Stewart, C.E., R.F. Follett, E.G. Pruessner, G.E. Varvel, K.P. Vogel and R.B. Mitchell. 2014. Nitrogen and harvest effects on soil properties under rainfed switchgrass and notill corn over 9 years: Implications for soil quality. Global Change Biology Bioenergy. doi:10.1111/gcbb.12142.

Swinton, S.M., B.A. Babcock, L.K. James and V. Bandaru. 2011. Higher US crop prices trigger little area expansion so marginal land for biofuel crops is limited. Energy Policy 39: 5254-5258.

Tan, K.T., K.T. Lee and A.R. Mohamed. 2008. Role of energy policy in renewable energy accomplishment: The case of second-generation bioethanol. Energy Policy 36: 3360-3365.

Tiwari, J., A. Kumar and N. Kumar. 2017. Phytoremediation potential of industrially important and biofuel plants: *Azadirachta indica* and *Acacia nilotica*. *In*: Phytoremediation Potential of Bioenergy Plants, Springer, Singapore.

Tolber, V.R., D.E. Todd, L.K. Mann, et al. 2002. Changes in soil quality and belowground carbon storage with conversion of traditional agricultural crop lands to bioenergy crop production. Environ. Poll. 116: 97-106.

Whitaker, J., J.L. Field, C.J. Bernacchi, C.E. Cerri, R. Ceulemans, C.A. Davies, E.H. DeLucia, I.S. Donnison, J.P. McCalmont, K. Paustian and R.L. Rowe. 2018. Consensus, uncertainties and challenges for perennial bioenergy crops and land use. GCB Bioenergy 10: 150-164.

Wyman, C.E. 2007. What is (and is not) vital to advancing cellulosic ethanol. Trends. Biotechnol. 25: 153-157.

Xiao, B., Q. Wang, J. Wu, C. Huang and D. Yu. 2010. Protective function of narrow grass hedges on soil and water loss on sloping croplands in Northern China. Agr. Ecosyst. Environ. 139: 653-664.

Zabaniotou, A.A., E.K. Kantarelis and D.C. Theodoropoulos. 2008. Sunflower shells utilization for energetic purposes in an integrated approach of energy crops: Laboratory study pyrolysis and kinetics. Bioresour. Technol. 99: 3174-3181.

Zhang, X., B. Gao and H. Xia. 2014. Effect of cadmium on growth, photosynthesis, mineral nutrition and metal accumulation of bana grass and vetiver grass. Ecotoxicol. Environ. Saf. 106: 102-108.

CHAPTER

12

Bioenergy through Bioelectricity in an Integrated Bioreactor-Bioelectricity Generator System for Phenol Biodegradation

Anoop M.* and Gayathri Premkumar

Dept. of Biotechnology and Research, KVM College of Science & Technology, Cherthala, Kerala, India

12.1 Introduction

Rapid industrialization activities strike the environment of the coastal regions of India and cause several serious problems. Clusters along the backwaters of India, especially Kerala, provides a coastal front for small-scale and large-scale industries like ice plants, food processing units, cement manufacturing companies, pottery making units, paper mills, organic fertilizer manufacturing industries, rice mills, oil mills, dye units, sugar mills, coir retting units, log sets, textile units etc. Among the small-scale industries, coir retting industries rank first (Nirmala et al. 2002). Coir fiber can be considered as the powerful natural fiber. Being a coconut by-product, it is an ideal and eco-friendly one for many applications which are nature friendly in environment. The coir was extracted from the coconut by retting. Coconut husk processing was practiced as an indigenous method in Kerala and was prevalent from early times. Degradable plant components are converted to simple compounds by these microbiological processes (Leena and Viveka 2009). India accounts for more than 85% of the total world coir production. Majority of the producers are from rural populations, especially women, who are engaged in the coir production (Nirmala et al. 2002). Indian coir industry has Coimbatore district as its major player in current scenario. The decrease in raw material and decline in the manpower requirement has affected many spaces even under such constraints, the coir industries managed its production units well.

Kerala, Tamil Nadu, Karnataka, Andhra Pradesh, Maharashtra, Goa, and Lakshadweep governs the Indian coir industry. Among these states, Kerala tops the

Corresponding author: mailmeanp86@gmail.com

list as the high quality and quantity producer of coir fiber and its products. Industrial coir sectors have increased their production from 7000 tons to 30000 tons in the year 2015-16 and are expected to rise a lot in future. Kerala government has already announced Rs112 crores from its budget for coir sector in 2021-22. A special scheme of small scale development was also marked in the present scenario.

The conventional method of immersing the husk in water initiates the natural process. Each retting yard usually contains about 1000 husks, either tied into a bundle or placed inside coir nets; it was termed as ‘Mallis’ (Meenatchisundaram 1975). They float in the retting area until it is completely soaked. The process requires 10-12 months of anaerobic fermentation to remove the dissolved substance. Husk retting follows the mentioned phases:

- Fast solubilisation and acidification of major organic compounds.
- Lingering removal of phenol substance will cause browning of the water bodies until the last step.
- Slower solubilisation of the remaining substance and production of gases.
- The workers crushing the fiber using a wooden hammer, enhancing the speed of retting.

The natural process is a time-consuming and slow rate process (Meenatchisunderam 1975).

The natural process of retting leads to the release of several chemicals into the environment. The initial stages show swell-up of husk and carbohydrates, tannins, and nitrogenous compounds, and the polyphenols are leached out. Polyphenols present in coconut husk are tannic acid, protocatechuic acid, catechol, pyrogallol etc. Microorganisms may act upon these chemicals and may produce intermediate compounds, which are less toxic or non-toxic to the nature. The pectin decomposition alone can be done on the biological process, and the disintegration of phenol components is also observed. The polyphenols are constantly liberating into the water bodies during the coir retting (Bijoy 1997). In bioremediation, biodegradation is an advanced branch that deals with an enormous potential of the microorganisms to remove toxic organic compounds present in the environment (Mishra and Kumar 2017). By providing the basic sustainable conditions for microbes such as changing pH, the redox reaction, nutritional requirements for pollutant absorption from contaminated localities will be enhanced (Omena and Olubukola 2017). Various microorganisms utilize phenol as the sole source of carbon and energy for their growth. The enzymatic actions of various fungi and bacteria will degrade the phenol by breaking down the aromatic ring. It shows that the microorganisms are capable of degrading phenol through variety of enzymatic actions. These enzymes include oxygenase, hydroxylase, peroxidases, and oxidases. The crucial moment of the aromatic compound metabolism is the breakage of the benzoin ring by fission and hydroxylation, and it occurs in the aerobic system by dioxygenase-catalyzed reactions. The ring fission is catalyzed by either ortho cleavage enzyme or meta cleavage enzyme. Catechols are cleaved either by ortho-fission or by a meta-fission.

Kerala coastal regions are witnessing the process of coir retting for ages, considering it as an indigenous practice for coconut husk processing. The backwaters of Kerala’s southern coasts provide all facilities for the coir industrial sectors. The over-exploitation of these backwaters for coir retting have caused a pernicious effect on the ecosystem. Fishing is the basic sustenance of the people who live in coastal areas of Kerala backwaters. As an impact of coir retting, water bodies near coir retting yards presented a drastic reduction in fish wealth and health because of low oxygen concentration by the activity of organic materials. Increases of toxic ingredients such as acids, alkalies, suspended solids, polyphenols, pectin, tannin etc. will occur as a result of waste disposal of retting industries. Tolerant fish varieties like Tilapia sp. were dominant in the retting yards as compared to the commercially important fishes. The Molluscan fishery has plummeted in the area because retting yards will not provide a suitable ground for the attachment for those species. Day by day, the diversity among the fish community is getting depleted because of the low pH, and the low concentration of oxygen provided by the coir industries. Retting leads to the release of enormous number of phenolic compounds, which brings out the revolting smells and also affect the growth of aquatic flora and fauna in that retting areas. In retting sites, heavy contamination and aquatic pollution occurred due to the presence of organic wastes like lignin, tannin, and polyphenols. The physicochemical quality of the water in the retting yard shows the prevalence of nil dissolved oxygen, very high BOD, electrical conductivity, hydrogen sulfide, and nutrients (Coir Vast Scope for Growth 2011). The concentration of phytoplankton in the coir retting yards was found to be more due to rich nutrient content and also shows the less abundance and diversity of zooplankton organisms. Polychaete species were prevalent on the bottom regional fauna of water bodies, which are favored by the organic enrichment in the area (Remani et al. 1989). The enhanced production of these organic enrichments represents crystal clear piece of evidence of pollution through organic contaminations. Drinking water sources are getting polluted due to industrial, domestic, and agricultural wastes. Small-scale industries like coir retting, log setting, food processing, and clay making pollute the water bodies. Coir retting industries discharge effluents like lignin, tannin, and polyphenols into the backwater system. Poisonous gases like hydrogen sulfide and methane pollute the air. The coir retting industry pours out a lot of phenols and polyphenols, which may alter the drinking water standards. In a study conducted on groundwater contamination of Kadinamkulam estuary, it was noticed that recorded values of color, pH, phosphate, calcium, magnesium, and hydrogen sulphide were above permissible limits, according to the Bureau of Indian Standards (1991). During monsoon days, the phosphate and nitrate contents from the retting yards will be wiped out into the bore wells and dug wells. Color, taste, and odor of the drinking water sources were found to be displeasing due to chemical release from the coconut husks during retting process (Bijoy 1997).

Many of the bacterial strain, like *Paenibacillus* sp. (AY952466) and *Bacillus cereus* (DQ002384), presented the capacity for degrading lignin and pentachlorophenol. At standard and optimum conditions, both the species, *Paenibacillus* sp. and *B. cereus* and its mixed culture degraded phenol (500 mg/l) up to 53.86%, 91.63% and 67.76% within 168 h of incubation, respectively, which was reported in many studies. Spectrophotometric predictions for toxic compound degradations were confirmed through HPLC, FTIR, and LCMS studies (Singh et al. 2009). *Paenibacillus* was considered as a potential candidate in the biodegradation of toxic aromatic compound phenol with our research studies. The natural method of retting coconut husk for extracting out fiber takes a long period of 6-12 months, which may release aromatic compounds like phenols and polyphenols to the environment that may harness the aquatic ecosystem and eliminate flora and fauna. In order to degrade the phenolic content in the water bodies, a bioreactor has been developed. It takes a year of practice for natural retting of coir for the release of organic compounds. During this period, the pectinolytic activity of microorganisms, especially bacteria, fungi, and yeasts, degrades the fiber binding material of the husks and liberates large quantities of organics and chemicals into the environment including pectin, pentosans, and polyphenols. That will affect the aquatic ecosystem and eliminate flora and fauna. In order to degrade the phenolic content in the water bodies and control the pollution evolved by these toxic compounds, a bioreactor has been developed for solving the problem at a minute level. The fabrication of the bioreactor was based on the combination of phyto degradation and the biodegradation process. The great advantage of this reactor is that it is an intrinsic bioreactor that generates bioelectricity by the combined action of both degradation processes. We can produce electricity in an alternative way by utilizing phenolic wastewater that will facilitate the control of severe problems associated with pollution.

With the energy crisis of 1970, along with the 1973 oil crisis and 1979 energy crisis, people soon became interested in the production of fuels using waste biomass through combustion and microbial fuel cells. Several researches have occurred as an effort to reduce production cost and the improvement of technology of conversions. The urban settlement of our nation being a decentralized settlement needs the usage of biomass as their energy source for power utilization. Bioenergy related projects have been recognized every year in the nation and are awarded approximately 9000 million dollars every year, thereby leading to an electricity generation of 4000-5000 million units approximately. As per data in accordance with the Ministry of Foreign Affairs, Denmark, more than 70% of country population depends on biomass as its energy sources. Even with such an experience in the biomass sector, there may be several gaps existing in the bioenergy sector which normally affect the supply chains. The collection and mobilization of raw materials is still found to be a challenging task. The need for an improved design for proper management of waste and utilizing it for energy generations was the prime focus of research. Hence, the need for conducting feasibility studies and proper development of research must be the main target for a consistent development in our bioenergy sectors. In response to the existing gaps, the present design that is proposed here will be a fruitful one by all the means.

12.2 Methodology

12.2.1 Analysis of biodegradation of selected polyphenols in a bioreactor

The selected polyphenols, gallic acid, tannic acid, and catechol, were assayed for biodegradation from 24 to 72 hours on a definite time interval of 24 hours.

12.2.2 Growth curve of the bacteria *Paenibacillus* sp strain GA2KVM on selected polyphenols

The bacterial strain *Paenibacillus* sp. was inoculated on luria Bertani broth and kept for incubation on an orbital shaker incubator. The overnight culture was centrifuged 10000 rpm for 10 minutes for collection of the pellets. The pellets were washed with 0.85% NaCl saline solution for the preparation of inoculums. 3% of the inoculums were inoculated on the minimal salt medium containing polyphenols like tannic acid, gallic acid, and catechol. The samples were kept in an orbital shaking incubator. The optical density was calculated at each 24 hrs of time interval from 24-72 hrs on 600 nm (Anoop and Jayachandran 2018).

12.2.3 Assay for polyphenols in wastewater taken from coir retting contaminated sources

Based on the rectitude behavior of bacterial strain than other organisms, the rate of biodegradation of some polyphenolic compounds such as tannic acid, gallic acid, and catechol is analyzed. A defined concentration of 22 μM of all the three compounds is added to 100 ml of each minimal salt medium. The bacterial strain was allowed to grow on Luria bertani and broth was centrifuged at 10000 rpm for 10 minutes, after 16 hours of incubation for harvesting the cells. The pellets collected after discarding the supernatant are washed with 0.85% NaCl saline solution. The saline washing was repeated and the inoculums were prepared. 3% of the inoculums was added to minimal salt medium containing the polyphenols. The minimal salt medium was kept for incubation on an orbital shaker incubator. The samples were collected and assayed for biodegradation studies by modified 4-amino antipyrine method (Anoop et al. 2018) at a regular interval of 24 hours from 24-72 hours of incubation.

12.2.4 Extraction of polyphenols from wastewater collected

The 40 ml wastewater collected from the coir retting place is measured and added to an extraction funnel. An equal amount of solvent hexane was added and and was rigorously shaken until the two phases combines together. The same was allowed to stand stable for some time at room temperature to get the lipids out from the water sample. The sample was collected from the bottom phase. The upper phase contains lipids that are washed out; the collected sample was centrifuged at 4000 rpm for 15 mins. The phases were viewed as separated. The water containing phase was collected and a repeated centrifugation was performed. Finally, sample collected after the final centrifugation was measured and added to the extraction funnel. An equal amount of ethyl acetate was added and the two phases were mixed together .The same were allowed to stand stable at room temperature. The bottom phase was collected and performed centrifugation for 10 mins at 4000 rpm. The water phase will be collected after centrifugation, and the same extraction procedure was repeated and followed by centrifugation. The bottom phase was kept under –20°C.

12.2.5 Study of total polyphenolic content in the wastewater

For the study of total polyphenol content in the wastewater, the sample will be analyzed by adding 2.5 ml of 0.2 N Folin Ciocalteu reagent to 0.5 ml of ethyl acetate fraction sample. The reaction was conducted in the dark for 5 min. After 5 min of incubation in dark, 2 ml of sodium carbonate was added onto it and was incubated in dark for 1 hour. After the incubation time, the optical density value was taken in 765 nm. The gallic acid reference standard graph was used to estimate the total polyphenol content (Molina-Cortés et al. 2020).

12.2.6 Absorption with powdered peels of watermelon

Watermelon was used as a potent organic constituent in removing phenol and polyphenol contaminations in wastewater (Ashiq 2018). The peels were washed with distilled water to remove water-soluble impurities. The peels were trimmed to smaller pieces and made as crispy in texture by placing them in sunlight for two days. The sample was powdered using a grinder, and made up to fine particles of approximately 500-900-micron sizes with the help of a molecular sieve. The powder was dried. The powder was again washed with 80-degree hot water and was dried in a hot air oven for 1 hour at 70 degrees of temperature. This processed powder is selected for the removal of toxic aromatic phenol by absorption. 5 g of the powder was added to 100 ml of minimal salt phenol medium (pH 6), and was incubated in a rotary shaker at 120 rpm for 1 hour, 3 hours, 5 hours, and 8 hours. The medium is then centrifuged at 5000 rpm for 15 min and the powdered fractions were removed to obtain the supernatant and then subjected to spectrometric assay for analysis.

12.2.7 Designing of integrated bioreactor/bioelectricity generator

In this study, bioreactor design is comparatively a multiplex engineering endeavor, in the treatment of polyphenolic waste water and generation of bioelectricity from this treatment. It is a multi-stage integrated reactor design with a fuel cell coupled bioelectricity generator (Fig. 12.1).

Figure 12.1: Integrated bioreactor/bioelectricity generator system for the treatment of phenol and polyphenols.

12.2.7.1 The first stage

The present stage consists of a reservoir, where the sample for the treatment is added, which passes through the bed prepared with fruit peels followed by an outlet for the collection of the treated samples and a connecting valve to the second stage. The first stage reservoir contains sample inlet, one outlet for sample collection and one flow controller for the next stage. The diameter of the inlet pipe of the reservoir is 5 cm for the easy adding of the samples. The storage capacity of the reservoir was 1000 ml, having 11 cm in diameter and 15.5 cm in height from the ground in size.

12.2.7.2 The second stage

In the second stage, the samples pass through the bed and are treated with the inoculum. The second stage will also contain an outlet for collection of the second stage treated samples for the analysis and a connection valve to the third phase. The second stage reservoir has the same storage capacity and height as the first phase. It also possesses an outlet for the sample collection and flow controller for the next stage. When the sample reaches half of the reservoir, the inoculums were added to the small opening in the side of the reservoir through a sterile syringe. Phenol assay were performed using the modified 4-amino anti pyrine method, and FTIR was performed for the rapid detection of pathway intermediates of biodegradation.

12.2.7.3 The third stage

It has a sparger for proper mixing with both direct current (DC) attached with 12V battery and alternating current (AC) power supply. The third phase of the reactor is a bioelectricity generator; the samples obtained after the two-stage different treatments were collected in reservoirs, which contain copper electrodes and a passage between two reservoirs, and also have a semi-permeable biological membrane for only the movements of the electrons. The potential difference between these two reservoirs and the production of electrons and their movements will generate a minute formation of voltage that will read through the digital millimeter connecting those copper electrodes. The whole reactor system is arranged in three different height level stands.

12.3 Results

12.3.1 Analysis of percentage degradation of polyphenols

The coir retting process may release several polyphenol compounds to a water source nearby. The bacteria *Paenibacillus* sp. strain GAKVM was subjected to. Furthermore, subjected to treatment with different polyphenol compounds like gallic acid, tannic acid, and catechol, the bacteria presented remarkable results during biodegradation. At 24 hours, the bacteria *Paenibacillus* sp. degraded 34.86% of gallic acid, 46% of tannic acid, and 46.4% catechol (Fig. 12.2).

12.3.2 Assay for bacterial growth curve

The growth of the bacteria in different incubation hours shows that the bacteria

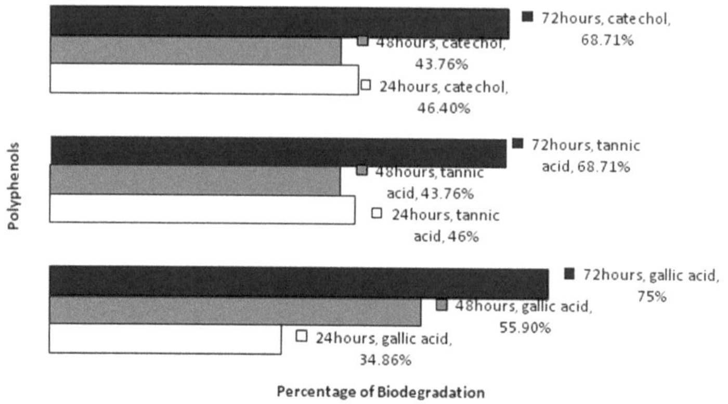

Figure 12.2: Percentage of biodegradation of three different polyphenols by *Paenibacillus* sp. strains GA2KVM on different incubation hours from 24-72 hours with a definite time interval of 24 hours.

multiplied rapidly between 24 and 48 hours for the three selected polyphenols as the sole source of carbon and energy. The bacteria presented a rapid multiplication, especially in 48 hours, which is found to be diminishing 72 hours of the incubation period (Fig. 12.3).

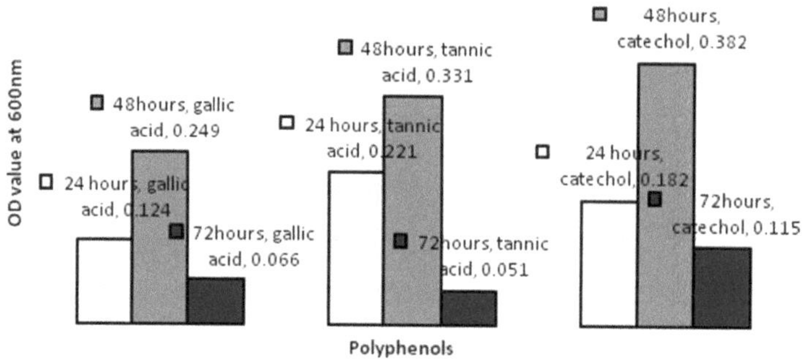

Figure 12.3: Assay of the growth of the bacteria *Paenibacillus* sp. strain GA2KVM on polyphenols.

12.3.3 Assay for polyphenols

The total polyphenols were assayed and values were recorded (Table 12.1).

Table 12.1: Assay of total polyphenolic content in wastewater from coir retting contaminated sources.

Total polyphenolic content (Optical Density value at 765 nm)		
0.316	0.348	0.229
Trial I	Trial II	Trial III

The concentration was calculated using the formulae (Control-Sample/ Control) * 100. The value was calculated as 67.52 ± 6.5%.

12.3.4 Analysis of the rate of absorption of phenol using powdered peels of watermelon

An attempt is also made to evaluate the possibility of removing a high concentration of phenol through dried powdered processed peels of watermelon. The experiment has been conducted at various incubation periods of 1 hour, 3 hours, 5 hours and 8 hours. The phenol removal by the processed powdered watermelon peels was speedy and could get more than 55% removal in 1 hour of treatment. However, the maximum removal at 8 hours was only 79.2% (Table 12.2).

Table 12.2: Analysis of absorption rate of phenol by powdered peels of watermelon on different incubation periods

Average percentage absorption of phenol in various incubation periods.			
1 hour	3 hours	5 hours	8 hours
60.28 %	64.64%	76.93 %	%

12.3.5 Treatment of the phenol containing synthetic wastewater using integrated bioreactor-bioelectricity generator

Treatment of phenol containing wastewater is performed with synthetic phenol-containing wastewater using the newly designed Integrated Bioreactor. Treatment of the phenol containing wastewater is carried out with a newly designed bioreactor at prolonged flow rate of 1 ml/min. The experiment is conducted at various incubation periods with a definite time interval of 2 hours.

12.3.6 Percentage of degradation of phenol in the wastewater samples collected after the treatment with the bioreactor system at the definite interval of time

Treatment of the phenol wastewater was done in the integrated bioreactor by the combined action of processed watermelon peels and the selected organism *Paenibacillus* sp. The maximum percentage of reduction of the phenol is 84.03% at 8 hours and also shows some slight variations in the multimeter for the reading of voltage formation; it should be noted as the lowest production of bioelectricity using this integrated bioreactor/bioelectricity generator (Fig. 12.4).

12.4 Discussion

Sustainable energy management is the key necessity and plays the major role in many energy industries. The largest contributor to such renewable and sustainable energy is bioenergy. Intensive researches are getting rapid attention in the field of bioenergy and the major event in the design of novel bioreactors for waste management. Bioreactors play significant roles in molecular biomass development and production through energy conversion from waste compounds. A representation of a new design of bioreactor for the conversion of toxic aromatic phenol and polyphenols

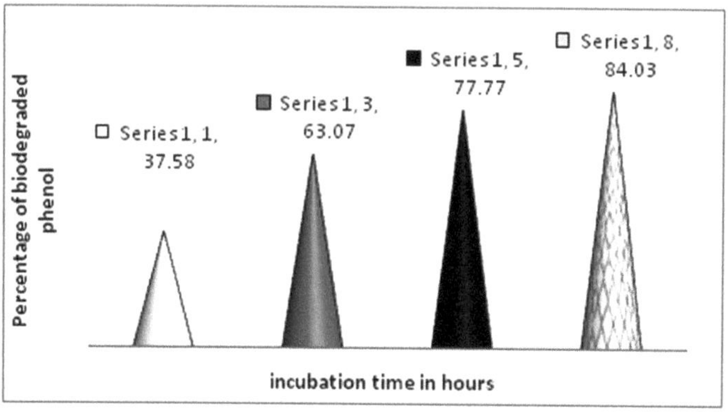

Figure 12.4: Phenol degradation percentage of various incubation periods taken from a sample collected from an integrated bioreactor/bioelectricity generator system.

into the generation of a bioelectricity was designed and explained in the present research study.

The reactor is designed with PVC pipes connected in series to 1000 ml glass beakers (Fig. 12.7). This type of design amounts to a continuous mode of treatment. The entire system is provided with a top inlet for sample addition and outlets at every junction for sample collections. The design holds three major stages where the first two stages of the bioreactor carried degradation, and a third stage collaborates a bioelectricity generator in the form of a microbial fuel cell. The results of polyphenol biodegradation (Fig. 12.2) indicated the ability of the strain *Peanibacillus* sp. as a potential candidate in the biodegradation of toxic aromatic compounds. The bacterial growth curve presented a clear indication of the significance of 48 hours of incubation (Fig. 12.3), as the bacteria multiplies rapidly from 24-48 hours of incubation, which showed that the polyphenols were utilized as sole source of carbon and energy by the bacteria as the incubation period got extended from 24-48 hours.

Absorption of chemicals by plants, which was another ingredient used in the research studies of development of bioreactor-bioelectricity generator, showed remarkable results (Table 12.2) as within 8 hours, more than 70% was absorbed. The phenol wastewater in the integrated bioreactor-bioelectricity generator got degraded by the synergistic action of the inoculums and plant material (Fig. 12.4) The multimeter deflections suggested the production of current from the waste material, thereby extrapolating bioenergy generation. Bioenergy generation based designs in bioreactors can be utilized for sustainable energy resources and can be commercialized. More and more research must be performed with these designs so as to enhance its performances and reliability for a large scale energy generation.

12.5 Summary and Conclusion

Design of a comprehensive strategy to biodegrade phenol is presented with the design of a novel bioreactor-bioelectricity generator. The sample used was phenol-containing wastewater. The composition of industrial effluents may vary periodically, and this may affect the performance of a newly designed bioreactor.

The sample was allowed to pass through first-stage watermelon beds that presented an effective absorption of toxic compounds. The inflow rate is arranged too slow which leads to an efficient treatment system utilizing time. The effluent was continued to second-stage, which was already occupied with an inline agitator system. The treatment is done at various flow rates of 1 ml/minute, 5 ml/minute and 10 ml/minute. The efficiency of treatment of phenol alone with watermelon beds, bacteria and in combination was evaluated through degradation assay (Mordocco et al. 1999). The decrease in phenol concentration was confirmed, and the sample was provided for FTIR assay for further research works.

The bioreactor system revealed a promising strategy for the treatment of synthetic wastewater containing phenol. The newly designed system could remove phenol and generate bioelectricity utilizing the toxic aromatic organic compounds. Appropriate technology development in the future may compile the scale-up studies of the present system, which was found to be a promising one that combines a bioreactor and bioelectricity generator.

The bacteria *Paenibacillus* sp. was found to be degrading the selected polyphenols very effectively. It showed a maximum degradation of 75% for gallic acid at 72 hours and minimum degradation of 34.86% at 24 hours. Wide metabolic diversity and powerful adaptability features were exhibited by microorganisms with which they are capable of degrading any naturally occurring or compounds derived out of any naturally occurring compounds. This reliability of the microorganisms is considered to be instrumental in the decontamination of many toxic organic compounds like phenol. The present attempt is one such step towards bringing the principles of biodegradation pathway analysis through a systems approach with the design of a bioreactor system with a view to degrade toxic contaminants that create severe environmental contamination problems. The bioreactor/bioelectricity generator system will be a handy tool in the degradation of phenol and polyphenols through an efficient interpretation of values obtained through a convergent system of chromatography and spectroscopy. The observations made in the present study will definitely pave the way for new trends in the field of biodegradation.

References

Anoop, M. and K. Jayachandran. 2018. Diauxic growth of *Raoultella* sp. SBS2 favouring enhanced degradation of phenol in a defined medium. European J. Biomed. Pharm. 5: 312-315.

Anoop, M., C.N. Indu and K. Jayachandran. 2018. Growth kinetics of *Alcaligenes* sp. d2n the biodegradation of phenol in mineral salt phenol medium. World J. Pharm. Res. 7(6): 1081-1092.

Ashiq, S.M. 2018. Microbial degradation of Aniline and its application in the biological treatment of industrial effluents. PhD. Thesis. Mahatma Gandhi University, Kerala, India.

Bijoy, S.N. 1997. Retting of coconut husk – A unique case of water pollution on the South West Coast of India. Int. J. Environ. Sci. 52(4): 335-355.

Coir Vast Scope for Growth. 2011. Agri Business, The Hindu.

Leena, B.G and S. Viveka. 2009. Effect of cocount husk retting on three backwater regions along the southwest coast of Kerala. Terr. Aquat. Environ. Toxicol. 3: 62-64.

Meenatchisunderam, R.I. 1975. Retting of coir – A review. Ceyion Cocon. Plrs. Rev. 7: 20-28.

Mishra, V.K. and N. Kumar. 2017. Microbial degradation of phenol: A review. J. Water Pollut. Purif. Res. 17-22.

Molina-Cortés., T.A., T. Sánchez-Motta, F. Tobar-Tosse and M. Quimbaya. 2020. Spectrophotometric estimation of total phenolic content and antioxidant capacity of molasses and vinasses generated from the sugarcane industry. Waste Biomass Valorization 11: 3453-3463.

Mordocco, A., C. Kuek and R. Jenkins. 1999. Continuous degradation of phenol at low concentration using immobilized Pseudomonas putida. Enzyme Microb. Technol. 25(6): 530-536.

Nirmala, E., T.K. Jalaja and K.N. Remani. 2002. Pollution hazards on the people and ecosystem of selected coir retting yards in the backwards of Calicut district. Final report, Centre for Water Resources Development and Management, Kozhikode, Kerala, India.

Omena, B.O. and O.B. Olubukola. 2017. Microbial and plant-assisted bioremediation of heavy metal polluted environments: A review. Int. J. Env. Res. Pub. 14: 1504.

Remani, K.N., E. Nirmala and S.R. Nair. 1989. Pollution due to coir retting and its effect on estuarine flora and fauna.

Singh, S., B.B. Singh, R. Chandra, D.K. Patel and V. Rai. 2009. Synergistic biodegradation of pentachlorophenol by *Bacillus cereus* (DQ002384), *Serratia marcescens* (AY927692) and *Serratia marcescens* (DQ002385). World J. Microbiol. Biotechnol. 25(10): 1821-1828.

CHAPTER

13

Phytoextraction of Potentially Toxic Elements from Wastewater Solids and Willow Ash – Experiences with Energy Willow (*Salix triandra x S. viminalis* 'Inger')

László Simon[1]*, Zsuzsanna Uri[1], Szabolcs Vigh[1], Katalin Irinyiné Oláh[1], Marianna Makádi[2] and György Vincze[1]

[1] University of Nyíregyháza, Institute of Engineering and Agricultural Sciences, Department of Agricultural Sciences and Environmental Management, H-4400 Nyíregyháza, Sóstói Str. 31/b, Hungary

[2] University of Debrecen, IAREF, Research Institute of Nyíregyháza, H-4400 Nyíregyháza, Westsik Vilmos Str. 4-6, Hungary

13.1 Introduction

Energy and mineral consumption by man is the main cause of trace element pollution in the biosphere. During the last century, as a consequence of mining, metal processing, industrialization, transport, burning of fossil fuels, disposal of wastes, etc., soil and water resources were contaminated with metals and certain metalloids all over the world (Kabata-Pendias 2011, Simon 2014). Trace elements mean the elements present at low concentrations (mg kg^{-1} or less) in the agroecosystems. Some trace elements, including copper (Cu), zinc (Zn), or manganese (Mn) are essential to plant growth and are called micronutrients. These elements are also potentially harmful and may be toxic to plants at high concentrations. Trace elements (metals and metalloids) such as cadmium (Cd), lead (Pb), chromium (Cr), nickel (Ni), mercury (Hg), arsenic (As), and partially barium (Ba) have toxic effects on living organisms and are often considered as contaminants (Kabata-Pendias 2011, Simon 2014).

Phytoremediation is the use of plants and their associated microbes for environmental clean-up. This technology makes use of the naturally occurring processes by which plants and their microbial rhizosphere flora degrade and sequester

Corresponding author: simon.laszlo@nye.hu

organic and inorganic pollutants (Arthur et al. 2005, Pilon-Smits 2005). Plants have a natural ability to uptake inorganic chemicals (including metals) from soil and sediment, and accumulate them in their tissues. Phytoextraction is the ability of plants to remove inorganic contaminants (primarily metals) from the contaminated matrix by transferring them to harvestable parts of the plant. Several factors like the extent of contamination, metal bioavailability, and the plant's ability to intercept, absorb, and accumulate metals contribute to the success of phytoextraction (Arthur et al. 2005). The harvested (above ground) plant material can be used for non-food purposes (e.g. wood, cardboard), or can be treated to decrease its volume and weight (by composting, compacting, drying, or thermal decomposition) (Pilon-Smits 2005, Šyc et al. 2012, Šuman et al. 2018).

One of the basic strategies for phytoextraction of potentially toxic elements (PTEs) from contaminated soil is cultivating fast-growing plant species with high biomass production (Šuman et al. 2018, Sameena and Puthur 2021). Bioenergy plants have the potential to adapt well in polluted lands and have the capacity to produce high biomass along with good energy potential (Jha et al. 2017). The heavy metals accumulated in the shoot biomass are usually below the standard toxicity levels in the case of a majority of the bioenergy plants. Therefore, the shoot biomass of bioenergy plants utilized for phytoremediation can be safely used for producing bioenergy (Jha et al. 2017, Sameena and Puthur 2021). According to Maxted et al. (2007), *Salix*-based phytoextraction may be applied to arable soils, which require only minor 'polishing' in order to meet arable soil standards or produce metal concentrations in edible crops which do not violate risk-based hazard quotients.

Among perennial energy crops, which are cultivated for their high aboveground biomass, the high-yielding, rapidly growing willow (*Salix*) species are very promising all over the world. Their harvestable shoot wet biomass can achieve 20 or 25 tons per hectare (or 10-12 tons of dry matter) annually (Merilo et al. 2006, Kulig et al. 2019). The willow sprouts well; its 2-6-meter-long shoots (whips) can be harvested every 1, 2, or 3 yr. Since short-rotation coppice (SRC) energy plantations can be cultivated for 15-20 yr in the same field, balanced and regular nutrient supply is required in the soil for access good aboveground biomass yields (Gyuricza et al. 2008, Smart and Cameron 2012, Simon et al. 2016).

Biomass yield of *Salix* spp., grown as an energy crop, can be stimulated by application of various inorganic or organic fertilizers and additives in soil, including biosolids (e.g. municipal sewage sludge), biochar or biomass ash (Pulford et al. 2002, Park et al. 2005, Dimitriou et al. 2006, Gyuricza et al. 2008, Smart and Cameron 2012, Sevel et al. 2014, Hytönen 2016, Saletnik and Puchalski 2019). Application of various soil amendments (e.g. municipal sewage sludge or wood ash) can enhance, however, not only the uptake rate of beneficial elements (e.g. nitrogen or potassium) but also the rate of PTE accumulation in the willow organs (Park et al. 2005, Dimitriou et al. 2006, Maxted et al. 2007, Gyuricza et al. 2008, Jama and Nowak 2012, Smart and Cameron 2012, Simon et al. 2013, Salam et al. 2019, Labrecque et al. 2020, Saletnik et al. 2020, Wójcik et al. 2020). This may have an impact on the toxic metal concentration of harvested shoots, and this way also on the toxic metal concentration of ash after biomass burning.

It is well documented that Cd accumulation or Zn uptake rates of *Salix* spp. are high, as compared to other plant species and other trace elements (Vysloužilová et al. 2003, Berndes et al. 2004, Dickinson and Pulford 2005, Dos Santos Utmazian et al. 2007, Maxted et al. 2007, Simon et al. 2017a, b, 2018, 2019). Resistance or tolerance of willows to other metals (Cr, Cu, Ni, Pb), and accumulation of elevated levels of As, Cu, Fe, Ni, Mn, and Pb (Meers et al. 2007, Gyuricza et al. 2008, Mleczek et al. 2013, Vandecasteele et al. 2015, Tőzsér et al. 2018, Salam et al. 2019, Labrecque et al. 2020) was also observed in the organs of various *Salix* spp. Therefore, from a phytoremediation point of view, the elevated level of certain toxic elements could be advantageous in the shoots of SRC willow to remove metal pollutants from the soil (Greger and Landberg 1999, Pulford and Watson 2003).

Considering the above preliminaries, our aim was to investigate the uptake or accumulation of nine selected PTEs (as As, Ba, Cd, Cr, Cu, Mn, Ni, Pb, and Zn) in aboveground organs of energy willow (*Salix triandra* × *S. viminalis* 'Inger'), grown in a long-term experiment. It was assumed that repeated soil application of wastewater solids (municipal sewage sludge compost – MSSC, municipal sewage sediment – MSS) or willow ash (WA) will change the depth distribution and plant availability of PTEs in soil, and this will influence the accumulation rate of PTEs in willow organs differently. Special attention was paid to the repeated WA soil application combined with MSSC or MSS and its long-term impact on PTEs' accumulation in willow since relatively few investigations (Dimitriou et al. 2006, Adler et al. 2008, Lazdina et al. 2011) focused on this formerly. It was supposed that WA reduces the accumulation of PTEs in willow when applied altogether with MSSC and MSS.

13.2 Materials and Methods

13.2.1 Long-term open-field experiment with willow

Open-field small plot long-term experiment was set up with energy willow (*Salix triandra* × *S. viminalis* 'Inger') during 2011. The research area is located in parallel to Westsik Street in Nyíregyháza city (Hungary) in the experimental field of Research Institute of Nyíregyháza, University of Debrecen, Centre of Agricultural Sciences. Research plots are located at 103.4-meter height above Baltic Sea level, and are delimited with geographical coordinates 47°58′42.48059″ N, 21°41′58.96964″ E; 47°58′40.72898″ N, 21°41′59.04370″ E; 47°58′42.59343″ N, 21°42′02.34824″ E; and 47°58′40.87105″ N, 21°42′02.46138″ E. The total area of the long-term experiment is 3800 m^2. The experiment was set up with a randomized block design with ten treatments in four replications, and hence there were 40 small plots (Fig. 13.1). The period of soil treatments was between April 2011 – June 2018 (Figs. 13.1-13.2) and period of presented soil and plant observations was between June 2018 – September 2020.

Willows were planted during April 2011; cuttings originated from Holland-Alma Ltd., Piricse, Hungary (license holder of the studied willow species is Lantmännen Agroenergi AB, Sweden). In one 27 m^2 experimental plot, 40 willow bushes are grown with 0.75 m line spacing and 0.6 m between plants. In every small plot, plants are grown in two twin rows with 1.5 meters spacing.

Before (during April 2011) first soil treatments of the long-term experiment, the basic characteristics of the uncontaminated Cambisol (brown forest soil with clay stripes) were the following at 0-25 cm depth: loamy sand texture; pH-H_2O: 8.10; pH-KCl: 7.52; total salt (m m^{-1} %): <0.02; $CaCO_3$ (m m^{-1} %): 4.80; humus (m m^{-1} %): 1.51%; CEC (cmolc kg^{-1}): 10.4 (Simon et al. 2018); P–621, K–2918, Ca–16307, Mg–4603; As–9.60, Ba–57.5, Cd–0.21, Cr–13.7, Cu–9.18, Mn–372, Ni–14.0, Pb–9.89, and Zn–35.5 mg kg^{-1}; as determined from cc. HNO_3– cc. H_2O_2 extract, following the instructions of a Hungarian Standard MSZ 21470-50 2006.

The top 0-25 cm layer of the soil was treated in the spring (during April, May or June) of 2011, 2013, and 2016 yr with municipal green waste compost, municipal sewage sludge compost, willow ash, and rhyolite tuff (Fig. 13.1). Every spring (during May or June) of the yr 2011-2014 ammonium nitrate, between 2014-2015 urea, then in 2016 and 2017 urea fertilizer with sulfur were applied to the soil as top dressings (Fig. 13.1) (Simon et al. 2016, 2017a, b, 2018).

Municipal sewage sludge compost (MSSC; producer Nyírségvíz Ltd., Nyíregyháza, Hungary) was applied to the top layer of the soil three times (during June 2011, May 2013, and May 2016) in 15 t ha^{-1} (wet weight with 48-56% dry matter) dose each yr; its PTEs' concentrations during 2016 were in cc. HNO_3 – cc. H_2O_2 extract (Hungarian Standard MSZ 21470-50 2006); As–12.2, Ba–212, Cd–0.55, Cr–19.3, Cu–79.0; Mn–318, Ni–15.1, Pb–22.0, and Zn–357 mg kg^{-1}. Other basic physical and chemical characteristics, and plant nutrient content of MSSC (Simon et al. 2018) were the following: pH-H_2O 5.93; pH-KCl 5.91, $CaCO_3$ (m m^{-1} %):0; total salt content (m m^{-1} %):3.34; total C (m m^{-1} %):10.4; total N (m m^{-1} %):1.84; NH_4-N (mg kg^{-1}):169; NO_3-N (mg kg^{-1}):42.3; P–18876, K–3424, Ca–39294, Mg–4479, Fe–17149 mg kg^{-1} in cc. HNO_3–cc. H_2O_2 extract (Hungarian Standard MSZ 21470-50 2006).

Willow ash (WA) was prepared with the burning of leafless twigs of the willows, grown formerly in the experimental plots. The topsoil of the plots was treated three times with WA, in June 2011 and May 2013, with 600 kg ha^{-1} doses, respectively, and in May 2016 with 300 kg ha^{-1} dose. PTE concentrations of WA (99% dry matter) were in 2016 the following: As–10.2, Ba–267, Cd–2.38, Cr–9.66, Cu–133, Mn–553, Ni–10.9, Pb–12.1, and Zn–1757 mg kg^{-1}, as determined from cc. HNO_3 – cc.H_2O_2 extract (Hungarian Standard MSZ 21470-50 2006). Applied WA can be characterized with the following basic characteristics (Simon et al. 2018): pH-H_2O 10.9; pH-KCl 10.7, total salt content (m m^{-1} %):1.17; NH_4-N (mg kg^{-1}):0; NO_3-N (mg kg^{-1}):0; P–6472, K–16508, Ca–43074, Mg–7991, Fe–17045 mg kg^{-1} in cc. HNO_3–cc. H_2O_2 extract (Hungarian Standard MSZ 21470-50 2006).

Above amendments and fertilizers (immediately rotated to upper 0-25 cm layer of the soil) were applied to the soil between 2011-2017 yr also in various combinations, as illustrated in Fig. 13.1. Control plots remained untreated throughout the experiment.

During March and April 2018, the shoots of all willow bushes were harvested (this was the 3rd harvest after 2013 and 2016 yr).

By 15 June 2018, the soil of plots formerly treated three times with MSSC was amended with air dry, unscreened 7.5 t ha^{-1} dose of MSS in four replications (Fig. 13.2). MSS originated from Lovász-zug suburban area of Debrecen city, Hungary,

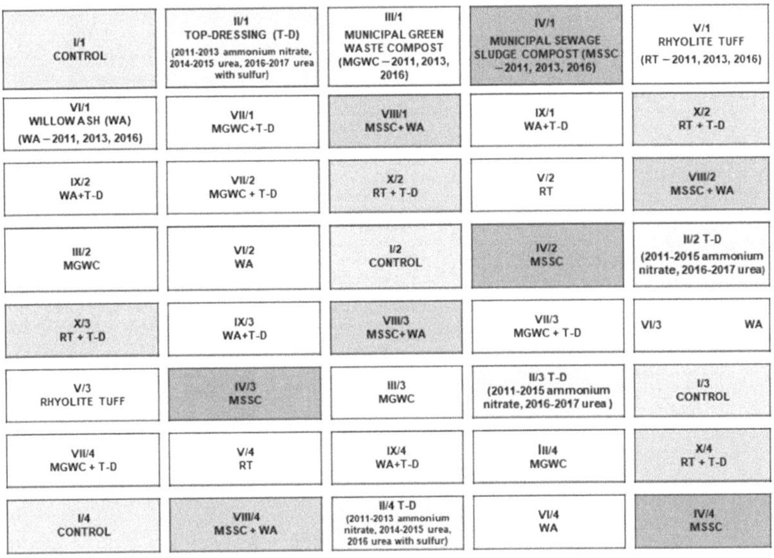

Figure 13.1: Scheme of the long-term experiment with energy willow (*Salix triandra* × *S. viminalis* 'Inger'), soil treatments between 2011-2017 yr in random block layout with four replications (Nyíregyháza, Hungary).

47°29′07″ N, 21°35′46″ E, where former sewage settling ponds were operated as a secondary biological purification unit (Tőzsér et al. 2018). MSS samples were collected from one of this recultivated sewage settling pond, where MSS was located under artificial soil cover in a 70-110 cm depth. Approximately 280-320 kg wet MSS samples were then spread in a 10-15 cm layer, regularly rotated, shredded, and air-dried in a covered, aerated building of the University of Nyíregyháza (Hungary). After two months, MSS dried to air-dry state. Four composite samples were taken from the air-dry MSS for chemical analysis. One composite sample with 1.0-1.5 kg total mass arises from combining 25 subsamples. Thoroughly mixed composite samples were passed through a 5 mm diameter sieve before analysis.

Willow ash was prepared at the beginning of June 2018 with the burning of dry leafless twigs of the willows from the 2016 harvest. WA was passed through 8-mm sieve before its soil application and sampled for chemical analysis, as described above for MSS. By 15 June 2018, the soil of plots formerly treated three times with WA was again amended with a completely dry, 300 kg/ha dose of freshly prepared willow ash in four replications. Control plots remained untreated (Fig. 13.2). MSSC and WA or MSS and WA were also applied to the soil in combinations during all treatments (Figs. 13.1-13.2).

13.2.1.1 Plant sampling

The first sampling of willow leaves was done five weeks after applying of MSS, WA, or MSS+WA soil amendments, by 25 July 2018. Willow leaves were sampled from 10 plants per plot, in four replicates. Five sampled plants were located in the middle section of the 2nd row, and five were found in the middle area of the 3rd row of a given

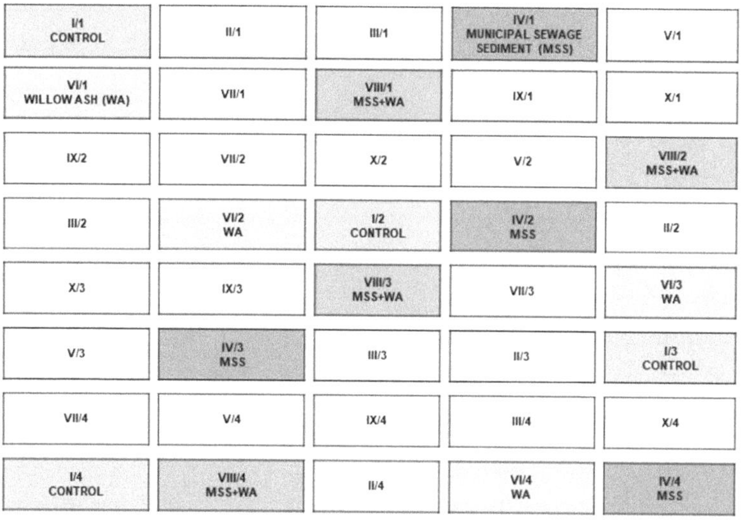

Figure 13.2: Scheme of the long-term experiment with energy willow (*Salix triandra* × *S. viminalis* 'Inger'), soil treatments during 2018 (Nyíregyháza, Hungary).

plot. Ten fully developed leaves per plant were collected from 10-20 cm uppermost section of the shoots. From one plot 100 leaves, from one treatment 400 leaves were collected, with 79-gram total fresh weight per plot on average. The second sampling of willow leaves was conducted by 24 June 2019, 53 wk after the last soil treatments. The sampling method was identical with the previous yr; from one plot, on average 46-gram leaves were collected. The third willow leaf sampling was done on 101 wk (by 27 May 2020) after the last soil treatments. The sampling protocol was the same as in 2018 and 2019; in 2020 yr 35-gram was the average fresh weight of the 100 collected willow leaves from one experimental plot. Besides three leaf samplings, also twigs of willows (without leaves) were sampled by 17 March 2020. The 80–100 cm height section of leafless young shoots of 10 plants per plot was cut out with stainless clippers. Sampled plants were positioned in the middle of the plots. All 16 plots included in the experiment (Fig. 13.2) were sampled in this way. The total fresh weight of twigs was 75-gram per plot on average.

Immediately after sampling, leaves or twigs were thoroughly washed in flowing tap water in the laboratory. The tap water was rinsed from the samples in two-times-changed distilled water. Samples were dried until constant loss of weight in drying oven (Mytron, Germany) at 70 °C for 10 hr (leaves) or 13 hr (twigs). Dry samples were ground to particles <1 mm in an ultra-centrifugal mill (Retsch ZEM 200, Germany).

13.2.1.2 Soil sampling

To check the impacts of 3 times (2011, 2013, 2016 yr) applied MSSC, WA, or MSSC+WA on the concentrations of PTEs in soil, samples were taken by 7 June 2018 from 16 plots (Fig. 13.2), including controls. Approximately 1.2-1.5 kg composite soil samples per experimental plot were collected, drilling 10 cm far from the stems of 25 willow bushes. Twenty-five subsamples per plot were taken from 0-25 cm depth with the help of a standard gouge auger (Eikelkamp, The Netherlands). To investigate the depth distribution of PTEs in plots, composite soil samples were taken by 25 September 2020 with the help of Edelman auger (Eikelkamp, The Netherlands) from 0-30 cm and 30-60 cm depth, in four replicates.

Immediately after sampling, all soil samples were taken to the laboratory. After removing foreign substances, the soil was homogenized and spread on plastic plates in a thin layer. After 14 d of drying at room temperature, the thoroughly mixed air-dry samples were passed through a 2-mm sieve.

13.2.2 Elemental analysis of plant, soil and soil additive samples

From the dried and ground to particles (<0.1 mm) plant samples, 0.50 g was loaded into the pressure-proof bombs of the microwave digester (Milestone Ethos Plus, Italy). To all samples, 5 ml of distilled cc. HNO_3 and 3 ml 30% (vv^{-1}) H_2O_2 (Scharlau, Spain) was added. The digestion was performed by the Application Note 076 program of the microwave digester, as follows: 3 min to 85 °C; 9 min to 145°C; 4 min to 200°C; and 14 min at 200°C.

To determine the "pseudo-total" element content of the soil or soil additives, the Hungarian Standard MSZ 21470-50 2006 was followed with slight modification. From the prepared (dried and ground to particles <0.1 mm) soil and soil additive samples, 0.50 g was loaded into the pressure-proof bombs of the microwave digester. To all samples, 6 ml of distilled cc. HNO_3 and 2 ml 30% (vv^{-1}) H_2O_2 was added. For soil samples or soil additives, the digestion was performed by the Application Note 031 program of the microwave digester, as follows: 10 min to 200 °C followed by 15 min at 200 °C. After the digestion, all samples were washed in a 50 ml volumetric flask with distilled water, homogenized, and filtered (MN 640 W paper; Macherey-Nagel, Germany).

To determine the soluble ("plant available") element content of the soil, the Hungarian Standard MSZ 20135 (1999) was followed. From the prepared (dried and ground to particles <0.1 mm) soil samples, 0.50 g was loaded into shakers, and 50 ml Lakanen-Erviö (LE) solution (0.02 M H_4-EDTA in 0.5 M ammonium acetate buffer and 0.5 M acetic acid, pH 4.65; Lakanen and Erviö 1971) was added. The shakers were uploaded to rotary shaker for 1 hr. After shaking, the samples were filtered through Munktell & Filtrak 292 ash-free filter paper (Germany).

Elemental analysis of all samples was conducted with Inductively Coupled Plasma-Optical Emission Spectrometry (ICP-OES) technique, applied on iCAP 7000 spectrophotometer (Thermo Fischer Scientific, USA). For the calibration, a multi-element standard solution (n=2) was used. All chemical analyses were done with three (soil additive samples) or four (plant and soil samples) replicates.

13.2.3 Bioconcentration factor

Bioconcentration factor (BCF) (Buscaroli 2017, Salam et al. 2019) was calculated by dividing the concentration of a PTE in a willow plant organ (leaf or twig) with "pseudo-total" or "plant available" concentration of a PTE in soil.

13.2.4 Statistical analysis

Statistical analysis of experimental data was conducted with SPSS 26.0 software using analysis of a variance (ANOVA) followed by treatment comparison using Tukey's b-test, and with Pearson's correlation.

13.3 Results and Discussion

Since the three times applied MSSC and WA contained more or less toxic elements in addition to the mineral nutrients that could be utilized by the plants, we assumed that surpluses of PTEs would be measurable in the soil of the experimental plots. To prove this, by 7 June 2018, we sampled the soil of the plots which were treated during 2011, 2013 or 2016 with MSSC, WA or MSSC+WA. Table 13.1 presents the concentrations of nine selected PTEs in the upper layer of the experimental soils.

Almost all treatments significantly enhanced the concentrations of all PTEs in experimental soils, as compared to control. The only exception was Cr in WA-treated soil. The highest As, Ba, Cd, Cu, Cr, Ni, and Pb concentrations were detected in MSSC-treated soil, while the most elevated Mn and Zn contents were present in WA-treated soil. Except for Ba, Mn, and Zn, the co-application of MSSC and WA resulted in statistically proven lower PTEs concentrations in soil than the application of MSSC alone. It can be supposed that only Mn and Zn were added in a significant amount from willow ash to soils treated with MSSC+WA (Table 13.1).

Table 13.1: Concentrations of potentially toxic elements in the topsoil (depth 0-25 cm) of the open-field long-term experiment set up with energy willows (*Salix triandra* × *S. viminalis* 'Inger') (Nyíregyháza, Hungary)

"Pseudo-total"* concentrations of potentially toxic element (mg kg^{-1})	Soil treatments (2011-2016)			
	Control	MSSC	WA	MSSC+WA
	Soil depth 0-25 cm (June 2018)			
As	8.45^{a}	22.3^{d}	13.4^{b}	19.8^{c}
Ba	97.8^{a}	130^{c}	117^{b}	127^{c}
Cd	0.291^{a}	0.805^{d}	0.470^{b}	0.642^{c}
Cr	9.51^{a}	18.0^{c}	9.58^{a}	12.0^{b}
Cu	11.3^{a}	15.5^{d}	12.4^{b}	13.7^{c}
Mn	263^{a}	429^{b}	481^{d}	454^{c}
Ni	11.3^{a}	14.9^{d}	12.6^{b}	13.9^{c}
Pb	15.4^{a}	35.3^{d}	16.6^{b}	23.1^{c}
Zn	40.3^{a}	51.4^{b}	55.6^{d}	54.0^{c}

*cc. HNO_3 + cc. H_2O_2 extract. MSSC = Municipal sewage sludge compost. WA = Willow ash. Means within the rows followed by the same letter are not statistically significant at $P<0.05$.

Our results confirm the well-known phenomenon that single or repeated application, long-term soil disposal of municipal sewage sludge or wood ash can considerably enhance the concentrations of PTEs in topsoil (Pulford et al. 2002, Dimitriou et al. 2006, Maxted et al. 2007). Except As in MSSC-treated soil, the PTEs concentrations measured in soils of our experimental plots were, however, equally lower than the valid Hungarian threshold limits (As–15, Ba–250, Cd–1, Cr–75, Cu–75, Ni–40, Pb–100, and Zn–200 mg kg^{-1}) for soil pollution (Hungarian KvVM-EüM-FVM Joint Decree 2009).

By 15 June 2018, the soil of plots formerly treated three times with MSSC was amended with MSS. Chemical analysis of MSS revealed that this biowaste is rich in calcium (Ca; 34724 mg kg^{-1} dry matter), magnesium (Mg; 7049 mg kg^{-1} d.m.), phosphorus (P; 4695 mg kg^{-1} d.m.), and potassium (K; 3077 mg kg^{-1} d.m.). The WA applied at the same time, however, contained seventeen times more K (54248 mg kg^{-1} d.m), or approximately five times more Ca (187550 mg kg^{-1} dry matter), Mg (35348 mg kg^{-1} d.m) and P (25403 mg kg^{-1} d.m.) than MSS. These values are in agreement with observations of other authors (Park et al. 2005, Dimitriou et al. 2006, Lazdina et al. 2011, Saletnik and Puchalski 2019, Wójcik et al. 2020), who noticed that sewage sludge or wood ash is a rich source of Ca, Mg, P or K. The MSS contained significantly more Cr and Pb than WA, while WA contained more Zn, since the measured values were As–31.0, Ba–596, Cd–1.23, Cr–1142, Cu–198, Mn–520, Ni–62.8, Pb–278, Zn–978 mg kg^{-1} in MSS, and As–18.5, Ba–403, Cd–0.60, Cr–9.10, Cu–130, Mn–670, Ni–14.2, Pb–26.7, and Zn–1853 mg kg^{-1} in WA (unpublished data). The concentrations of PTEs in WA are in the range observed by Saletnik and Puchalski (2019), Zając et al. (2019), and Wójcik et al. (2020) in willow ash or wood ash.

In Hungary, the 50/2001 Government Decree controls the agricultural utilization of sewage sludge. The Cr content of the studied MSS exceeded the 1000 mg kg^{-1} limit value for total chromium in sewage sludge (Hungarian Government Decree 2001). In Hungarian 36/2006 Decree of the Ministry of Agriculture and Rural Development for various crop-enhancing substances (mineral fertilizers, inorganic soil improvers) containing waste, the limits are 10 mg kg^{-1} for As, 2 mg kg^{-1} for Cd, 100 mg kg^{-1} for Cr, 100 mg kg^{-1} for Cu, 50 mg kg^{-1} for Ni, and 100 mg kg^{-1} for Pb (Hungarian FVM Decree 2006). In the studied WA, As and Cu concentrations were above these regulatory limits.

Soil sampling was repeated by 24 September 2020, which was 116 wk after the last soil treatments, done by 15 June 2018. That time soil samples were collected not only from the topsoil (0-30 cm) but also from the 30-60 cm subsoil layer. "Pseudo-total" and "plant available" concentrations of the PTEs were determined from cc. HNO_3+cc. H_2O_2 and H_4-EDTA in ammonium acetate buffer+acetic acid (Lakanen and Erviö 1971) and soil extracts, respectively. According to Meers et al. (2007) complexation by EDTA and acetic acid simulates complex-forming behavior of root exudates, whereas NH_4^+ is capable of desorbing the exchangeable soil fraction, and the pH simulates rhizosphere acidity. Results are shown in Table 13.2.

2020 yr results (Table 13.2) confirm our observations from yr 2018 (Table 13.1), that all soil additives, to a varying degree but uniformly, significantly increased the "pseudo-total" concentrations of the studied group of PTEs in topsoil, as compared to untreated control. Comparing the results of 2018 (Table 13.1) with those of 2020 (Table 13.2), it can also be stated that the relative concentrations of As and Cd

Table 13.2: "Pseudo-total" and "plant available" concentrations of potentially toxic elements in the topsoil (depth 0-30 cm) and subsoil (depth 30-60 cm) of the open-field long-term experiment set up with energy willows (*Salix triandra* × *S. viminalis* 'Inger') (Nyíregyháza, Hungary)

Concentrations of potentially toxic elements (mg kg^{-1})	Soil treatments (2011-2018)			
	Control	**MSSC+MSS**	**WA**	**MSSC+MSS+WA**
"Pseudo-total" *	**Soil depth 0-30 cm (September 2020)**			
As	9.19^{a}	35.6^{d}	27.3^{b}	30.4^{c}
Ba	56.0^{a}	67.3^{d}	59.6^{b}	63.5^{c}
Cd	0.214^{a}	0.625^{d}	0.391^{b}	0.494^{c}
Cr	12.6^{a}	17.5^{c}	13.1^{a}	15.8^{b}
Cu	9.47^{a}	12.0^{d}	10.1^{b}	11.2^{c}
Mn	368^{a}	482^{b}	529^{b}	500^{b}
Ni	13.5^{a}	14.1^{b}	14.9^{c}	14.7^{c}
Pb	10.1^{a}	13.7^{d}	11.4^{b}	12.4^{c}
Zn	35.3^{a}	40.2^{b}	43.2^{b}	41.9^{b}
	Soil depth 30-60 cm (September 2020)			
As	8.56^{a}	30.4^{d}	23.0^{b}	26.0^{c}
Ba	52.5^{a}	62.4^{c}	53.4^{a}	59.1^{b}
Cd	0.179^{a}	0.448^{c}	0.313^{b}	0.360^{b}
Cr	12.0^{a}	13.6^{b}	12.7^{a}	13.9^{b}
Cu	7.46^{a}	9.43^{c}	7.80^{a}	8.43^{b}
Mn	324^{a}	440^{b}	491^{c}	469^{c}
Ni	12.7^{a}	13.7^{b}	13.8^{b}	13.6^{b}
Pb	8.74^{a}	9.90^{d}	9.09^{b}	9.42^{c}
Zn	30.3^{a}	37.5^{b}	39.0^{b}	38.7^{b}
"Plant available" **	**Soil depth 0-30 cm (September 2020)**			
As	0.41^{a}	1.66^{d}	1.30^{b}	1.47^{c}
Ba	34.3^{a}	42.1^{d}	38.6^{b}	40.5^{c}
Cd	0.048^{a}	0.118^{d}	0.072^{b}	0.097^{c}
Cr	0.245^{a}	0.338^{d}	0.257^{b}	0.304^{c}
Cu	3.15^{a}	3.86^{d}	3.58^{b}	3.71^{c}
Mn	309^{a}	403^{b}	455^{d}	435^{c}
Ni	2.85^{a}	2.98^{b}	3.17^{b}	3.09^{c}
Pb	1.44^{a}	1.93^{d}	1.65^{b}	1.82^{c}
Zn	3.78^{a}	4.39^{b}	4.93^{d}	4.64^{c}

	Soil depth 30-60 cm (September 2020)			
As	0.40[a]	1.49[d]	1.04[b]	1.26[c]
Ba	32.1[a]	38.0[d]	33.4[b]	35.9[c]
Cd	0.037[a]	0.086[d]	0.057[b]	0.067[c]
Cr	0.228[a]	0.254[c]	0.245[b]	0.264[d]
Cu	2.08[a]	3.17[d]	2.36[b]	2.73[c]
Mn	279[a]	365[b]	411[d]	391[c]
Ni	2.65[a]	2.89[b]	2.93[b]	2.90[b]
Pb	1.25[a]	1.39[c]	1.31[ab]	1.33[bc]
Zn	3.16[a]	3.87[b]	4.02[c]	3.96[bc]

*cc. HNO_3+cc. H_2O_2 soil extract, **H_4-EDTA in ammonium acetate buffer+acetic acid soil extract. MSSC = Municipal sewage sludge compost. MSS = Municipal sewage sediment. WA = Willow ash. Means within the rows followed by the same letter are not statistically significant at $P<0.05$.

increased, while those of Ba, Cu, Mn, Ni, Pb, and Zn decreased during 2020, in all treated topsoils, as compared to untreated control soil. These changes can be attributed to the various rate of downward migration (leaching) of PTEs, to the different rates of binding of PTEs to soil colloids (Table 13.2), or to uptake or accumulation of PTEs in willow organs (Table 13.3).

Similar to 2018 yr results (Table 13.1), it was recognized that co-application of MSSC+MSS and WA resulted in significantly lower "pseudo-total" As, Ba, Cd, Cr, Cu, and Pb concentrations in topsoil and lower As, Ba, Cd, Cu, and Pb concentrations or subsoil than the application of MSSC+MSS alone (Table 13.2). In other respects, it can be again concluded that only Mn and Zn were added from willow ash in significant amounts to soils treated with MSSC+MSC+WA, since the Mn and Zn concentrations in MSSC+MSC+WA were higher than in MSSC+MSC soil treatments.

During September 2020, in 0-30 cm layer of MSSC+MSS, WA or MSSC+MSS+WA treated soils, 197–287% more As, 6–20% more Ba, 83–192% more Cd, 4–39% more Cr, 7–27% more Cu, 31–44% more Mn, 4–10% more Ni, 13–36% more Pb, and 14–22% more Zn were detected in HNO_3–H_2O_2 extracts (Table 13.2), as compared to PTE concentrations in untreated control soil. The most significant increment was observed in the amount of As and Cd in treated soils. As a general trend, lower concentrations of all observed PTEs were recorded in 30-60 cm subsoil than in 0-30 cm topsoil layer (Table 13.2). MSSC+MSS or MSSC+MSS+WA soil treatments significantly increased the concentrations of all PTEs in the subsoil. In the WA-treated subsoil, however, the Ba, Cu, and Cr concentrations were not statistically different from control values (Table 13.2).

Table 13.2 also presents the "plant available" concentrations of the PTEs in topsoil or subsoil. Similar to "pseudo-total" concentrations, all soil treatments significantly enhanced the PTEs' "plant available" concentrations both in topsoil and in the subsoil, as compared to untreated control. The only exception was Pb in 30-60 cm layer of subsoil, where WA-treatments resulted in similar Pb concentration than in control. Special attention was paid to concentrations of cadmium, since this toxic element is very mobile in the soil-plant system (Kabata-Pendias 2011, Simon 2014), and willows efficiently phytoextract this metal from contaminated soil (Pulford et al. 2002, Hammer et al. 2003, Vysloužilová et al. 2003, Berndes et al. 2004, Dickinson and Pulford 2005, Maxted et al. 2007, Unterbrunner et al. 2007). Comparing "plant available" Cd concentrations to "pseudo-total", it was found that in topsoil 22.4%–18.9%–18.4%–19.6% of Cd is present in "plant available" form in control, MSSC+MSS, WA or MSSC+MSS+WA-treated soils, respectively. This ratio was 20.7%–19.2%–18.2%–18.6% in subsoil. Our results support that in untreated control soil, higher ratio of Cd is present in the "plant available" form than in treated soils. This phenomenon suggests that willow accumulated more "plant available" Cd from treated than from control soil.

Besides Cd, willows are good phytoextractors of Zn from contaminated soils (Pulford et al. 2002, Hammer et al. 2003, Vysloužilová et al. 2003, Maxted et al. 2007, Unterbrunner et al. 2007, Tőzsér et al. 2018). The highest "pseudo-total" or "plant available" Zn concentrations were measured in 0-30 cm or 30-60 cm layers when WA was applied to the soil. These values were significantly 23-30% higher than in control soils (Table 13.2). In contrast to the 18.2-19.2% ratio of Cd in the "plant available" fraction of treated topsoils, the Zn was present in "plant available" extracts only in 8.8-11.1% ratio, as related to "pseudo-total" concentrations. In subsoils, this ratio for Zn changed between 9.8-10.4% (Table 13.2).

Table 13.3 shows the concentrations of selected PTEs in the leaves and twigs of willows grown in soils repeatedly treated with MSSC, MSS, or WA. It can be generally declared that all former soil treatments significantly enhanced the uptake or accumulation of PTEs in the leaves or twigs of willows (Table 13.3). Elevated levels of As, Cd, Cr, Cu, Ni, or Pb in aboveground organs of willow are, however, in the normal range (Kabata-Pendias 2011, Simon 2014). Only Zn concentrations (160-183 mg kg^{-1} d.m.) measured during 2018 in leaves can be considered excessive, considering that 100-400 mg kg^{-1} of zinc in mature leaf tissue is excessive or toxic (Kabata-Pendias 2011, Simon 2014).

Comparing the concentrations of various PTEs in the leaves of willows during 2018, 2019, or 2020 yr, it can be observed that overall accumulation rate was higher just after soil treatments during 2018 than during 2019 or 2020. The highest Ba, Cd, Cu, Mn, Ni, and Zn concentrations were measured in the leaves 5 wk after soil treatments during 2018 yr. Later, during 2019 or 2020, 53 or 101 wk after the last soil treatments, the accumulation rate of PTEs in leaves decreased (Table 13.3). During May 2020, however, in leaves of MSSC+MSS, WA or MSSC+MSS+WA treated willows, still 78–124% more As, 7–18% more Ba, 30–125% more Cr, 7–15% more Cu, 29–44% more Mn, 202–695% more Pb, or 11–35% more Zn was detected than in untreated control. During 2020 yr, only MSSC+MSS treatments enhanced the Cd accumulation significantly in the leaves by 14%, while former WA treatments reduced it by 12%, as compared to control (Table 13.3). This is in agreement with our observation that during 2020 yr the "pseudo-total" Cd concentrations in treated soils were lower (Table 13.2) than in 2018 yr (Table 13.1), and during 2020 the "plant available" concentrations of Cd were higher in MSSC+MSS than in WA-treated soil (Table 13.2).

Table 13.3: Effects of soil treatments on the concentrations of potentially toxic elements in the aboveground organs of energy willows (*Salix triandra* × *S. viminalis* 'Inger') grown in a long-term open-field experiment (Nyíregyháza, Hungary)

Concentrations of potentially toxic elements (mg kg^{-1})	Soil treatments (2011-2018)			
	Control	MSSC+MSS	WA	MSSC+MSS+WA
Willow leaves (July 2018)				
As	0.174^{a}	0.276^{d}	0.218^{b}	0.250^{c}
Ba	6.18^{a}	9.12^{c}	8.56^{b}	8.89^{bc}
Cd	0.806^{a}	1.111^{d}	0.974^{b}	1.041^{c}
Cr	0.231^{a}	0.473^{d}	0.323^{b}	0.399^{c}
Cu	8.71^{a}	14.8^{c}	9.08^{a}	10.9^{b}
Mn	44.6^{a}	51.0^{b}	71.2^{d}	60.8^{c}
Ni	1.08^{a}	2.28^{d}	1.76^{b}	2.03^{c}
Pb	0.156^{a}	0.568^{d}	0.180^{b}	0.372^{c}
Zn	123^{a}	160^{b}	183^{d}	173^{c}
Willow leaves (June 2019)				
As	0.176^{a}	0.295^{d}	0.200^{b}	0.243^{c}
Ba	4.66^{b}	4.22^{a}	6.89^{d}	5.47^{c}
Cd	0.847^{b}	0.913^{d}	0.792^{a}	0.883^{c}
Cr	0.266^{a}	0.543^{d}	0.348^{b}	0.461^{c}
Cu	8.04^{a}	8.99^{c}	8.40^{b}	8.54^{b}
Mn	35.8^{a}	34.3^{a}	41.1^{c}	37.9^{b}
Ni	1.06^{b}	1.57^{d}	0.97^{c}	1.21^{c}
Pb	0.097^{a}	0.308^{c}	0.114^{a}	0.220^{b}
Zn	79.7^{c}	67.9^{a}	79.9^{c}	73.0^{b}
Willow leaves (May 2020)				
As	0.194^{a}	0.439^{c}	0.349^{b}	0.392^{bc}
Ba	5.45^{a}	5.83^{a}	6.46^{b}	6.34^{b}
Cd	0.872^{b}	0.994^{c}	0.768^{a}	0.868^{b}
Cr	0.304^{a}	0.683^{d}	0.395^{b}	0.516^{c}
Cu	8.13^{a}	9.39^{d}	8.66^{b}	9.01^{c}
Mn	39.0^{a}	50.5^{b}	56.2^{c}	53.7^{c}
Ni	1.02^{b}	1.33^{c}	0.93^{a}	1.06^{b}
Pb	0.110^{a}	0.874^{d}	0.332^{b}	0.587^{c}
Zn	98.5^{a}	133^{c}	109^{b}	117^{b}
Willow twigs without leaves (March 2020)				
As	0.303^{a}	2.10^{d}	1.58^{b}	1.86^{c}
Ba	5.90^{a}	6.67^{a}	8.97^{c}	7.73^{b}

(Contd.)

Table 13.3: *(Contd.)*

Concentrations of potentially toxic elements (mg kg^{-1})	Soil treatments (2011-2018)			
	Control	MSSC+MSS	WA	MSSC+MSS+WA
Cd	0.636^{a}	0.852^{b}	1.517^{d}	1.083^{c}
Cr	0.329^{a}	0.953^{d}	0.712^{b}	0.813^{c}
Cu	5.09^{a}	6.55^{d}	5.56^{b}	5.97^{c}
Mn	11.7^{a}	17.0^{b}	21.0^{d}	19.5^{c}
Ni	0.171^{a}	0.434^{d}	0.229^{b}	0.323^{c}
Pb	0.139^{a}	0.660^{d}	0.228^{b}	0.452^{c}
Zn	61.8^{a}	98.3^{b}	93.3^{b}	95.7^{b}

Means within the rows followed by the same letter are not statistically significant at $P<0.05$.

During March 2020, in leafless twigs of MSSC+MSS, WA or MSSC+MSS+WA treated willow cultures, 421–593% more As, 13–52% more Ba, 36–138% more Cd, 116–190% more Cr, 9–29% more Cu, 45–79% more Mn, 34–153% more Ni, 13–36% more Pb, and 50–59% more Zn were measured, as compared to control. Our results (Table 13.3) are in partial agreement with Unterbrunner et al. (2007), who generally found lower Cd and Zn concentrations in the wood of investigated *Salix* spp. than in leaves. In our case, this was true only for Zn.

It is striking that willows developing on treated soils accumulated definitely more As in their twigs than in their leaves (Table 13.3). Contrary to our observation, Purdy and Smart (2008) found in their hydroponic experiment with willows that As concentrations were generally greater in leaves than in stems. Tlustoš et al. (2007) also found more arsenic in the leaves of various willow clones grown on contaminated soils than in twigs and considered the amount accumulated in aboveground biomass negligible. The amount of As detected by us in willow leaves or twigs is not high either (Table 13.3), but it is clear that it reflects the surplus of As present in WA or MSSC+MSS, applied to our experimental soil (Table 13.2).

In soils, we observed an interesting phenomenon that MSSC+MSS+WA-treatments resulted in more "pseudo-total" and "plant available" Mn and Zn in topsoil or subsoil than MSSC+MSS soil application by itself (Tables 13.1-13.2). This was reflected in leaves or twigs of willows where, in most cases, significantly higher Mn or Zn concentrations were measured in MSSC+MSS+WA than in MSSC+MSS treatments (Table 13.3).

Bioconcentration factors (BCF) can be used to estimate a plant's potential for phytoremediation, and phytoextraction purposes (Salam et al. 2019). According to Buscaroli (2017), an element's pseudo-total and actually available fractions are preferable when assessing the bioconcentration ability of plants. Therefore, we calculated BCFs not only as a ratio of a PTE concentration in the willow leaves or twigs to PTE "pseudo-total" concentration but also to "plant available" concentration in soil. Bioconcentration factors calculated from 2020 yr values of accumulated PTEs in leaves of willows (Table 13.3) and from "pseudo-total" concentrations of PTEs in soils (Table 13.2) were in this order: Zn>Cd>Cu>Mn>Ba>Ni>Pb>Cr>As, while for willow twigs the order was Cd>Zn>Cu>Ba>As>Cr>Mn>Pb>Ni. The highest BCF for Zn (3.31) in leaves was found in MSSC+MSS+WA soil treatment, while the highest BCF for Cd (3.88) in twigs was found in WA soil treatment during the yr 2020. Besides BCFs for Zn and Cd, it is remarkable that the highest BCF for Cu in leaves was 0.86 in the case of WA soil treatments, while in the twigs, the highest BCF value of 0.55 for Cu was also found in the case of WA soil treatments. In the case of Cd, the above BCF is in the range established by Cloutier-Hurteau et al. (2014) for willow. In the case of Zn or Cu, however, our BCFs are higher than those calculated by Cloutier-Hurteau et al. (2014). In contrast to Kacálková et al. (2014), we found definitively higher BCFs for Cu in willow leaves or twigs.

With the help of Pearson's correlation analysis, a statistically significant correlation ($P<0.05$; $r = 0.537$) was found between Cd concentrations in leaves and "pseudo-total" Cd concentrations in topsoils, while no correlation was found for Zn. In twigs, the trend was the opposite: significant correlation ($P<0.01$; $r = 0.763$) was found between Zn concentrations measured in plants and "pseudo-total" Zn concentrations in topsoils, while no correlation was found for Cd.

If BAFs were calculated on the basis of "plant available" soil PTE concentrations (Table 13.2), their order was Zn>Cd>Cu>Cr>Ni>As>Pb>Ba>Mn for leaves, and Zn>Cd>Cr>Cu>As>Ba>Pb>Ni>Mn for twigs. In the case of leaves, the highest BCF for Zn (30.3) was found for MSSC+MSS soil applications, while in the case of twigs, the highest BCF for Zn (22.4) was also found for MSSC+MSS soil treatments. Pearson's correlation analysis revealed that that is not a significant correlation between Zn concentrations measured in the leaves of willows (Table 13.3) and "plant available" Zn concentrations in topsoils (Table 13.3). A statistically significant correlation ($P<0.05$; $r = 0.578$) was found, however, between Cd concentrations in leaves and "plant available" Cd concentrations in topsoils. In twigs, an opposite tendency was observed; a significant correlation ($P<0.01$; $r = 0.640$) was found between Zn concentrations measured in willows and "plant available" Zn concentrations in topsoils, while no correlation was found for Cd.

Trace element concentrations in plants reflect, in most cases, their abundance in growth media. Generally, Cr is very slightly soluble in soil solution and is not easily taken up by plants; As and Pb are relatively strongly sorbed to soil particles and are not readily transported to aboveground parts of plants; Cu, Mn, and Ni are mobile in soil and readily taken up by plants, while Cd and Zn are very mobile in soil, and are easily bioaccumulated by plants (Kabata-Pendias 2011, Simon 2014). These general observations are primarily supported by our calculated BCFs for willow aboveground organs.

13.4 Conclusions

Repeated application of wastewater solids (MSSC, MSS) and wood ash (WA) significantly enhanced the amounts of PTEs (As, Ba, Cd, Cu, Mn, Pb, and Zn) in topsoil or subsoil of willows, grown for energetical purposes in a long-term experiment. Besides "pseudo-total", also the "plant available" PTEs' concentrations were changed in treated soils. Experimental soils, however, became only mildly contaminated with PTEs, since only the concentration of As was higher than the regulatory threshold limit. MSSC+MSS application in combination with WA resulted in significantly higher Mn and Zn, and lower As, Ba, Cd, Cr, and Pb concentrations in topsoil or subsoil, than MSSC+MSS treatment of soil without WA.

All soil treatments enhanced the uptake or accumulation of PTEs in leaves or harvested twigs of willows significantly. Significantly higher Mn or Zn concentrations were measured in MSSC+MSS+WA than in MSSC+MSS treatments. It was confirmed that willows are effective phytoextractors of Cd and Zn.

It can be concluded that with regular harvesting of willow shoots, significant amounts of Cd and Zn, and moderate or low amounts of Cu, Ba, As, Cr, Mn, Pb, and Ni can be phytoextracted from a soil, which is mildly contaminated with PTEs through regular application of wastewater solids or wood ash.

Acknowledgments

This research was supported in the frame of Hungarian GINOP 2.2.1-15-2017-00042 R&D Competitiveness and Excellence Cooperation by the "Genetic utilization of plants in the Pannon region" project. The authors appreciate the precise service of Dr. Tünde Pusztahelyi and her team in the plant and soil analysis from the Agricultural Instrument Center, University of Debrecen, Debrecen, Hungary.

References

Adler, A., I. Dimitriou, P. Aronsson, T. Verwijst and M. Weih. 2008. Wood fuel quality of two *Salix viminalis* stands fertilised with sludge, ash and sludge–ash mixtures. Biomass Bioenerg. 32: 914-925.

Arthur, E.L., P.J. Rice, P.J. Rice, T.A. Anderson, S.M. Baladi, K.L.D. Henderson and J.R. Coats. 2005. Phytoremediation – An overview. Crit. Rev. Plant. Sci. 24: 109-122.

Berndes, F., F. Fredrikson and P. Börjesson. 2004. Cadmium accumulation and *Salix*-based phytoextraction on arable land in Sweden. Agr. Ecosyst. Environ. 103: 207-223.

Buscaroli, A. 2017. An overview of indexes to evaluate terrestrial plants for phytoremediation purposes (review). Ecol. Indic. 82: 367-380.

Cloutier-Hurteau, B., M.-C. Turmel, C. Mercier and F. Courchesne. 2014. The sequestration of trace elements by willow (*Salix purpurea*) – Which soil properties favor uptake and accumulation? Environ. Sci. Pollut. R. 21: 4759-4771.

Dickinson, N.M. and I.D. Pulford. 2005. Cadmium phytoextraction using short-rotation coppice *Salix*: The evidence trail. Environ. Int. 31: 609-613.

Dimitriou, I., J. Eriksson, A. Adler, P. Aronsson and T. Verwijst. 2006. Fate of heavy metals after application of sewage sludge and wood-ash mixtures to short-rotation willow coppice. Environ. Pollut. 142: 160-169.

Dos Santos Utmazian, M.N., G. Wieshammer, R. Vega and W.W. Wenzel. 2007. Hydroponic screening for metal resistance and accumulation of cadmium and zinc in twenty clones of willows and poplars. Environ. Pollut. 148: 155-165.

Greger, M. and T. Landberg. 1999. Use of willow in phytoextraction. Int. J. Phytoremediat. 1: 115-123.

Gyuricza, C., L. Nagy, A. Ujj, P. Mikó and L. Alexa. 2008. The impact of composts on the heavy metal content of the soil and plants in energy willow plantations (*Salix* sp.). Cereal Res. Commun. 36: 279-282.
Hammer, D., A. Kayser and C. Keller. 2003. Phytoextraction of Cd and Zn with *Salix viminalis* in field trials. Soil Use Manage. 19: 187-192.
Hungarian FVM Decree No. 36/2006 (V.18) of the Ministry of Agriculture and Rural Development on the licensing, storage, distribution and use of crop-enhancing substances. https://net.jogtar.hu/jogszabaly?docid=a0600036.fvm (in Hungarian).
Hungarian Government Decree No. 50/2001 (IV.3.) about the rules of agricultural utilization and treatment of sewage and sewage sludge. https://net.jogtar.hu/jogszabaly?docid=a0100050.kor (in Hungarian).
Hungarian KvVM-EüM-FVM Joint Decree No. 6/2009 (IV.14.) of the Ministry of Environment and Water Management, Ministry of Health, Ministry of Agriculture on the threshold limits and measurement of pollutants necessary to protect the geological medium and groundwater against pollution. https://net.jogtar.hu/jogszabaly?docid=a0900006.kvv (in Hungarian).
Hungarian Standard MSZ 20135.1999. Determination of the soluble nutrient element content of the soil. Hungarian Standards Board, Budapest (in Hungarian).
Hungarian Standard MSZ 21470-50. 2006. Environmental testing of soils. Determination of total and soluble toxic element, heavy metal and chromium(VI) content. Hungarian Standards Board, Budapest (in Hungarian).
Hytönen, J. 2016. Wood ash fertilisation increases biomass production and improves nutrient concentrations in birches and willows on two cutaway peats. Balt. For. 22: 98-106.
Jama, A. and W. Nowak. 2012. Willow (*Salix viminalis* L.) in purifying sewage sludge treated soils. Pol. J. Agron. 9: 3-6.
Jha, A.B., A.N. Misra and P. Sharma. 2017. Phytoremediation of heavy metal-contaminated soil using bioenergy crops. pp. 63-96. *In:* K. Bauddh, B. Singh and J. Korstad [eds.]. Phytoremediation Potential of Bioenergy Plants. Springer, Singapore.
Kabata-Pendias, A. 2011. Trace Elements in Soils and Plants. Fourth Edition. CRC Press. Taylor & Francis Group, Boca Raton, FL.
Kacálková, L., P. Tlustoš and J. Száková. 2014. Phytoextraction of risk elements by willow and poplar trees. Int. J. Phytoremediat. 17: 414-421.
Kulig, B., E. Gacek, R. Wojciechowski, A. Oleksy, M. Kołodziejczyk, W. Szewczyk and A. Klimek-Kopyra. 2019. Biomass yield and energy efficiency of willow depending on cultivar, harvesting frequency and planting density. Plant Soil Environ. 65: 377-386.
Labrecque, M., Y. Hu, G. Vincent and K. Shang. 2020. The use of willow microcuttings for phytoremediation in a copper, zinc and lead contaminated field trial in Shanghai, China. Int. J. Phytoremediat. 22: 1331-1337.
Lakanen, E. and R. Erviö. 1971. A comparison of eight extractants for determination of plant available micronutrients in soil. Acta Agr. Fenn. 123: 223-232.
Lazdina, D., A. Bardule, A. Lazdins and J. Stola. 2011. Use of waste water sludge and wood ash as fertiliser for *Salix* cultivation in acid peat soils. Agron. Res. 9: 305-314.
Maxted, A.P., C.R. Black, H.M. West, N.M.J. Crout, S.P. McGrath and S.D. Young. 2007. Phytoextraction of cadmium and zinc by *Salix* from soil historically amended with sewage sludge. Plant Soil 290: 157-172.
Meers, E., B. Vandecasteele, A. Ruttens, J. Vangronsveld and F.M.G. Tack. 2007. Potential of five willow species (*Salix* spp.) for phytoextraction of heavy metals. Environ. Exp. Bot. 60: 57-68.
Merilo, E., K. Heinsoo, O. Kull, I. Söderbergh, T. Lundmark and A. Koppel. 2006. Leaf photosynthetic properties in a willow (*Salix viminalis* and *Salix dasyclados*) plantation in response to fertilization. Eur. J. Forest Res. 125: 93-100.

Mleczek, M., M. Gąsecka, K. Drzewiecka, P. Goliński, Z. Magdziak and T. Chadzinikolau. 2013. Copper phytoextraction with willow (*Salix viminalis* L.) under various Ca/Mg ratios. Part 1. Copper accumulation and plant morphology changes. Acta Physiol. Plant 35: 3251-3259.

Park, B.B., R.D. Yanai, J.M. Sahm, D.K. Lee and L.P. Abrahamson. 2005. Wood ash effects on plant and soil in a willow bioenergy plantation. Biomass Bioenerg. 28: 355-365.

Pilon-Smits, E. 2005. Phytoremediation. Annu. Rev. Plant Biol. 56: 15-39.

Pulford, I.D., D. Riddell-Black and C. Stewart. 2002. Heavy metal uptake by willow clones from sewage sludge-treated soil: The potential for phytoremediation. Int. J. Phytoremediat. 4: 59-72.

Pulford, I.D. and C. Watson. 2003. Phytoremediation of heavy metal contaminated land by trees – A review. Environ. Int. 29: 529-540.

Purdy, J.J. and L.B. Smart. 2008. Hydroponic screening of shrub willow (*Salix* spp.) for arsenic tolerance and uptake. Int. J. Phytoremediat. 10: 515-528.

Salam, M.M.A., M. Mohsin, E. Kaipiainen, A. Villa, S. Kuittinen, P. Pulkkinen, P. Pelkonen and A. Pappinen. 2019. Biomass growth variation and phytoextraction potential of four *Salix* varieties grown in contaminated soil amended with lime and wood ash. Int. J. Phytoremediat. 21: 1329-1340.

Saletnik, B. and C. Puchalski. 2019. Suitability of biochar and biomass ash in basket willow (*Salix viminalis* L.) cultivation. Agronomy 9: 577. https://doi.org/10.3390/agronomy9100577

Saletnik, B., G. Zaguła, A. Saletnik, M. Bajcar and C. Puchalski. 2020. Biochar and ash fertilization alter the chemical properties of basket willow (*Salix viminalis* L.). Agronomy 10: 660. doi:10.3390/agronomy10050660

Sameena, P.P. and Jos T. Puthur. 2021. Heavy metal phytoremediation by bioenergy plants and associated tolerance mechanisms. Soil Sediment Contam. 30: 253-274.

Sevel, L., T. Nord-Larsen, M. Ingerslev, U. Jørgensen and K. Raulund-Rasmussen. 2014. Fertilization of SRC willow, I: Biomass production response. Bioenerg. Res. 7: 319-328.

Simon, L., B. Szabó, M. Szabó, Gy. Vincze, Cs. Varga, Zs. Uri and J. Koncz. 2013. Effect of various soil amendments on the mineral nutrition of *Salix viminalis* and *Arundo donax* energy crops. Eur. Chem. Bull. 2: 18-21.

Simon, L. 2014. Potentially harmful elements in agricultural soils. pp. 85-137, 142-150. *In*: C. Bini and J. Bech [eds.]. PHEs, Environment and Human Health. Potentially Harmful Elements in the Environment and the Impact on Human Health. Springer, Dordrecht, Heidelberg, New York, London.

Simon L., G. Vincze., Z. Uri., K. Irinyiné Oláh, S. Vígh., M. Makádi., T. Aranyos and L. Zsombik. 2016. Long-term open-field fertilisation experiment with energy willow (*Salix* sp.) – Observations of the first five years. Növénytermelés 65: 59-76 (in Hungarian).

Simon, L., G. Vincze, Z. Uri, K. Irinyiné Oláh, M. Makádi, T. Aranyos, L. Zsombik and S. Vigh. 2017a. Long-term open-field fertilization experiment with energy willow (*Salix* sp.) – Experiences of the 2016 year. p. 23. *In*: M. Makádi [ed.]. Proceedings of Abstracts. International Conference on the Long-term Field Experiments (LOTEX 2017). September 27-28, 2017. Nyíregyháza, Hungary. University of Debrecen. ISBN 978-963-473-973-9.

Simon, L., Z. Uri, G. Vincze, K. Irinyiné Oláh and S. Vigh. 2017b. Long-term effects in a field experiment set up with willow (*Salix* sp.), grown for energy purposes. pp. 449-457. *In*: J. Kátai and Z. Sándor [eds.]. Soil Protection, 2017 – Special Issue. ISSN 1216-9560. http://talaj.hu/wp-content/uploads/2015/02/Talajv%C3%A9delem_K%C3%BCl%C3%B6nsz%C3%A1m_2016.pdf (in Hungarian).

Simon, L., M. Makádi, G. Vincze, Z. Uri, K. Irinyiné Oláh, L. Zsombik, S. Vígh & B. Szabó. 2018. Long-term field fertilization experiment with energy willow (*Salix* sp.) – Elemental composition and chlorophyll fluorescence in the leaves. Agrokem Talajtan 67: 91-103.

Simon, L., S. Vigh, Z. Uri, K. Irinyiné Oláh, M. Makádi and G. Vincze. 2019. Accumulation of toxic elements from sewage sediment and wood ash in energy willow grown in a long-term experiment. p. 122. *In*: M. Makádi [ed.]. Book of Proceedings. 2nd International Conference on Long-term Field Experiments (LOTEX 2019). November 21-22, 2019. Nyíregyháza, Hungary. University of Debrecen. ISBN: 978-963-490-149-8.

Smart, L.B. and K.D. Cameron. 2012. Shrub willow. pp. 687-708. *In*: C. Kole, C.P. Joshi and D.R. Shonnard [eds.]. Handbook of Bioenergy Crop Plants. CRC Press, Boca Raton, London, New York.

Šuman, J., O. Uhlik, J. Viktorova and T. Macek. 2018. Phytoextraction of heavy metals: A promising tool for clean-up of polluted environment? Front. Plant Sci. 9: 1476. doi:10.3389/fpls.2018.01476

Šyc, M., M. Pohořelý, P. Kameníková, J. Habart, K. Svoboda and M. Punčochář. 2012. Willow trees from heavy metals phytoextraction as energy crops. Biomass Bioenerg. 37: 106-113.

Tlustoš, P., J. Száková, M. Vysloužilová, D. Pavlíková, J. Weger and H. Javorská. 2007. Variation in the uptake of arsenic, cadmium, lead, and zinc by different species of willows *Salix* spp. grown in contaminated soils. Cent. Eur. J. Biol. 2: 254-275.

Tőzsér, D., S. Harangi, E. Baranyai, G. Lakatos, Z. Fülöp, B. Tóthmérész, and E. Simon. 2018. Phytoextraction with *Salix viminalis* in a moderately to strongly contaminated area. Environ. Sci. Pollut. R. 25: 3275-3290.

Unterbrunner, R., M. Puschenreiter, P. Sommer, G. Wieshammer, P. Tlustoš, M. Zupan and W.W. Wenzel. 2007. Heavy metal accumulation in trees growing on contaminated sites in Central Europe. Environ. Pollut. 148: 107-114.

Vandecasteele, B., P. Quataert, F. Piesschaert, S. Lettens, B. De Vos and G. DuLaing. 2015. Translocation of Cd and Mn from bark to leaves in willows on contaminated sediments: Delayed budburst is related to high Mn concentrations. Land 4: 255-280. doi:10.3390/land4020255

Vysloužilová, M., P. Tlustoš and J. Száková. 2003. Cadmium and zinc phytoextraction of seven clones of *Salix* spp. planted on heavy metal contaminated soils. Plant Soil Environ. 49: 542-547.

Wójcik, M., F. Stachowicz and A. Masłoń. 2020. The use of wood biomass ash in sewage sludge treatment in terms of its agricultural utilization. Waste Biomass Valori 11: 753-768.

Zając, G., J. Szyszlak-Bargłowicz and M. Szczepanik, 2019. Influence of biomass incineration temperature on the content of selected heavy metals in the ash used for fertilizing purposes. Appl. Sci. 9: 1790. doi:10.3390/app9091790

CHAPTER
14

Bioremediation of Wasteland Using Hydrocarbon Yielding Plant: A Case Study

Shobha Johari[1]* and Ashwani Kumar[2]

[1] Department of Botany, L.B.S. Government College, Kotputli – 303108, Jaipur
[2] Alexander von Humboldt Fellow (Germany); Former Head, Department of Botany, University of Rajasthan, Jaipur – 302004

14.1 Introduction

Climate change impacts such as higher temperatures, extreme weather, drought, increasing levels of carbon dioxide and sea level rise are jeopardizing the quality and quantity of our food supplies (Kumar 2011, Kumar et al. 2018a, 2020). The negative impact of salinity on plant growth and survival has been studied recently (Shekhawat et al. 2006a,b, Ebadi et al. 2018). Influence of salinity stress has also been reported on growth and productivity of various plants (Shekhawat and Kumar 2004, Shekhawat et al. 2006, Garg and Kumar 2011, Vijayvargiya and Kumar 2011, Kumar and Roy 2018, Kumar et al. 2018b). Yensen (2006) has also described uses of halophytes for 21st century. *Salicornia* is a halophyte and can be grown in coastal areas on marshy soils extending thousands of km worldwide (Kumar et al. 2018b). *Salicornia* was found a promising alternative for biomass from saline and semiarid and arid conditions as a source of biofuels (Kumar and Kumar 2011, Kumar et al. 2018b, Johari and Kumar 2016). This and other halophytes can be utilized for bioremediation of saline soils of semi-arid and arid regions (Kumar et al. 2018b). Plant tissue culture technique has been used to develop salt resistant plants (Shekhawat and Kumar 2006). Soil salinity negatively impacts plant growth and the survival of rhizosphere biota (Ebadi et al. 2018).

Rajasthan is the largest state of India in terms of area. It is situated between 23°3′ northern latitude to 30° 12′ northern latitude and 69° 30′ eastern meridians to 70° 17′ eastern latitude. Major part of Rajasthan lies on northern side of line of cancer. On the western side of Aravalli ranges, major area is sandy having annual rainfall less than 300 mm. It is Thar Desert of Rajasthan. It is the 17th largest desert of world.

Corresponding author: shobhajohari96@gmail.com

Sandy desert area of Rajasthan is almost 85%. Out of the total desert area of India, 62% lies in Rajasthan state. Major landscape is sand-dunes and sandy arid plains. Western part of Thar Desert has sand dune areas covering 59-60% area. Besides sandy soil, saline lakes Sambhar, Pachpadra and Deedwana are found in Thar Desert of Rajasthan. Soil of these areas is saline due to groundwater salinity.

Although the conventional food crops cannot be grown in these lands, a large number of hydrocarbon yielding plants are able to grow under such adverse conditions in semi-arid and arid regions. These hydrocarbon yielding plants can be developed into energy crops for liquid fuels (Calvin 1979). This avoids food versus fuel competition (Kumar 2018a). Besides providing hydrocarbons, these plantations shall also check soil erosion due to wind and water (Kumar 2014). In this way, solar energy can be harvested on renewable basis by growing hydrocarbon yielding plants on wastelands. The wider use of biomass for development offers minimal ecological imbalance and provides means to recycle nutrients and carbon dioxide from the atmosphere (Dayal 1986, Vimal 1986, Kumar et al. 2018b, 2020).

When the possibility of obtaining liquid fuels from plants by direct extraction was introduced, the emphasis of bioenergy research, the presence of whole plant oils called biocrude on one hand and hydrocarbon yielding biomass on the other, became an important criterion in the selection of potential energy crops (Buchanan et al. 1978a,b,c, Bagby et al. 1981, Kumar 2014). This shift of emphasis led to the proposal of extensive, rather than intensive, agricultural systems as an alternative for arid land energy farms (Lipinsky et al. 1980, Kumar 2013). Implementing salt-tolerant tree plantations while utilizing saline drainage or groundwater represents a promising alternative of using abandoned lands (Qadir et al. 2008).

The current primary energy consumption is dominated by conventional fossil fuels including coal, oil and gas leading to sustainability problems such as a declining amount of fossil fuels, environmental impacts and huge price fluctuations (Kumar 2008, 2013, Agarwal and Kumar 2018, Kumar 2018a,b, 2020a, Lijó et al. 2019, Khoo et al. 2020, Kumar and Gadhwal 2020). Greenhouse gas emissions, global climate change as well as intense energy demand have driven a number of professionals to develop novel solutions to replace fossil fuels. Among the alternative energy sources, biomass accounts for around 80% of the energy produced by global renewable energy carriers (Strzalka et al. 2017, Kang and Lee 2015, Khoo et al. 2020). Today, bioenergy contributes about 10% of the global energy demand (International Energy Agency (IEA) 2014). According to the envisioned roadmap for fuel production by 2050, 27% of the global transport fuel is devoted to biofuels in the future. It has now been projected that the lion's share of biofuel use will increase until 2050 (International Energy Agency (IEA) 2011).

Non-food feedstocks like lignocellulosic and microalgae biomass are second- and third-generation feedstocks used to produce bioenergy, respectively (Kumar et al. 2018b, 2021). Bioalcohol derived from corn, wheat, sugar beet and sugarcane as well as biodiesel produced by transesterification of oils extracted from rapeseed, palm, soybean and sunflower are examples of bioenergy generated from first-generation feedstock (Kumar and Gupta 2018). Furthermore, anaerobic digestion has been employed commercially to produce biogas, which is also a type of bioenergy. The biogas produced has been utilized to generate heat and electricity. It plays an important role to provide the necessary energy to rural areas for cooking and lighting (Li and Khanal 2016). In addition, bioenergy can be produced by microbial fuel cells (MFCs), which uses naturally occurring microorganisms with biological electricity generation ability. Despite having numerous scientific breakthroughs, there are still various technical barriers to tackle for bioenergy production so that it can compete with fossil fuels (Kumar 2020b).

Presence of salts into the soil also causes physiological drought. Osmotic pressure of soil solution increases due to saline soil solution. It causes a harmful effect on growth of plants. pH value of soil also increases because of the presence of sodium. This causes imbalance in available nutrients for the plants. Hence, land of this area becomes a 'wasteland'. Still, many xerophytic plants grow in Thar Desert. Among herbs are *Calotropis procera, C. gigantea, Argyeria, Cryptostegia gradiflora, Balliospermum montanum, Catharanthus, Leptadaenia* and species of *Euphorbia.* Latex found in many species of *Euphorbia* is an emulsion of oil and water, roughly one third oil in water viz. 30 percent of the weight of the latex is oil (Shukla and Crishna-Murti 1971, Calvin 1984).

Milkweeds (Asclepiadaceae) have also been proposed as a renewable source of fuel chemicals and also chemical feedstocks (Buchanan et al. 1978a,b, Gaertner 1979, Erdman and Erdman 1981, Dehgan and Wang 1983). *Catharanthus roseus* contains more than 100 alkaloids. This species performs successfully at a salinity of 10 $dS.m^{-1}$ without decline in biomass production (Anwar et al. 1988). Several laticiferous plants have been screened in India for large scale cultivation, as a source of energy and hydrocarbon (Marimuthu et al. 1989). First efforts to cultivate hydrocarbon producing plant *Euphorbia abyssinica* J.F. Gmel. for fuel production were made by Italians in Ethiopia (1935-36) and French in Morocco (1940). In recent years, Calvin (1979) and his collaborators have revived the idea again and have advocated study of petrocrops as a possible feed stock for petroleum like materials. The use of salt-affected land and saline water resources by crop diversification could also be an alternative for bioremediation of waste lands. In recent years, it appears that one of the promising options for producing liquid fuels is the direct extraction of low molecular weight, nonpolar constituents (biocrude) from laticiferous plants followed by catalytic upgrading (Nielsen et al. 1977, Buchanan et al. 1978c, Weisz et al. 1979, Calvin 1980, Haag et al. 1980, Hinman et al. 1980, Nemethy et al. 1981, McLaughlin et al. 1983). Conversion of jojoba oil, castor oil, corn oil, *Hevea brasiliensis* latex (over HZSM catalyst in the presence of hydrogen) (Weisz et al. 1979) and of biocrudes from *Asclepias speciosa* Torr., *E. lathyris, Grindelia squarrosa* (Pursh.) Dunal (Mobils ZSM-5 Zeolite catalyst) (Haag et al. 1980, Adams et al. 1984) into hydrogen fuels, particularly in the gasoline range, has already been demonstrated by previous workers (Bhatia et al. 1989). In a typical run, products obtained from the latex processed biocrude were gases (6.61), naphtha (23.85), kerosene (21.97), gas oil (18.22), heavier (26.47), water (0.95) and cake and losses (1.93). In case of biomass processed biocrude, the products were gases (6.46), naphtha (16.17), kerosene (28.26), gas oil (16.17), heavier (12.93), water (0.97) and cake and losses (19.04). Over 90 percent of the products have been identified.

According to Srivastava (1985), there are at least 85 families that represent members which yield hydrocarbons. Important resin and latex bearing families found in Rajasthan are Asclepiadaceae, Euphorbiaceae, Sapotaceae, Convolvulaceae, Moraceae (Urticaceae) and Apocynaceae. There are about 386 laticiferous plants placed under families Euphorbiaceae (187), Asclepiadaceae (73), Apocynaceae (51), Urticaceae (36), Sapotaceae (15) and Convolvulaceae (4). Out of these, 45 plant species have been identified as potential hydrocarbon yielding plants. The important ones are *Euphorbia antisyphilitica* Zucc.(Johari and Kumar 2013a), *Baliospermum montanum* Muell. Agr., *Calotropis procera* (Ait.) R. Br. (Kumar 2018b), *C. gigantea* (Linn.) R. Br., *Cryptostegia grandiflora* (Roxb.) R. Br., *Euphorbia antiquorum* Linn.,*E. caducifolia* Haines, *E. tirucalli* Linn., *E. trigona* Haw., *Argyreia nervosa* Boj., *Pedilanthus tithymaloides* (L.) Poit., *Euphorbia lathyris* L. (Garg and Kumar 2012a,b). Kotiya et al. (2018) reviewed the potential biofuel plants for semi-arid and arid regions. *Euphorbia* sp have other uses in traditional medicine, due to its antifungal activity (Johari and Kumar 2020a,b). Johari and Kumar (2020c) studied anatomical features of such plants (*Euphorbia antisyphilitica*).

The bioenergy crops can be genetically improved for producing more biomass, being more resilient to stresses like drought conditions and pathogens. Besides, they can synthesize cell wall materials with reduced recalcitrance towards deconstruction processes (Bhansali and Kumar 2018). Kumar et al. (2020) reviewed application of synthetic biology for obtaining biofuels through synthetic routes. According to Mortimer (2019) to date, much synthetic biology research has focused on microbial engineering.

Number of latex producing species, mainly belonging to families Euphorbiaceae and Asclepiadaceae, were screened for assessing their suitability as a source of low molecular weight and non-polar petroleum like hydrocarbons. Calvin (1979) concluded that exploration of hydrocarbon yielding plants gives rise to two practical approaches: first, to use the hydrocarbon as it comes from the plants itself as a crude oil, refine it, remove the sterols which it contains, crack the rest of the compounds to ethylene, propylene etc. and then reconstruct other chemicals from those products and second, to learn how the molecular weight is controlled and to manipulate the plant to construct materials of the desired molecular weight, an approach which will be longer and more complex, using plants as the collecting and constructing vehicles. *Euphorbia* latex has received increasing attention because it contains a mixture of light hydrocarbons, which have a molecular weight of the order of 20,000 instead of 2 million. Hence, after the removal of water from the latex, the resulting material is liquid oil. The hydrocarbons from *Euphorbia* are primarily a blend of C_{15}, C_{20} or C_{30} compounds (Nielsen et al. 1977, 1979) that, when subjected to catalytic cracking, yield various products virtually identical to those obtained by cracking naphtha (Maugh 1979), a high quality petroleum fraction that is one of the principal raw materials used in the chemical industries. Latex remains in special cells and/or vessels, called laticifers, and is a byproduct of photosynthetic conversion of solar energy into biochemical energy. Because of greater economic value of reactive chemical intermediates which can be obtained from latex yielding plants, the production of chemical feedstocks is now considered to be an attractive short term goal in the development of bioenergy crops (Lipinsky1981, Palsson et al. 1981). Latex can be collected from tree species by following methods: i) Tapping – Lateral channels are opened on the trunk at an angle of about 45° to the axis at intervals of 20 to 30 cm along with entire bole and main branches and joined by a vertical incision. ii) Leaf extraction – This technique was established to practice the leaves with a solvent or by processing them mechanically. iii) Felling and girdling – After felling the tree, larger branches were opened at intervals of 30 to 50 cm around entire length of bole, and latex is collected.

An extensive and thorough trial for *E. lathyris* was conducted in Spain, examining planting densities, harvest date, water and fertilizer requirements and yields of dry matter, sugars and oils. It has been concluded that low plant density, lack of irrigation and prolonged growth of *E. lathyris* result in optimum hydrocarbon yields (Ayerbe et al. 1984, Garg and Kumar 2012a). In Kenya, a 100 ha project with *E. tirucalli* was started in 1980. Over 300 species from six genera of latex bearing plants were screened before *E. tirucalli* was selected. This species was widely grown in arid and semi-arid regions of Africa for live fencing for stock and around houses. The trials were conducted to obtain a techno-economic evaluation of the potential to produce liquid and solid fuels without irrigation and fertilizers. After 18-24 months, growth yields of up to 20 dry t/ha/year have been reported (Hall 1985). Establishment of *E. tirucalli* plantations appeared relatively easy and regrowth after first cutting was possible. Need of long term trials were felt to establish fertilizer requirement, especially with the removal of so much plant matter by harvesting. Outcome of these trials will be an important milestone in establishing validity of hydrocarbon plants in dry environments.

In *E. tirucalli,* lower levels of salinity were found to promote growth and hexane extractables, although inhibition was noted at higher salinity conditions. In a separate experiment, 100 percent field capacity irrigation produced maximum biomass but hexane extractables were best in 50 percent field capacity irrigation. These reports show effect of agronomical factors on laticiferous plants. The production of liquid fuels from desert plants was originally envisioned as a near term possibility based on the rising costs of crude oil (Calvin 1979). Search for alternative sources of energy has led to the suggestion that it will be necessary to make use of bioenergy, which also provides chemical feedstocks and have several commercial applications.

One potential source of biocrude is *Euphorbia,* a large genus of Euphorbiaceae, a family of laticiferous herbs, shrubs and small trees, distributed in the tropical and warm temperature regions of the world. Many of the species are succulent and inhabit dry places; they resemble cacti in appearance but are distinguished from them by the presence of milky latex. About 3000 species have been reported throughout the world, chiefly in tropical regions (Bhatia et al. 1989).

Studies have been performed to find out the influence of environment (Nemethy et al. 1981, Vasudevan and Giridhar 1986), water stress (Sachs et al. 1981a, Kingsolver 1982), nutritional factor (Sachs et al. 1981b, Kingsolver 1982, Tenorio et al. 1984, Kumar and Kumar 1985) and hormonal influence (Garg and Kumar 1987a) on *E. lathyris.* Environmental influence has also been studied on *C. procera* (Rani et al. 1989). Besides, studies on the effect of soil type (Johari et al. 1989, Roy 1989), growth regulators (Johari and Kumar 2013b) and salinity (Garg and Kumar 1989b) have also been reported on various members of Euphorbiaceae.

Stock plants of *E. antisyphilitica* and *E. tirucalli* were raised from the cutting obtained from National Botanical Research Institute (NBRI), Lucknow. *E. lathyris* was raised from the seeds obtained from Prof. Melvin Calvin, USA. *E. caducifolia* (Udaipur and Achrol), *E. nivulia* (Galtaji, Jaipur), *E. neriifolia* (Bharatpur) and *E. hirta* (Jaipur) were collected from different parts of Rajasthan.

Acetone extractables were maximum in *E. antisyphilitica,* followed by *E. neriifolia, E. lathyris, E. caducifolia, E. nivulia, E. tirucalli* and *E. hirta.* Benzene extractables were maximum in *E. antisyphilitica* followed by *E. caducifolia.* Minimum benzene extractables were recorded in *E. hirta.* Total extractives (acetone + benzene) were maximum in *E. antisyphilitica,* followed by *E. neriifolia* and minimum in *E. hirta* (Table 14.1).

Maximum hexane extractables were recorded in *E. antisyphilitica* followed by *E. caducifolia, E. nivulia, E. neriifolia, E. lathyris, E. tirucalli* and *E. hirta.* Methanol extractables were maximum in *E. lathyris* followed by *E. nivulia. E. hirta* was poorest among all the species in methanol extractables. Hexane + Methanol extractables were maximum in *E. lathyris* followed by *E. antisyphilitica* (Table 14.1). Maximum dry weight production was recorded in *E. lathyris* followed by *E. hirta.* Extractables, using non-polar solvent hexane, were maximum in *E. antisyphilitica;* therefore, it was selected for detailed investigations.

Table 14.1: Hydrocarbon yield of aboveground parts of different *Euphorbia* species

Name of the plant	Dry Wt. (%)	Acetone Ex. (%)	Benzene Ex.(%)	Acetone + Benzene Ex. (%)	Hexane Ex. (%)	Methanol Ex. (%)	Hexane + Methanol Ex. (%)
E. antisyphilitica	10.00	11.53	1.32	12.85	7.00	11.50	18.50
E. lathyris	22.63	9.45	0.49	9.94	5.57	21.56	27.13
E. tirucalli	8.80	4.85	0.91	5.76	3.48	6.31	9.79
E. caducifolia	13.31	8.83	0.98	9.81	6.60	11.36	17.96
E. nivulia	11.30	8.61	0.55	9.16	6.40	12.00	18.40
E. neriifolia	11.59	10.82	0.65	11.47	6.31	7.13	13.44
E. hirta	20.00	4.75	0.17	4.92	2.12	4.50	6.62

Ex. = Extractables

Euphorbia antisyphilitica (Fig. 14.1) does not occur in wild form in India but had been introduced in India in 1965. Now it grows well in arid and semi-arid regions of India. This species is also dendroid and profuse amount of latex remains present in its stem. This plant possesses an important place not only in the field of botanochemical substitutes of petroleum, but also in the commercial field of waxes and traditional medicines (Johari and Kumar 2020a,b). It has been used in Mexico for its waxes for at least last 50 years (Campos-Lopez and Roman-Alemany 1980).It is also being utilized as remedy for venereal diseases, in Mexico. Sternburg and Rodriguez (1982) identified $C_{29}H_{60}$, $C_{31}H_{64}$ and $C_{33}H_{68}$ compounds in a Gas liquid chromatogram, from wax samples of *E. antisyphilitica.* Its anatomical studies reveal that stem is rich in anastomosing laticifers (Johari and Kumar 2020c).

Figure 14.1: *Euphorbia antisyphiliticastock* plants.

Studies on growth and productivity of plant are essential to estimate its future potential in relevant field. Growth requirement of plants vary with species to species. When introducing a new plant in commercial field, knowledge about the growth period and corresponding productivity is necessary to estimate inputs and outputs. Many anabolic reactions involved in the growth occur at optimum rate in specific physical and chemical conditions. It has been a general experience that survival of a plant and its productivity varies with the changes in climatic conditions. Annual yield of *E. lathyris* L. is reported to vary from 7.5 bbl/ha (Kingsolver 1982), 11.0 bbl/ha (Sachs et al. 1981b, Ayerbe et al. 1984) to 15-25 bbl/ha (Calvin 1978). Experiments were conducted to determine morphological and physiological factors promoting propagation through cuttings (Table 14.2).

Table 14.2: Survival and sprouting in different portions of stem

Portion of stem	Survival rate of cutting (%)	Period of sprouting in the cuttings (days)
Apical	100	12
Middle	100	7
Basal (woody stem base)	0	-

Observations indicate that apical part of stem was best suitable for further vegetative multiplication of the plant.

The size of cuttings has been reported to influence propagation of laticiferous plants. Experiment was conducted to determine optimum size of cutting for *E. antisyphilitica* (Table 14.3). Growth index was found best in 15.0 cm size of cutting. 10.0 cm sized cutting were next in giving better results. Beyond 15.0 cm size growth index was found decreasing.

Table 14.3: Correlation of size on growth index of cuttings

Initial size of cutting (cm)	Initial Fresh Wt. (g)	Final Fresh Wt. (g)	Root Length (cm)	Growth Index
10.00	0.575	2.40	2.70	3.17
15.00	1.440	6.50	5.92	3.51
20.00	2.340	8.12	3.86	2.47
25.00	3.520	10.40	3.66	1.95
30.00	4.980	11.34	3.88	1.27
35.00	5.700	15.86	4.00	1.78
40.00	7.400	19.84	5.42	1.68
45.00	9.020	16.84	5.72	0.86

In order to evaluate the sprouting ability of the cutting with the duration of time and possible effect of latex on the rooting or plant establishment, the cuttings were obtained everyday upto ten days and were planted on tenth day. Ten replicates were taken for each day and the experiment was replicated thrice. Uniform cuttings measuring 15.0 cm and weighing 2.5 to 3.0 g were taken for the present investigation. Cuttings were harvested after two months of planting and results are presented in Table 14.4. Maximum percent survival and root length was observed in the cuttings, which were planted immediately after excision from the stock plant (standing time 0 day). In contrast to this, minimum percent survival and root length was observed in cuttings planted after nine days. However, keeping the cutting from 0 to 9 days improved their sprouting ability and the cuttings which were stored for nine days gave earliest sprouting from apex only. Freshly harvested cuttings took longest period for first sprouting, but the root length was maximum in them. Apparently latex oozing out from cuttings improved their survival rate.

Table 14.4: Effect of duration of time on survival, sprouting and rooting of cuttings

Standing time after cutting (days)	Survival of cutting (%)	Number of days required for first sprouting	Final root length (cm)
0	100.00	28.00	6.20
1	86.67	20.00	5.10
2	80.00	20.00	4.80
3	66.67	20.00	3.80
4	66.67	16.00	3.80
5	53.34	12.00	3.70
6	53.34	11.00	3.20
7	46.67	10.00	3.20
8	40.00	9.00	3.00
9	33.34	8.00	2.70

Out of the total land area of India (329 m ha), a vast expanse is covered by wastelands (180 m ha) of which around 7 m ha is salt affected (Abrol and Bhumbla

1971). Majority of east-coast belt is characterized by long and continuous stretches of saline and saline-alkaline soils. This continuous belt is occasionally interrupted by coastal alluvial soils. Salt affected lands in southern states are confined mostly to the black cotton soil tracts in the semi-arid part of the plateau. Here saline soils are mostly black clays with high SP (Sodium percentage) value. Saline soils also occur along with sea coast, which is about 56,000 km long, skirting west.

The problem of wastelands is most serious in Rajasthan, not only in magnitude but intensity as well. A considerable part of Rajasthan desert is comprised of unproductive land in the form of saline lands, covered by wind-blown sand. Pali, Bhilwara, Bharatpur, Ajmer, Alwar, Jaipur and Chittorgarh are salt affected districts of Rajasthan. A significant part of Rajasthan canal command area (0.16 m ha) is affected by salinity. Quality of underground water is poor in semi-arid and arid regions of India. About 54 percent of various districts of Rajasthan have EC value of underground water more than 2250.0 micro mho/cm, of which 28.4 percent have EC value >5000 micro mho/cm. Use of saline water in agriculture has shown that, in general, if one irrigation with saline water was followed by two canal water irrigations, the crop yield was improved significantly as compared to that with saline water alone (Yadav 1975). Total of approximately 45 m ha of land are irrigated in our country, out of which 16 m ha are from groundwater sources, which are mostly saline in Rajasthan, and 29 m ha employ surface waters.

Salinity reduces growth and yield of plants, before visible toxicity occurs. Salinity affects plant growth (Rai 1977) and induces changes in anatomy and morphology of stem and leaves (Strongonov 1962). Jojoba plants, when grown on different levels of salinity, showed decreased stem diameter and smaller vessel elements (Yermanos et al. 1967). In case of *Euphorbia lathyris* L., 100 percent germination was recorded upto 300 ppm saline concentration but higher concentrations inhibited germination in a linear fashion (Kingsolver 1982). Similarly, growth of roots, shoots and leaf area were all substantially reduced by salt stress in *Phaseolus vulgaris* L. (Prisco and O'Leary 1973). Biomass production of *Acanthus ilicifolius* L. decreased at higher salinity levels, but at moderate levels (ECe 8.30 dsm^{-1}) it produced biomass exactly double to that of control (Bal and Dutt 1987). Effect of salinity has also been studied on grain crops such as wheat (Reddy and Vora 1986), rice (Yeo et al. 1985) and maize (Patel and Vora 1985). Reports of such study are also available on other crops like sugar beet (Nunes et al. 1984), grapes (Divate and Pandey 1981) and guar (Datia and Dayal 1988).

Studies were undertaken with an objective to find the influence of saline irrigation on biomass and latex productivity of *E. antisyphilitica.*

14.2 Materials and Methods

14.2.1 Preparation of saline water

Saline water was prepared artificially for irrigating the plants. Sodium chloride was added in water to make saline concentrations ranging from 0.125, 0.25, 0.50, 1.0 and 2.0 percent (w/v). The control was devoid of added sodium chloride. Ten replicates were taken for each set of experiment.

A set of 60 pots were made and sandy soil was filled. Two-month old plants were transferred to the pots used for experimentation. Saline water was given in each set for irrigation. The control plants were irrigated with water. Saline irrigation was given upto four months at fifteen days interval, i.e. total eight such irrigations were given. Electrical conductivity (EC) of each saline concentration was determined (Table 14.5) and in each pot saline irrigation was given upto 100 percent level of field capacity.

Table 14.5: EC value of different concentrations of saline water

Saline concentration (%)	EC (micro mho/cm)
Control	770
0.125	1840
0.25	3000
0.50	5800
1.0	10700
2.0	19000

Plants were harvested after fifteen days of last irrigation and their aboveground and underground length, fresh and dry weights, percent hexane extractables, chlorophyll and sugars were estimated.

14.3 Results

14.3.1 Saline irrigation in sand

Lower concentrations (upto 0.25%) improved plant growth but higher concentrations (0.5, 1.0, 2.0%) inhibited further increase in growth (Fig. 14.2). Aboveground as well as underground dry weight was also maximum in 0.25% saline concentration but aboveground percent dry matter yield was best in 0.125% salinity (Table 14.6, Fig. 14.3). Highest saline concentration (2.0%) given in present experiment was found inhibitory for growth as well as productivity of various parameters like chlorophyll and sugars.

Lower saline concentrations (0.25% and below), which favoured biomass yield, also showed gradual increase in percent hexane extractables (Fig. 14.4) from control to 0.25% salinity (Table 14.6) and declined linearly with further increasing saline concentration but still were better than control. Sugar contents also decreased gradually with increasing percentages of salinity, maximum being in control plants (Fig. 14.5). Total chlorophylls tended to decrease from 0. 125% to 2.0% saline irrigation (Fig. 14.6). It was observed that after two saline irrigations with 2.0% salinity, leaves became yellow and later fell down but stem did not show any visible adverse effects.

Figure 14.2: Effect of different concentrations of salinity (0.125 to 2.0%) of irrigation water on the growth of *E. antisyphilitica.* Con = Control (without saline irrigation).

Table 14.6: Effect of different percentages of saline irrigation on plant growth, hexane extractables, sugars and chlorophyll contents of *E. antisyphilitica*

		Saline Conc. (%)						SEm	CD at 5%	CD at 1%
		Con.	**0.125**	**0.25**	**0.50**	**1.0**	**2.0**			
Length (cm)	AG	39.73	40.97	54.65	53.76	44.43	37.73	±4.69	9.80	13.36
	UG	22.16	24.25	32.45	21.30	26.45	15.15	±1.02	2.14	2.92
Fresh Wt. (g)	AG	31.83	46.10	59.53	50.30	42.60	17.40	±5.25	11.44	16.04
	UG	7.40	12.82	14.35	12.33	12.65	4.45	±3.01	6.57	9.21
Dry Wt. (g)	AG	2.68	4.49	5.38	4.53	3.73	1.30	NS		
	UG	1.06	2.21	2.36	1.80	2.05	0.65	NS		
Dry Wt. (%)	AG	8.42	9.74	9.04	9.00	8.76	7.47			
	UG	14.32	7.24	16.45	14.60	16.21	14.61			
HE%		7.0	8.9	9.2	8.3	8.3	7.1			
Sugars (mg/g)		60.0	59.5	57.5	57.5	53.0	52.7			
Chlorophylls	a	0.514	0.571	0.573	0.350	0.341	0.211			
	b	0.276	0.222	0.080	0.192	0.114	0.192			
Total (mg/g)		0.790	0.793	0.653	0.542	0.455	0.403			

NS = Non significant

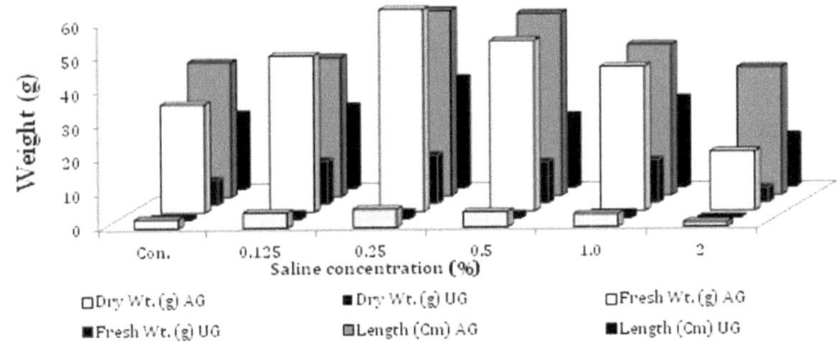

Figure 14.3: Effect of different concentrations of salinity (0.125 to 2.0%) of irrigation water on aboveground and underground length, fresh weight and dry weight of *E. antisyphilitica*. Con – Control (without saline irrigation)

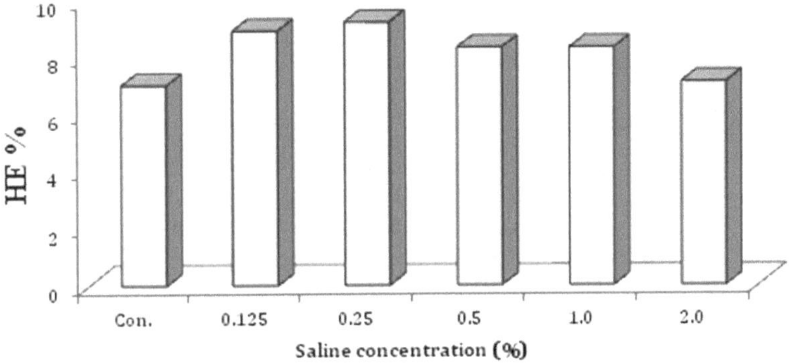

Figure 14.4: Effect of different concentrations of salinity (0.125 to 2.0%) of irrigation water on hexane extractables (percent dry weight) of aboveground parts of *E. antisyphilitica*. Con – Control (without saline irrigation). HE = Hexane Extractables

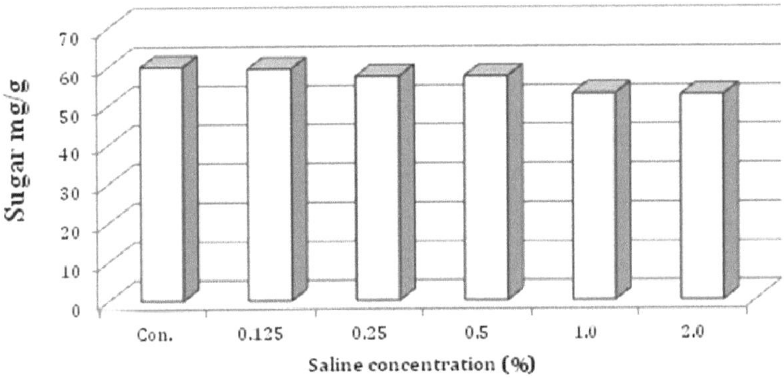

Figure 14.5: Effect of different concentrations of salinity (0.125 to 2.0%) of irrigation water on sugar contents (mg/g dry weight) of aboveground parts of *E. antisyphilitica*. Con – Control (without saline irrigation)

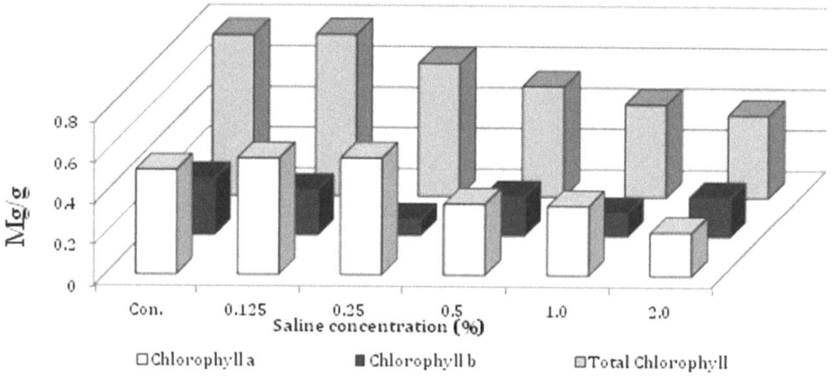

Figure 14.6: Effect of different concentrations of salinity (0.125 to 2.0%) of irrigation water on chlorophyll a, chlorophyll b and total cholorphyll contents (mg/g fresh weight) of aboveground parts of *E. antisyphilitica.* Con – Control (without saline irrigation)

14.4 Discussion

The energetics and economics of biocrude production scenarios appear to be greatly improved by the incorporation of a cogeneration facility directly into the processing plant (McLaughlin et al. 1983). Plants contain a variety of natural products that are readily extractable with non-polar organic solvents. They contain triglycerides, oils, waxes, terpenes, phytosterols and other modified isoprenoid compounds. These extracts have been considered as a potential source of fuels and chemicals. In contrast to the food and fibre crop plants, the work on energy plants is in its infancy and has been carried by only a few groups in a scattered manner (Kumar 1984,1989, Srivastava 1986, Garg and Kumar 1989a). For tropical plants, there are no reports such as there are for temperate plant species (Bagby et al. 1981, Roth et al. 1982, 1984, Carr 1985, Carr and Bagby 1987, Carr et al. 1986).

The entire biocrude mixture of several oil yielding xerophytic species could be cracked to produce chemical feedstocks and gasoline mixtures (Haag et al. 1980). Seed oils may also have potential as a diesel fuel substitute (Morgan and Shultz 1981). Detailed analysis of the major chloroform methanol extractable components of *Asclepias linaria* Cav. (Asclepiadaceae) and *Ilex verticilliata* L. (Aquifoliaceae), two potential hydrocarbon yielding crops, has been reported (Abbott et al. 1990). *A. linaria*, a desert milkweed, contains total 30.3 percent and *I. verticilliata* contains 41.5 percent extractable materials on dry weight basis. Chloroform and methanol extracts were 20.8 and 26.0 percent, respectively, in *A. linaria* and *I. verticilliata.* Hexane solubles of the plants, which correspond to the oil and hydrogen fractions, were also reported. Although fatty acid triglycerides were the major component in hexane extracts of seeds, aboveground plant material from *A. linaria* has relatively small amounts of triglycerides. Refinement of the crude extract into a natural biodegradable surfactant could be possible. Studies have been performed to find out the influence of environment (Nemethy et al. 1981, Vasudevan and Giridhar 1986), water stress (Sachs et al. 1981a, Kingsolver 1982), nutritional factor (Sachs et al.

1981a, Kingsolver 1982, Tenorio et al. 1984, Kumar and Kumar 1985) and hormonal influence (Garg and Kumar 1987a) on *E. lathyris*. Environmental influence has also been studied on *C. procera* (Rani et al. 1989). Besides, studies on the effect of soil type (Johari et al. 1989, Roy 1989) and salinity (Garg 1987, Garg and Kumar 1989b) have also been reported on various members of Euphorbiaceae. The latex of some plants are very rich in reduced photosynthetic materials (polyisoprenes and sterols) comprising upto 80 percent of dry weight, making these species bright, prospective candidates for future petrochemical plantations (Nielsen et al. 1977).

The practical way is to make use of land resources that are unsuitable for conventional agriculture because such latex yielding plants have been found to grow successfully on marginal lands (Kumar 2013, Kumar et al. 2020). The economic development of hydrocarbon from plant materials will ultimately depend upon agronomy and conversion costs (Weisz and Marshall 1979). In India, there is a vast scope for harvesting solar energy because in most of the days in a year full sunshine occurs in North-West and South India. Indian arid zone, having wasteland and full sunshine throughout the year, occupies an area of about 0.32 million km^2 of hot desert, out of which 12.65 percent is arid and 29.57 percent is semi-arid. Rajasthan comprises 61 percent arid zone and 13 percent semi-arid zone, while Gujarat state has 19.6 percent arid and 9 percent semi-arid lands (Sharma et al. 1988). Energy farm must be developed specifically for arid and semi-arid lands (Bassham 1977). Developing biocrude farming on marginal lands is desirable because they would not compete with food and fibre crops (Johnson and Hinman1980). Soil deterioration of wastelands can also be improved by green plants (Cunningham and Berti1993). Soils containing excess neutral soluble salts dominated by chlorides and sulphates are saline soils. Saline soils are widely distributed in arid, semi-arid and in reasonably dry coastal areas. Introduction of irrigation in arid and semi-arid areas without adequate provision of drainage leads to salt salinization within a few years. The main contributing factors are high salt content in the profile, saline groundwater and high water-table, seepage from canal and adjacent irrigated areas, restricted surface and subsurface drainage, irrigation with water of high salt content and ingress of sea-water. Excess salinity in most soils is due to the dominance of sodium chloride (NaCl) in India. In certain areas of U.P., Delhi, Haryana, Punjab, Rajasthan and also in South India, saline soils occur with a highly mineralized water table, well within the critical limit to cause salinization. Sodium often constitutes more than half of the total soluble cations. Excess of sodium on the soil exchange complex and/or soluble salts in the soil profile reduces the productivity of an estimated seven million hectare of cropland in India (Abrol and Bhumbla 1971). Present study shows that *E. antisyphilitica* can be successfully raised in wastelands of the Rajasthan state.

14.5. Conclusion

E. Antisyphilitica is a potential plant candidate that maintained the biomass production and sugar biosynthesis in extreme salinity toxicity. The present study shows that *E. antisyphilitica* can be successfully raised in wastelands of the Rajasthan state.

References

Abbott, T.P., R.E. Peterson, L.W. Tjarks, D.M. Palmer and M.O. Bagby. 1990. Major extractable components in *Asclepiaslinaria* (Asclepiadaceae) and *Ilex verticilliata* (Aquifoliaceae), two potential hydrocarbon crops. Econ. Bot. 44: 278-284.

Abrol, J.P. and D.R. Bhumbla.1971. Saline and alkaline soils in India: Their occurrence and management. World Soil Resources, FAO Report 41: 42-51.

Adams, R.P., M.F. Balandrin and J.R. Martineau.1984. The showy milkweed *Asclepias speciosa*: A potential new semi-arid land crop for energy and chemicals. Biomass 4: 81-104.

Agarwal, S. and A. Kumar. 2018. Historical development of biofuels.pp. 17-46. *In*: Kumar, A., S. Ogita and Y.Y. Yau [eds.]. Biofuels: Greenhouse Gas Mitigation and Global Warming Next Generation Biofuels and Role of Biotechnology. Springer, Heidelberg, Germany.

Anwar, M., D.V. Singh and K. Subrahmanyam. 1988. Safe limits of salinity for three important medicinal plants. Int. J. Trop. Agric. 6: 125-128.

Ayerbe, L., E. Funes, J.L. Tenorio, P. Ventas and L. Mellado. 1984. *Euphorbia lathyris* as an energy crop II: Hydrocarbon and sugar productivity. Biomass 5: 37-42.

Bagby, M.O., R.A. Buchanan and F.H. Otey. 1981. Multi-use crops and botanochemical production. Am. Chem. Soc. Symp. 144: 125-136.

Bal, A.R. and S.K. Dutt. 1987. Salt tolerance mechanism in *Acanthus ilicifolius* L. Indian J. Pl. Physiol. 30: 170-175.

Bassham, J.A. 1977. Increasing crop production through more controlled photosynthesis. Science 197: 630.

Bhansali, S. and A. Kumar. 2018. Synthetic and semisynthetic metabolic pathways for biofuel production. pp. 421-432. *In*: Kumar, A., S. Ogita and Y.Y. Yau [eds.]. Biofuels: Greenhouse Gas Mitigation and Global Warming Next Generation Biofuels and Role of Biotechnology. Springer, Heidelberg, Germany.

Bhatia, V.K., K.G. Mittal, P.R. Mehrotra and M. Mehrotra. 1989. Hydrocarbon fuels from biomass. Fuel 68: 475-479.

Buchanan, R.A., I.M. Cull, F.H. Otey and C.R. Russel. 1978a. Hydrocarbon and rubber producing crops. Evaluation of U.S. plant species. Econ. Bot. 32: 131-145.

Buchanan, R.A., I.M. Cull, F.H. Otey and C.R. Russel. 1978b. Hydrocarbon and rubber producing crops. Evaluation of 100 U.S. plant species. Econ. Bot. 32: 146-153.

Buchanan, R.A., I.M. Cull, F.H. Otey and C.R. Russel. 1978c. Whole plant oils, potential new industrial raw materials. J. Am. Oil Chem. Soc. 55: 657-662.

Calvin, M. 1978. Chemistry, population, resources. Pure Appl. Chem. 50: 407-426.

Calvin, M. 1979. Petroleum plantations for fuel and materials. Bioscience 29: 533-537.

Calvin, M. 1980. Hydrocarbons from plants: Analytical methods and observations. Die Naturwissen 67: 525-533.

Calvin, M. 1984. Renewable fuels for the future. J. Appl. Biochem. 6: 3-18.

Campos-Lopez, E. and Roman-Alemany. 1980. Organic chemicals from the Chihuahuan Desert. J. Agric. Food Chem. 28: 171-183.

Carr, M.E. 1985. Plant species evaluated for new crop potential. Econ. Bot. 39:336-345.

Carr, M.E. and M.O. Bagby. 1987. Tennesse plant species screened for renewable energy sources. Econ. Bot. 41: 78-85.

Carr, M.E., M.O. Bagby and W.B. Roth. 1986. High oil and polyphenol producing species of the North-West. J. Am. Oil Chem. Soc. 63: 1460-1464.

Cunningham, D.S. and R.W. Berti. 1993. Remediation of contaminated soils with green plants: An overview. In Vitro Cell Dev. Biol. 29: 207-212.

Datia, K.S. and J. Dayal. 1988. Effect of salinity on germination and early seedling growth of Guar (*Cyamopsis tetragonoloba* L.). Indian J. Plant Physiol. 31: 357-363.

Dayal, M. 1986. Production and utilization of petrocrops. pp. 2-7. *In*: Proc. Workshop on Petrocrops. Department of Non-Conventional Energy Sources, New Delhi.

Dehgan, B. and S.C. Wang. 1983. Evaluation of hydrocarbon plants suitable for cultivation in Florida. Proc. Soil Sci. Florida 42: 17-19.

Divate, M.R. and R.M. Pandey. 1981. Salt tolerance in grapes. III. Effect of salinity on chlorophyll, photosynthesis and respiration. Indian J. Plant Physiol. 24: 74-79.

Ebadi, A., N.A. Khoshkholgh Sima, M. Olamaee, M. Hashemi and R. Ghorbani Nasrabadi. 2018. Remediation of saline soils contaminated with crude oil using the halophyte *Salicorniapersica* in conjunction with hydrocarbon-degrading bacteria. J. Environ. Manage. 219: 260-268.

Erdman, M.D. and B.A. Erdman. 1981. *Calotropis procera* as a source of plant hydrocarbons. Econ. Bot. 35: 467-472.

Garg, J. 1987. Studies on growth and physiology of some latex yielding plants. Ph.D. Thesis. University of Rajasthan, Jaipur. pp. 159.

Garg, J. and A. Kumar. 1987a. Effect of growth regulators on the growth, chlorophyll development and productivity of *Euphorbia lathyris* L., a hydrocarbon yielding plant. pp. 403-406. *In*: Biggins, J. [ed.]. Progress in Photosynthesis Research; Proceedings of the VIIth International Congress on Photosynthesis Providence. Vol. 4. Nijhoff: Dordrecht, Netherlands Elsevier Applied Science, London.

Garg, J. and A. Kumar. 1987b. Improving growth and hydrocarbon yield of *Euphorbia lathyris* L. pp. 93-97. *In*: Sharma, R.N., O.P. Vimal and A.N. Mathur [eds.]. Proc. Bioenergy Society IV Convention and Symposium. Bioenergy Society of India, New Delhi.

Garg, J. and A. Kumar. 1989a. Potential petro-crops for Rajasthan. J. Indian Bot. Soc. 68: 199-200.

Garg, J. and A. Kumar. 1989b. Influence of salinity on growth and hydrocarbon yield of *Euphorbia lathyris*. J. Indian Bot. Soc. 68: 201-204.

Garg, J. and A. Kumar. 2011. Hydrocarbon from plants as renewable source of energy. Bioherald 1: 31-35.

Garg, J. and A. Kumar. 2012a. Effect of various organic nutrients on growth, biomass yield and hydrocarbon production of *Euphorbia lathyris* L.: A hydrocarbon yielding plant. Intl. Natl. J. Life Science Pharma. Res. 2: 83-89.

Garg, J. and A. Kumar. 2012b. Analysis of hexane extracts of *Euphorbia lathyris*, a potential source of biofuels. Bioherald: International J. Biodiver. Environ. 2: 62-64.

Gartner, E.E. 1979. The history and use of milkweed (*Asclepias syriaca* L.). Econ. Bot. 33: 119-123.

Haag, W.O., P.G. Rodewald and P.G. Weisz. 1980. Catalytic production of aromatics and olefins from plant materials. pp. 63-76. *In*: 2nd Chemical Congress of American Society. Las Vagas, Nev.

Hall, D.O. 1985. Plant hydrocarbon resources in arid and semi-arid lands. pp. 369-384. *In*: Wickens, G.E., J.R. Goodin and D.V. Field [eds.]. Plants for Arid Lands. Proc. Kew International Conference on Economic Plants for Arid Lands. George Allen and Unwin, London.

Hinman, C.W., J.J. Hoffmann, S.P. McLaughlin and T.R. Peoples. 1980. Hydrocarbon production from and plant species. Proc. Ann. Meet. Am. Sect. Int. Soil Energy Soc. 3: 110.

International Energy Agency (IEA). 2011. Technology roadmap (biofuels for transport).

International Energy Agency (IEA). 2014. Key world energy statistics. http://www.iea.org/publications/freepublications/publication/keyworld2014.pdf

Johari, S. and A. Kumar. 2013a. Improving growth and productivity of *Euphorbia antisyphilitica*: A biofuel plant for semi-arid regions. Int. J. Life Sci. and Pharma. Res. 3(4): 20-24.

Johari, S. and A. Kumar. 2013b. *Euphorbia antisyphilitica:* Effect of growth regulators in improving growth and productivity of hydrocarbon yielding plant. Int. J. Life Sci. and Pharma. Res. 3(4): 25-28.

Johari, S. and A. Kumar. 2016. Climate change and greenhouse gas mitigation by biofuels. pp. 292-312. *In*: Basu, S.K., P. Zandi and S.K. Chalaras [eds.]. Global Environmental Crises, Challenges and Sustainable Solutions from Multiple Perspectives.Haghshenass Publication, Teheran, Iran.

Johari, S. and A. Kumar. 2020a. *Euphorbia* spp. and their use in traditional medicines: A review. World J. Pharma. Res. 9(14): 1477-1485.

Johari, S. and A. Kumar. 2020b. Antifungal activities of *Euphorbia* spp. and its use in traditional medicines: A review. World J. Pharma. Res. 9(14): 1499-1506.

Johari, S. and A. Kumar. 2020c. Characterization of anatomical and biochemical characteristics of *Euphorbia antisyphilitica*: A plant of traditional medicine. World J. Pharma. Res. 9(14): 1507-1515.

Johari, S., S. Roy and A. Kumar. 1989. Influence of edaphic and nutritional factors on growth and hydrocarbon yield of *Euphorbia antisyphilitica* Zucc. *In*: Grassi, G., G. Gosse and G. dos Santos [eds.]. Biomass for Energy and Industry. Vol. I. 1.552-1.526. Elsevier Applied Science, London.

Johnson, J.D. and H. Hinman. 1980. Oils and rubber from land plants. Science 208: 460-464.

Kang, A. and T.S. Lee. 2015. Converting sugars to biofuels: Ethanol and beyond. Bioengineering 2: 184-203.

Khoo, K.S., W.Y. Chia, D.Y.Y. Tang, P.L. Show, K.W. Chew and W.H. Chen. 2020. Nanomaterials utilization in biomass for biofuel and bioenergy production. Energies 13: 892.

Kingsolver, B.E. 1982. *Euphorbia lathyris* reconsidered its potential as an energy crop for arid lands. Biomass 2: 281-298.

Kotiya, A., M.K. Sharma and A. Kumar. 2018. Potential biomass for biofuels from wastelands. pp. 59-80. *In*: Kumar, A., S. Ogita and Y.Y. Yau [eds.]. Biofuels: Greenhouse Gas Mitigation and Global Warming Next Generation Biofuels and Role of Biotechnology. Springer, Heidelberg, Germany.

Kumar, A. 1984. Hydrocarbon from plants in arid and semi-arid regions. *In*: Proceedings of the National Seminar on Applications on Science and Technology for Afforestation Act. 81-86. Jaipur.

Kumar, A. 2008. Bioengineering of crops for biofuels and bioenergy. pp. 346-360. *In*: Kumar, A. and S. Sopory [eds.]. Recent Advances in Plant Biotechnology. I.K. International, New Delhi.

Kumar, A. 2011. Biofuel resources for greenhouse gas mitigation and environment protection. pp. 221-246. *In*: Trivedi, P.C. [ed.]. Agriculture Biotechnology. Avishkar Publishers, Jaipur.

Kumar, A. 2013. Biofuels utilisation: An attempt to reduce GHGs and mitigate climate change. pp. 199-224. *In*: Nautiyal, S., K. Rao, H. Kaechele, K. Raju and R. Schaldach [eds.]. Knowledge Systems of Societies for Adaptation and Mitigation of Impacts of Climate Change. Environ.Sci. and Eng.Springer, Berlin, Heidelberg.

Kumar, A. 2014. Role of horticulture in biodiversity conservation. pp. 143-156. *In*: Nandwani D. [ed.]. Sustainable Horticultural systems: Issue Technology and Innovation. Springer, Germany.

Kumar, A. 2018a. Alternative biomass from semiarid and arid conditions as a biofuel source: *Calotropis procera* and its genomic characterization. pp. 241-270. *In*: Kumar, A., S. Ogita and Y.Y. Yau [eds.]. Biofuels: Greenhouse Gas Mitigation and Global Warming Next Generation Biofuels and Role of Biotechnology. Springer, Heidelberg, Germany.

Kumar, A. 2018b. Global warming, climate change and greenhouse gas mitigation. pp. 1-16. *In*: Kumar, A., S. Ogita and Y.Y. Yau [eds.].Biofuels: Greenhouse Gas Mitigation and Global Warming Next Generation Biofuels and Role of Biotechnology. Springer, Heidelberg, Germany.

Kumar, A. 2020a. Climate change: Challenges to reduce global warming and role of biofuels. pp. 13-54. *In*: Kumar, A., Y.Y. Yau, S. Ogita and R. Scheibe [eds.]. Climate Change, Photosynthesis and Advanced Biofuels. Springer, Singapore.

Kumar, A. 2020b. Synthetic biology and future production of biofuels and high-value products. pp. 271-302. *In*: Kumar, A., Y.Y. Yau, S. Ogita and R. Scheibe [eds.]. Climate Change, Photosynthesis and Advanced Biofuels. Springer, Singapore.

Kumar, A. and P. Kumar. 1985. Agronomic studies on growth of *Euphorbia lathyris*. pp. 170-175. *In*: Egneus, H. and H. Ellegard [eds.]. Bioenergy 84. Vol. II. Biomass Resources. Elsevier Applied Science Publishers, Amsterdam.

Kumar, A. and V.R. Kumar. 2011. New and renewable source of energy: *Salicornia*spp.: A halophyte has potential for biomass yield in coastal axes of the world. pp. 686-688. *In*: Proceedings of 19th European Biomass Conference, Berlin, Germany.

Kumar, A. and N. Gupta. 2018. Potential of lignocellulosic materials for production of ethanol. pp. 271-290. *In*: Kumar, A., S. Ogita and Y.Y. Yau [eds.]. Biofuels: Greenhouse Gas Mitigation and Global Warming Next Generation Biofuels and Role of Biotechnology. Springer, Heidelberg, Germany.

Kumar, A. and S. Roy. 2018. Agrotechnology, production, and demonstration of high-quality planting material for biofuels in arid and semiarid regions. pp. 205-228. *In*: Kumar, A., S. Ogita and Y.Y. Yau [eds.]. Biofuels: Greenhouse Gas Mitigation and GlobalWarming Next Generation Biofuels and Role of Biotechnology. Springer, Heidelberg, Germany.

Kumar, A., S. Ogita and Y.Y. Yau. 2018a. Biofuels: Greenhouse Gas Mitigation and Global Warming Next Generation Biofuels and Role of Biotechnology. Springer, Heidelberg, Germany. pp. 432.

Kumar, A., A. Ebin and A. Gupta. 2018b. Alternative biomass from saline and semiarid and arid conditions as a source of biofuels: *Salicornia*. pp. 229-240. *In*: Kumar, A., S. Ogita and Y.Y. Yau [eds.]. Biofuels: Greenhouse Gas Mitigation and Global Warming Next Generation Biofuels and Role of Biotechnology. Springer, Heidelberg, Germany.

Kumar, A. and N. Gadhwal. 2020. Global warming, social justice and role of community services in climate change. Soc. Service Practices Commun. Dev. J. 1: 20-26.

Kumar, A., Y.Y. Yau, S. Ogita and R. Scheibe. 2020. Introduction. pp. 490. *In*: Kumar, A., Y.Y. Yau, S. Ogita and R. Scheibe [eds.]. Climate Change, Photosynthesis and Advanced Biofuels. Springer, Singapore.

Li, Y. and S.K. Khanal. 2016. Bioenergy: Principles and Applications. Wiley: Hoboken, NJ, USA.

Lijó, L., S. González-García, D. Lovarelli, M.T. Moreira, G. Feijoo and J. Bacenetti. 2019. Life cycle assessment of renewable energy production from biomass. pp. 81-98. *In*: Basosi, R., M. Cellura, S. Longo and M.L. Parisi [eds.]. Life Cycle Assessment of Energy Systems and Sustainable Energy Technologies: The Italian Experience. Springer International Publishing: Cham, Switzerland.

Lipinsky, E.S. 1981. Chemicals from biomass – Petrochemical substitution options. Science.212:1465-1471.

Lipinsky, E.S., S. Kresovich and A. Scantland. 1980. Fuels from new crops. pp. 307. *In*: E. Campoz-Lopez [ed.]. Renewable Resources: A systematic Approach. Academic Press, New York.

Marimuthu, S., R.B. Subramanian, I.L. Kothari and J.A. Inamdar. 1989. Laticiferous taxa as a source of energy and hydrocarbon. Econ. Bot. 43: 255-261.

Maugh, T.H. 1979. Unlike money, diesel fuel grows on trees. Science 206: 436.

McLaughlin, S.P., B.E. Kingsolver and J.J. Hoffmann. 1983. Biocrude production in arid lands. Econ. Bot. 37: 150-158.

Morgan, R.P. and E.B. Shultz. 1981. Fuels and chemical from novel seed oils. Chem. Eng. News 7: 69.

Mortimer, J.C. 2019. Plant synthetic biology could drive a revolution in biofuels and medicine. Exp. Biol. Med. 244(4): 323-331.

Nemethy, E.K., J.W. Otovas and M. Calvin. 1981b. Natural production of high energy liquid fuels from plants. pp. 405. *In*: Klass, D.L. and G.H. Emert [eds.]. Fuel from Biomass and Wastes. Ann Arbor Science Publishers, United States.

Nielsen, P.E., H. Nishimura, Y. Liang and M. Calvin. 1979. Steroids from *Euphorbia* and other latex bearing plants. Phytochemistry 18: 103-104.

Nielsen, P.E., H. Nishimura, J.W. Otvos and M. Calcin. 1977. Plant crops as a source of fuel and hydrocarbon-like materials. Science 198: 942-944.

Nunes, M.A., M.A. Dias, M.M. Correia and M.M. Oliveira. 1984. Further studies on growth and osmoregulation of sugar beet leaves under low salinity conditions. J. Exp. Bot. 35: 322-331.

Palsson, B.O., S. Fathi-Afshar, D.F. Rudd and E.N. Lightfoot. 1981. Biomass as source of chemical feedstocks: An economic evaluation. Science 213: 513-517.

Patel, J.A.A. and A.B. Vora. 1985. Salinity induced changes in *Zea mays* L. (Ganga-2). J. Expt. Bot. 36: 49-54.

Prisco, J.T. and J.W. Oleary. 1973. The effect of humidity and cytokinin on growth and water relations of salt-stressed bean plants. Plant Soil 39: 263-276.

Qadir, M., A. Tubeileh, J. Akhtar, A. Larbi, P.S. Minhas and M.A. Khan. 2008. Productivity enhancement of salt-affected environments through crop diversification. Land Degrad. Develop. In press.

Rai, M. 1977. Varietal resistance to salinity in maize. Ind. J. Plant Physiol. 20: 100-104.

Rani, A., S. Roy and A. Kumar. 1989. Influence of morphological and environmental factors on growth and hydrocarbon yield of *Calotropis procera* (Ait.) R. Br. pp. 1.480-1.483. *In*: Grassi, G., G. Gosse and G. dos Santos [eds.]. Biomass for Energy and Industry. Vol.1. Elsevier Applied Science, London.

Reddy, M.P. and A.B. Vora. 1986. Salinity induced changes in pigment composition and chlorophyllase activity of wheat. Indian J. Plant Physiol. 29: 331-334.

Roth, W.B., I.M. Cull, R.A. Buchanan and M.O. Bagby. 1982. Whole plants as renewable energy sources: Checklist of 508 species analysed for hydrocarbon, oil, polyphenol and protein. Trans. Illinois. State Acad. Sci. 75: 217-231.

Roth, W.B., M.E. Carr, I.M. Cull, B.S. Phillips and M.O. Bagby. 1984. Evaluation of 107 legumes for renewable sources of energy. Econ. Bot. 38: 358-364.

Roy, S. 1989. Effect of various factors on growth and productivity of *Jatrophacurcas*: Ahydrocarbon yielding plant. pp. 1.484-1.488. *In*: Grassi, G., G. Gosse and G. dos Santos [eds.]. Proc. Biomass for Energy and Industry. Vol. 1. Elsevier Appl. Sci. London.

Sachs, R.M., C.B. Low, J.D. McDonald, A.R. Award and M.J. Sully. 1981a. *Euphorbia lathyris:* A potential source of petroleum like products. California Agricult. pp. 29-32.

Sachs, R.M., C.B. Low, J.D. McDonald, A.R. Award and M.J. Sully.1981b. Agronomic studies with *Euphorbia lathyris*: A potential source of petroleum like products. *In*: Energy from Biomass and Wastes, V. Institute of Gas Technology, U.S.A.

Sharma, R.D., R.C. Tiwari and S.K. Gupta. 1988. Greening of wastelands – An overview. *In*: Sharma, R.N., H.L. Sharma and H.S. Garcha [eds.]. Proc. Bioenergy Society V Convention and Symposium. 205212. Bioenergy Soc. of India, New Delhi.

Shekhawat, V.P.S. and A. Kumar. 2004. Effect of salinity on proline and reducing sugar content in selected halophytes and glycophyte species. Nature Env. Pollut. Technol. 3: 365-368.

Shekhawat, V.P.S. and A. Kumar. 2006. Somaclonal variants for salt tolerance and *in vitro* propagation of peanut. pp. 82-196. *In*: Kumar, A., S. Roy and S.K. Sopory [eds.]. Plant Biotechnology and its Application in Tissue Culture. I.K. International, New Delhi.

Shekhawat, V.P.S., A. Kumar and K.H. Neumann. 2006a. Bio-reclamation of secondary salinized soils using halophytes. pp. 147-154. *In*: Öztürk, M., Y. Waisel, M.A. Khan and G. Görk [eds.]. Biosaline Agriculture and Salinity Tolerance in Plants. Birkhäuser Basel https://doi.org/ 10.1007/3-7643-7610-4_16.

Shekhawat, V.P.S., A. Kumar and K.H. Neumann. 2006b. Effect of NaCl salinity on growth and ion accumulation in some chenopodiaceous halophytes. Commun. Soil Sci. Plant Anal. 37: 1933-1946.

Shukla, O.P. and C.R. Crishna-Murti. 1971. The biochemistry of plant latex. J. Sci. Industr. Res. 30: 640-642.

Srivastava, G.S. 1985. Availability and production of petrocrops in India. pp. 68-89. *In*: Behl, H.M. and O.P.Vimal [eds.]. Proc. Natl. Seminar-cum-workshop on Bioenergy Education. Rahul and Co., Ajmer.

Sternburg, C. and E. Rodriguez. 1982. Hydrocarbons from *Pedilanthus macrocarpus* (Euphorbiaceae) of Baja California and Sonora, Mexico. Am. J. Bot. 69: 214-218.

Srivastava, G.S. 1986. Petrocrops, their availability and cultivation. pp. 8-26. *In*: Proc. Petrocrop Workshop. Department of Non-Conventional Energy Sources, New Delhi.

Strongonov, B.P. 1962. Structure and function of plant cells in saline habitat. B. Gollek [ed.]. pp. 236. John Wiley, New York.

Strzalka, R., D. Schneider and U. Eicker. 2017. Current status of bioenergy technologies in Germany. Renew. Sustain. Energy Rev. 72: 801-820.

Tenorio, J.L., P. Ventas, E. Funes and L. Ayerbe. 1984. Biomass production from *Euphorbia lathyris*. pp. 176-180. *In*: Egneus, H. and E. Ellegard [eds.]. Bioenergy, 84 Vol. II Biomass Resources. Elsevier Applied Science, London.

Vasudevan, P. and G. Giridhar. 1986. *Calotropis procera* (Ait.) R. Br. A potential petrocrop. pp. 27-44. *In*: Proc. Workshop on Petrocrops. Department of Non-Conventional Energy Sources, New Delhi.

Vijayvargiya, S. and A. Kumar. 2011. Influence of Salinity Stress on Plant Growth and Productivity: Salinity Stress Influences on Plant Growth. Germany. Lambert Academic Publishers.170 pp.

Vimal, O.P. 1986. Strategies for use of petro crops. pp. 221-229. *In*: Proc. Workshop on Petrocrops. Department of Non-Conventional Energy Sources, New Delhi.

Weisz, P.B., W.O. Haag and P.G. Rodewald. 1979. Catalytic production of high-grade fuel (gasoline) from biomass compounds by shape-selective catalysis. Science 206: 57-58.

Weisz, P.B. and J.F. Marshall. 1979. High grade fuel from biomass farming: Potential and constrains. Science 206: 24.

Yadav, J.S.P. 1975. Improvement of saline alkali soils through biological methods. Indian For. 101: 385-395.

Yensen, N.P. 2006. Halophyte uses for the twenty-first century. *In*: Khan, M.A. and D.J. Weber [eds.]. Ecophysiology of High Salinity Tolerant Plants. Series: Tasks for Vegetation Sci. 40.

Yeo, A.R., S.J.M. Capron and T.J. Flowers. 1985. The effect of salinity upon photosynthesis in Rice (*Oryzasativa* L.): Gas exchange by individual leaves in relation to their salt content. J. Exp. Bot. 36: 1240-1248.

Yermanos, D.M., L.E. Francois and T. Tammadoni. 1967. Effect of soil salinity on the development of Jojoba. Econ. Bot. 21: 69-80.

Index

O

P

R

S

T

W

About the Editors

Dr. Jos T. Puthur, Professor in the Department of Botany, University of Calicut, Malappuram, Kerala, India is an eminent plant physiologist and expert in the area of 'osmoregulation, biochemical and molecular responses, as well as alterations in photosynthetic processes of plants exposed to abiotic stresses'. He was awarded with TWAS research grant for promising scientific research projects, BOYSCAST fellowship of DST, Govt. of India, UGC Fellowship under Indo-Hungarian Educational Exchange programme, and TWAS-UNESCO Associate ship at centers of Excellence in South (CEFOBI, Rosario, Argentina). He is a faculty member of the massive online open course (MOOC) of Consortium for Educational Communication (CEC), MHRD, Govt of India. He was selected as a visiting scientist under INSA-DFG programme of International Scientific Collaboration and Exchange of Scientists to Institut für Molekulare Physiologie und Biotechnologie der Pflanzen (IMBIO), University of Bonn, Germany. He was a Visiting Scientist at the Institute of Plant Biology, Biological Research Centre, Hungarian Academy of Sciences, Hungary. He has guided 10 students for Ph.D. and published 100 research publications in well reputed national and international journals.

Dr. Om Parkash Dhankher, Professor of Agriculture Biotechnology in the Stockbridge School of Agriculture, University of Massachusetts, Amherst (U.S.A) received his M.Sc. and M. Phil. in Botany from Kurukshetra University (India) and Ph.D. in Plant Molecular Biology from Durham University (United Kingdom). He was the recipient of the prestigious Commonwealth Scholarship by the Commonwealth Commission London. He developed the first transgenic plant-based approach for arsenic phytoremediation by combining the expression of two bacterial genes and translating this research from model plant Arabidopsis to high biomass non-food field crops. His major research focus is phytoremediation, bio-energy production, and developing climate-resilient crops. Along with this, his laboratory is developing arsenic-free and arsenic tolerant food crops in order to improve human health using both forward and reverse genetic approaches. Prof. Dhankher has published more than 100 referred publications and book chapters in high impact journals including *Nature, Nature Biotechnology, PNAS, Plant Cell, Plant Biotechnology, New Phytologist, Plant Physiology, Environmental Science & Technology*, etc., three edited books, guest edited five special issues for several journals, and six international patents were awarded to him. He is an elected fellow of Indian Society of Plant Physiology and a member of the executive committee of the American Society of

Plant Biologists (ASPB), and was elected as the Vice President for the International Society for Phytotechnologies (IPS, 2015-2022). Prof. Dhankher is also serving as the senior associate editor for the *International Journal of Phytoremediation*, editor for the *International Journal of Plant & Environment*; *Plant Physiology Reports,* and the associate editor for the *Crop Science*, *Plants*, and the *Food and Energy Security* journal, etc. He has supervised over two dozen Ph.D. and M.Sc. students, and postdoctoral research associates. Prof. Dhankher has established widespread national and international collaborations with researchers in Australia, India, China, Italy, Egypt, U.K., and USA.